VLSI
and
PARALLEL
COMPUTATION

Sponsoring Editor   Bruce Spatz
Production Editor   Sharon Montooth
Typesetter   Rosenlaui Publishing Services, Inc.
Cover Designer   Andrea Hendrick
Cover Mechanical   Victoria Philp
Copy Editor   Linda Medoff
Proofreader   Martha Ghent
Cover Image   Brian Wyvill

Library of Congress Cataloging-in-Publication Data
VLSI and parallel computation / edited by Robert Suaya and Graham
  Birtwistle.
      p.   cm.
   ISBN 0-934613-99-0
   1. Parallel processing (Electronic computers) 2. Integrated
circuits—very large scale integration. I. Suaya, Robert.
II. Birtwistle, G. M. (Graham M.)
QA76.5.V564   1990   89-49717
004′.35—dc20   CIP

Morgan Kaufmann Publishers, Inc.
Editorial Office:
    2929 Campus Drive
    San Mateo, CA 94403
Order from:
    P.O. Box 50490
    Palo Alto, CA 94303-9953

The material in Chapter 5 is based upon work supported by the Department of the
Navy, Office of the Chief of Naval Research under grant number N00014-89-J-1062.
Any opinions, findings, and conclusions or recommendations expressed in this
publication are those of the authors and do not necessarily reflect the views of the
Department of the Navy.

94   93   92   91   90   5   4   3   2   1

VLSI
and
PARALLEL
COMPUTATION

Edited by Robert Suaya and Graham Birtwistle

Morgan Kaufmann Publishers, Inc.
San Mateo, California

Preface

This book is an outgrowth of a winter school on very large scale integration (VLSI) and parallel computations held at the Banff Springs Hotel in Alberta. Leading practitioners in the field described recent contributions to the theory and practice of building highly concurrent systems, their communication requirements, concurrent programming techniques in practice, complexity of parallel computations, and a review of neural networks.

Why did we choose to couple VLSI so closely with parallel computation? Perhaps it helps to state what this book is not about. It is not concerned with progress in vector supercomputer technology. Neither is it concerned with the evolution of conventional computers into a few tens of processors. This book deals with some of the distinct issues that appear when one enters the world of highly parallel systems containing hundreds or thousands of processors. Vector supercomputers are arguably some of the most powerful machines today, but there are clear limits to their technology. Vector machines rely on the fastest technologies available today (emitter coupled logic (ECL) soon to be followed by gallium arsenide (GaAs)). The limiting factors are power dissipation and miniaturization. Vector supercomputers support a limited form of concurrency, typically done in expression evaluation and iterations at the innermost levels of a computation. Over the next decade the progress in supercomputer technology will primarily come from further use of parallelization. Clock rates on the other hand are unlikely to improve by more than one order of magnitude. For example, today's CRAY YMP has a six-nanosecond clock rate, and the CRAY 4, two generations from now, is expected to have a one-nanosecond clock rate.

In order to advance the state of the art in many scientific endeavors we will need improvement by several orders of magnitude over current architectures. Examples for this need abound in vision and speech research, emulation of biological systems, VLSI design verification, plasma confinement, structural mechanics, 3D animation in graphics, automated reasoning in system verification, theory of strong interactions (QCD), and turbulence.

With a large set of processors, VLSI and concurrency together bring an opportunity to tackle these fundamental problems; they call for a different way of reasoning about and structuring the work. The concerns associated with large-scale parallel multiprocessors stem directly from the high level of concurrency required, and are very different from those associated with sequential computation, or shared memory extensions into the parallel domain. These concurrency issues are relatively new to computer science, and are quite foreign to those of us who have spent our time designing and/or programming Von Neumann machines. On the other hand, concurrency has been used for millenia in biological systems. "Nature loves concurrency", as Chuck Seitz will remind the reader. By exploiting concurrency, the human brain does much better than our fastest supercomputers in many sensory activities, and it achieves this feat with a technology a million times slower than ECL. Recent developments in neural network research and analog computations, discussed in this book, provocatively suggest the possibility of obtaining human-sensory level computational throughput in the not too distant future. The medium that can make this paradigm a reality is VLSI. Today we identify VLSI with CMOS (complementary metal oxide semiconductors), but this need not be the case in the future.

In 1990 we can fabricate tens of millions of transistors on a single silicon chip, inexpensively and reliably, with the potential to increase this number to a billion transistors on a single packaged chip before the year 2000. Rapid advances in packaging technologies, such as packing multiple dies on the same substrate, indicate that another two orders of magnitude may be gained from these physically compacting modules.

Computing power measured in multiple trillions of floating point instructions per second (TFLOPs) could be achieved with relative ease with inexpensive and slow CMOS components. With our understanding of VLSI and its communication capabilities, and with new programming techniques that take advantage of the fact that highly complex systems can be fabricated on very compact modules, it is fair to say that radical changes will emerge in the way computations are performed. These are the issues on which we try to shed some light and this is why we couple VLSI with concurrency.

Perhaps a bit of personal history is in order here. After a good hiking trip in the Canadian Rockies, the authors had the idea of holding workshops to stimulate the brightest graduate students by exposing them to the breath and diversity of the fields to which VLSI contributes. We felt that it was very important to bring together leading researchers with the best North American graduate students in an environment conducive to the exchange of novel ideas. The choice of venue was easy: the Banff Springs Hotel is a true jewel set in the splendid Banff National Park. Similar meetings in other fields have a tradition of producing leaps of insight in their participants not normally experienced in conventional academic or professional settings. Simply put, something special happens when you mix young talent with leaders of the academic and professional world in an atmosphere free from the distractions of everyday life.

The second Frontiers workshop was held at Banff in March, 1988. The workshop was entitled *Massively Parallel Models of Computation* and attracted a select number of top graduate students from Canada and the USA to hear Yaser Abu-Mostafa (Caltech), Bryan Ackland (AT&T Bell Laboratory), Bill Dally (MIT), Lennart Johnsson (Yale), Dick Lyon (Apple), Ernst Mayr (Stanford), and Chuck Seitz (Caltech) lecture on their current research directions. These contributing talks were the raw material from which the the chapters in this book were hewn after fruitful follow-up discussions and substantial further writing, rewriting, honing and polishing by the authors.

- Chuck Seitz (Caltech) starts our book with a chapter entitled **Concurrent Architectures** in which he examines general questions on concurrency, routing automata, status of second generation multicomputers, and multicomputing programming techniques. Chuck starts by arguing that the AT^2 results (A: Area, T: Time) of VLSI complexity theory can be interpreted as showing that the cost AT of a computation can be expected to vary as $1/T$, and that computations that can be partitioned N ways are going to be $1/N$th as expensive to run concurrently than sequentially, if the same elapsed time is allowed for each. He then describes recent second-generation developments in multicomputers (message-passing concurrent computers), including their architecture, design, and programming. These second-generation systems have substantially higher node performance and a greatly improved relationship between communication and computation performance, in comparison with the Cosmic Cube and commercial first-generation multicomputers. Finally, he reviews computational models and programming techniques for existing multicomputers.

- Ernst Mayr (Stanford and J. W. Goethe Universities) surveys the theoretician's view of parallel computation in Chapter 2 entitled **Theoretical Aspects of Parallel Computation**. He first introduces various parallel machine models including Parallel Random Access Machine (PRAM), fixed interconnection networks, as well as cloning Turing machines and their relations. He then presents several fundamental parallel algorithms within the context of the PRAM model; the algorithms include doubling, path doubling, parallel prefix computations, list ranking, sorting, and the Euler contour path technique. His next topic is DTEP (the dynamic tree expression problem) which exemplifies $\mathcal{NC}$, the class of problems solvable in polylogarithmic time on a PRAM or Hypercube with a polynomial number of processors. Ernst's fourth topic is the complexity of parallel scheduling and $\mathcal{P}$-complete algorithms, and the implications that such algorithms and/or their underlying sequential programming paradigms cannot be efficiently parallelized. As a consequence, such problems pose serious challenges for the efficient use of parallel architectures. This leads into Ernst's final topic: seeking fast and efficient parallel algorithms giving good approximations for such problems. He studies several examples and shows that the situation in parallel computation very much parallels that for $\mathcal{P}$ versus $\mathcal{NP}$.

- Bill Dally (MIT) writes Chapter 3, **Network and Processor Architectures for Message-Driven Computers**. The four sections in this chapter touch on aspects of scheduling and resource management problems, both crucial in parallel computation. Bill first addresses the design of multicomputer communication networks that give minimum latency and maximum throughput for a given wiring density. He then proceeds to review topology, routing algorithms, and flow control mechanisms. This discussion supplements Seitz's treatment in Chapter 1 of a communication network for the Ametek 2010. Bill then describes methods for analyzing the performance of multicomputer communication networks and discusses the design of communication controllers. This is followed by an efficient implementation of scheduling mechanisms in the message-driven processor, and the operating system that supports the MIT J-Machine. Finally, Bill describes an approach to programming on fine-grain message-passing machines.

- Lennart Johnsson (Yale and Thinking Machines) writes Chapter 4 **Communication in Network Architectures**. He discusses efficient communication in parallel computer architectures with interconnection networks in the form of Boolean cubes or related networks. Algorithms are classified with respect to their communication structures such as multidimensional lattices, butterfly networks, data manipulator networks, hierarchical grids (pyramid networks), and various tree structures for copy, or reduction operators on sets. Optimum routing and scheduling algorithms are given for the emulation of these structures under various loading conditions of the network. He also discusses optimum routing algorithms for arbitrary communication patterns.

- Yaser S. Abu-Mostafa (Caltech) gives us a critical introduction to neural network models in Chapter 5 entitled **Neural Networks**.[1] He starts with examples such as perceptrons, the Hopfield model, and multilayered networks. He then proceeds to address the capacity, computability and limitations of neural nets. Finally, Yaser covers some applications of neural networks which involve a learning phase when the network is tuned by a learning algorithm and training examples to compute the desired function. Yaser exposes some significant limitations in using neural networks to find optimum solutions to hard optimization problems. Indeed, he shows that in trying to use a neural network to solve a problem that is inherently exponential, the size of the network has to be exponential.

- Dick Lyon (Apple Computer) contributes Chapter 6 entitled **VLSI and Machines that Hear**. Dick first supplements Chuck Seitz's treatment of the AT measure of VLSI complexity. He then gives us his insight on some novel RAM cells and analog circuits. Finally he presents an introduction to parallel analog computation in hearing: an electronic cochlea.

- Bryan Ackland (AT&T Bell Laboratories) examines two specific knowledge based design projects underway at Bell Laboratories in Chapter 7, entitled **Knowledge Based VLSI Design Synthesis**. The first system, *Cadre*, is being developed to synthesize automatically full custom layout from a hierarchical structural circuit description. It is being built as a collection of cooperating expert agents communicating via a central coordinating design manager. The second system, *Synapse*, is being developed to generate circuit structure and layout from a high-level behavioral description. Here, the emphasis is on formal design representations and transformations that will advance the state of a design in a provably correct fashion.

The special character and location of this meeting were triggering factors in generating a large amount of enthusiasm and excitement in the audience of 49 attendees representing the following universities: Berkeley, Caltech, MIT, Stanford, and Yale, plus Calgary, Concordia, Simon Fraser, McGill, Montreal, Victoria; together with researchers from AT&T Bell Labs, Bell Northern Research, the Canadian Microelectronics Corporation, HP Labs,

---

[1] This chapter was written and reviewed in cooperation with David Schweizer, also from Caltech.

Schlumberger Palo Alto Research, SRI International and SUN Microsystems.

Acknowledgements

First and foremost we thank the contributors—Yaser Abu-Mostafa, Bryan Ackland, Bill Dally, Lennart Johnsson, Dick Lyon, Ernst Mayr, and Chuck Seitz—for their patience, insight, and companionship both at the workshop and whilst the book was being compiled. Bruce Spatz, the Managing Editor from Morgan Kaufmann, deserves special credit for his diligence and unrelenting confidence in the end result.

The workshop was supported through a Strategic Grant from the Natural Sciences and Engineering Research Council of Canada, and would not have been possible without their generous and enlightened funding policy. Detailed organization and day to day running of the workshop was left in the very capable hands of Eileen Coe. The talks were recorded as they were given, transcribed by Bev Frangos, Francie Hill, and Lorraine Storey, all secretaries in the Computer Science Department at the University of Calgary, and rendered into LaTeX by Mark Brinsmead, Brian Graham, Mike Hermann, Cameron Patterson, Glen Stone, Konrad Slind, and Simon Williams, all graduate students in Computer Science at the University of Calgary. We thank them for their dedication and skills in getting the job done accurately and quickly. Last but not least, we want to thank Brian Wyvill from the University of Calgary (who is also a world class mountaineer in his spare time) for giving us a dazzling slide show of big wall climbs in Yosemite at the final banquet. He also gave us some of the fruits of his professional skills by rendering the cover of this book. It took massive amounts of sequential computer power on an Evans and Sutherland machine to render the images into their final form. Oh for the ideas in this book to be realised!

Robert Suaya   Graham Birtwistle
Computer Science Laboratory   Computer Science Department
SRI International   University of Calgary

Contents

Preface   **v**

1 Concurrent Architectures   **1**
Charles L. Seitz
     1.1   Introduction to Concurrency 1
     1.2   Multicomputers 21
     1.3   Concurrent Programming 48
     1.4   Application Programming 65
     Bibliography 81

2 Theoretical Aspects of Parallel Computation   **85**
Ernst W. Mayr
     Introduction 85
     2.1   Parallel-Machine Models and Complexity Classes 86
     2.2   Fundamental Parallel Algorithms 100
         2.2.1   Doubling 101
         2.2.2   Pointer Jumping or Path Doubling 104
         2.2.3   Parallel Prefix 105
         2.2.4   List Ranking 107
         2.2.5   Parallel Sorting 109

2.2.6   Euler Contour-Path Technique 110
2.3   The Dynamic Tree Expression Problem 111
2.3.1   The Generic Problem 111
2.3.2   Applications of DTEP........................115
2.3.3   Application to Algebriac Straightline
Programs 120
2.3.4   General Remarks About DTEP 125
2.4 $\mathcal{P}$-Complete Algorithms.................................125
2.5   Parallel Approximation Algorithms....................131
Bibliography 136

3 Network and Processor Architecture for Message-Driven Computers   140

William Dally

Introduction 140
3.1   Wire-Efficient Communication Networks for
Multicomputers...141
3.1.1   Topology.................................145
3.1.2   Routing...................................150
3.1.3   Flow Control..............................152
3.1.4   Deadlock Avoidance.........................158
3.1.5   Wire-Efficient Topology 165
3.1.6   Summary 169
3.2   Analysis of Multicomputer Communication Networks ... 169
3.2.1   Parameters 172
3.2.2   Latency Calculation..........................172
3.2.3   Latency of an Indirect Network...............177
3.2.4   Latency of a Direct Network 179
3.2.5   Summary 182
3.3   Design of Communication Controllers 184
3.3.1   Router Organization 184
3.3.2   Synchronization 190
3.3.3   Driver Circuit Design 196
3.3.4   Summary 199
3.4   The Message-Driven Processor (MDP) 199
3.4.1   Processors are Inexpensive 200
3.4.2   Fine-Grain Programs Expose More
Concurrency 204
3.4.3   The Message-Driven Processor (MDP)........206
3.4.4   JOSS: The Jellybean Operating System 214
3.4.5   Summary 218
Bibliography 219

4 Communication in Network Architectures   **223**
S. Lennart Johnsson

  Acknowledgement..223
  Introduction..223
  4.1   Communication Requirements in Scientific
          Computation..230
          4.1.1   An Explicit Navier-Stokes Compressible
                  Flow Solver..................................231
          4.1.2   Stress Analysis by the Finite Element
                  Method.......................................233
          4.1.3   Acoustic Field Computation by an
                  Alternating Direction Method.................235
          4.1.4   Lattice-Gauge Physics........................236
          4.1.5   Linear Algebra...............................241
          4.1.6   Permutations.................................242
          4.1.7   Summary......................................243
  4.2   The Value of Locality in Computation.................245
  4.3   Networks...247
  4.4   Address Maps...251
  4.5   Graph Notions..259
  4.6   Lattices...263
          4.6.1   One-Dimensional Arrays.......................264
          4.6.2   Multidimensional Arrays......................265
          4.6.3   Arbitrary Lattices...........................268
  4.7   Butterfly Network Emulation..........................276
  4.8   Tree Embeddings......................................281
          4.8.1   Spanning Graphs..............................285
          4.8.2   Spanning Graphs Composed of n Spanning
                  Trees..289
          4.8.3   Summary of Topological Characteristics
                  of the Communication Graphs..................297
          4.8.4   One-to-All Broadcasting......................297
          4.8.5   One-to-All Personalized Communication........299
          4.8.6   All-to-All Broadcasting......................301
          4.8.7   All-to-All Personalized Communication........303
          4.8.8   Summary......................................308
  4.9   Pyramid Embeddings...................................308
          4.9.1   Embedding a $\mathcal{P}(k,d)$ Hyper-pyramid in a
                  Boolean Cube with Minimal Expansion
                  and Dilation d.............................315
          4.9.2   Embedding Multiple Pyramids with
                  Unit Expansion...............................320
  4.10   Permutations...321

	4.10.1	Cyclic Shifts Along an Axis of Arrays Embedded by Gray Code Encoding	322
	4.10.2	Conversion Between Binary Encoding and Binary-Reflected Gray Code Encoding	322
	4.10.3	Matrix Transposition	325
	4.10.4	The Single Path Transpose (SPT) Algorithm	327
	4.10.5	The Dual Paths Transpose (DPT) Algorithm	328
	4.10.6	The Multiple Paths Transpose (MPT) Algorithm	328
	4.10.7	Dimension Permutations	333
	4.10.8	Extended-Cube Permutation Algorithms	352
4.11	Emulation with Wafer Scale Integration		352
	4.11.1	Host Architectures	353
	4.11.2	Emulation of Two-Dimensional Meshes	356
	4.11.3	Butterfly-Based Algorithms	358
	4.11.4	Summary	359
4.12	Shared Memory		362
	4.12.1	Shared Memory Emulation	364
	4.12.2	The Fluent Abstract Machine	369
	4.12.3	A Radix Sort by Multiprefix	371
	4.12.4	Functionality of the Switches	372
	4.12.5	Message Decoding and Scheduling	373
	4.12.6	Area Requirements for a Switch	375
	4.12.7	Global Architecture	376
	Bibliography		379

5 Neural Networks — **390**

Yaser Abu-Mostafa
David Schweizer

	Introduction	390
5.1	Networks and Neurons	390
5.2	Feedback Networks	391
5.3	Choosing the Stable States	394
5.4	Feedforward Networks	396
5.5	Back Error Propagation	400
5.6	Collective Computation	404
5.7	Nearest Neighbor Search	405
5.8	Traveling Salesman Problem	406
5.9	Limitations	411
	Bibliography	414

6 VLSI and Machines that Hear   416

Richard Lyon

6.1   VLSI Complexity and Area Cost 416
6.2   RAMs and Circuits 419
       6.2.1   Analog Circuits 419
       6.2.2   RAM Cells...................................420
6.3   Analog Computation in Hearing.......................427
       6.3.1   VLSI Implementation 428
       6.3.2   System Response............................433
Bibliography..441

7 Knowledge Based VLSI Design Synthesis   442

Bryan Ackland

Introduction..442
7.1   SYNAPSE–Objectives and Techniques 443
       7.1.1   Leaf Cell Synthesizer........................444
       7.1.2   Representations of Desired Behavior..........445
       7.1.3   Representation of the Circuit Technology......445
       7.1.4   Transformation Technique....................448
7.2   CADRE–Custom Layout Synthesis 449
       7.2.1   Cooperating Agents 450
       7.2.2   Floorplanner 452
       7.2.3   Leaf Cell Layout Agent......................456
       7.2.4   Design Manager 464
7.3   Future Directions 468
Bibliography..469

CHAPTER 1

CHARLES L. SEITZ
California Institute of Technology

Concurrent Architectures

1.1 Introduction to Concurrency

Roberto Suaya has asked me to devote this first section to an overview of why we are all here. I shall try to do so by examining why we might be interested in concurrency in computing.

I have been fascinated for many years by concurrency in computing, at levels ranging from circuits to programs. In all this time I have found only two reasons sufficiently fundamental and compelling to justify (and sometimes convey) this fascination to others. The first reason is aesthetic and is related to the observation that sequential programs are usually overspecified. The second reason is practical: A realistic accounting of the relationship between the cost and performance of computations will lead you to conclude that computations that can be formulated with concurrent (or parallel) threads are physically less costly to perform than computations that have only a sequential formulation.

Let me put this practical reason aside for a few minutes so that I can say enough about the aesthetic reason to start you thinking about it (Figure 1.1). We know from the theory of computation that if we are given a sequential program, the transformations that we can perform to convert that program into one in which there are opportunities for highly concurrent execution are quite limited; that is, the semantics-preserving

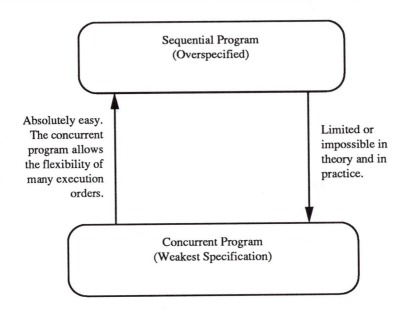

Figure 1.1   What we know for sure.

transformations available to us are weak. Nevertheless, people do translate sequential programs into concurrent programs; for example, vector supercomputers depend for their performance on compilers that extract the limited concurrencies in expression evaluations and iterations at the innermost levels of a computation. However, trying to transform the structure of an entire large program is not very fruitful. In general, we cannot prove the equivalence of two programs, such as a concurrent matrix-inversion program and a sequential matrix-inversion program, because they may use different algorithms. If it is an intractable task even to prove that the two programs produce equivalent results, then the likelihood that we can mechanically translate from the overspecified sequential program to the concurrent program is slight.

Instead of translating sequential programs, we must learn to write concurrent programs. In a concurrent program, we are aiming for a recipe for performing a given computation in which the concurrencies are exposed and sequencing is used only where actually required. Such a concurrent program is, of course, easy to run on a sequential computer just by binding the execution order. Running such a program on a concurrent computer, however, depends on a variety of complex factors. A simple analogy will clarify the problem.

We have all heard of "grain size," a vague term that can refer either to the size of the computing elements of a machine or to the size of the execution units in a concurrent program. A fine-grain program is required for a fine-grain machine, but a fine-grain program can also be executed without difficulty on either a medium-grain concurrent machine or a sequential computer. Because we have so much flexibility in so many different orders of execution, these highly concurrent, fine-grain programs adapt themselves like a liquid to the shape of the container, that is, to the shape of the particular target architecture. You get an essentially correct picture of the relationship between programs and machines if you think of concurrent programs as analogous to granular materials that can be poured into vessels of differing shapes and sizes.

The large sequential programs we have been writing for years are monolithic and will not pour at all. The moderately partitioned concurrent programs that we write for the medium-grain hypercube multicomputers are like gravel—gravel pours, but with some difficulty. The finest grain size we can achieve today in writing programs resembles a fine sand that pours like a liquid. The point of this analogy is that as a computing problem is expressed in terms of smaller execution units, it becomes feasible to run it on a wider variety of machines.

I believe that concurrency in computation is primarily an issue of program design and programming, of learning to express computations with the generality that results from exposing concurrency. It is only secondarily an issue of computer hardware, even though VLSI is an elegant and important medium for expressing highly concurrent computations. Expressing complex computations as highly concurrent programs is sometimes difficult and apparently becomes increasingly difficult at finer grain. So many real-life computing problems contain so much "natural" concurrency, however, that the difficulties of writing concurrent application programs have been overrated. Look on the bright side: After many years of writing sequential programs, you may become bored—at least I did. If you want a new challenge in programming, try writing some concurrent programs; it can restore your interest in programming.

Now let us turn to the second reason for being interested in concurrency that has struck me as being both fundamental and compelling, a rather hard-nosed practical reason involving the cost and performance of computations. Let me start by mentioning an ancillary argument that I cannot regard as fundamental, compelling, or even practical: Some research papers (a friend of mine even wrote a joke paper along this line) claim that because VLSI has taken the cost of a digital gate asymptotically to zero, we can use an arbitrary number of them; therefore, we should use parallel architectures. Does that sound familiar? In fact, however, a satisfying explanation of why concurrent computers might be preferred over sequential computers can be arrived at only by being tediously careful about relating the time and cost of a computation. My presentation of this argument will include some conjecture and a small amount of deliberate deception, which is why much of what I have to say is previously unpublished.

By my reading, the complexity theory for VLSI evolved in largest part from Clark Thompson's doctoral thesis [39] and has been confirmed and elaborated in the work of many other theoretical computer scientists. Thompson noted that for certain transitive computations (in which any output might potentially depend on any input), the chip area A and computing time T were related such that AT^2 must exceed a problem-dependent function of the problem size. The coefficients accommodate the characteristics of the technology. A is the actual area of the chip and could be expressed in square microns or in units normalized to the feature size of the process, such as λ^2, so that A is a measure of the complexity; the choice affects only the coefficients. T is the *latency* for completing the given computation—the time from when the inputs are applied until the outputs are produced. This is a time for a single problem instance; do not confuse T as being related in a simple way to throughput.

My argument hinges on this observation: Let us fix the technology for some constituent operation in a computation, for example, binary multiplication of 32-bit integers in a 1.2μm CMOS process. If $AT^2 \geq k$, we can re-express this relationship so that the area-time product $AT \geq k/T$. Now, this is very curious and interesting. Most of you could probably identify the area-time product as a cost; that is, a simple model of the cost of a rented apartment is that it varies linearly with the area and with how long you occupy it. If you use a part of a VLSI chip to do a binary multiplication, the cost of doing the multiplication varies as the area of the multiplier times the time required to get the multiplication performed. Let me emphasize that we are talking about the *cost per operation* of the *constituent* operations of a computation. We are not going to deal yet with other important issues that depend on the overall organization of the computation in which the multiplications are used, such as the duty factor with which the multiplier is used. We shall assume for now that we can use the multiplier continuously and repeatedly.

The area-time product goes as $1/T$; thus, this theory evidently claims that the cost of performing the multiplication operations varies as $1/T$. This result differs from the usual model that digital system engineers have applied for many years, but the traditional model does not quite jibe with reality. The AT^2 results do. Also, costs in the physical world are most often associated with energies, and we will examine some suggestive cases in which the energy required to perform a computation also varies as $1/T$.

A relationship that shows the cost of a computation decreasing with the time allowed should be plausible to you. As graduate students or postdoctoral fellows, you have learned that you can get as many computing cycles as you want if you are not fussy about when you get them or how fast they are. Computing cycles cost asymptotically nothing if you are not in a hurry for them. When we spend money for computers, we are actually buying response time, not cycles. We are buying the assurance that we can get an answer within certain bounded times. If the cost or energy per operation varies as $1/T$, we might call this the haste-makes-waste hypothesis. If you want computational results quicker, then you must pay more for them.

The obvious loophole in this relationship between cost and time is concurrency; but before getting to the loophole, let us examine some evidence in favor of the basic hypothesis.

In Chapter 2 of *Computational Aspects of VLSI* [40], Jeffrey Ullman presents an elegant, intuitive explanation for how these AT^2 complexity results come about. He describes three kinds of lower bounds (Figure 1.2). We are not going to be interested in the first two, but they provide a convenient way of sneaking up on the third. One kind of computation tends to be entirely limited by storage; that is, the computation might consist of a storage reference into a very large working set, or a name-value association. In that case, when you implement this computation in silicon, you end up with a very simple bound on the area. You are limited by how densely you can place storage cells on a chip. This is a simple area limit, and some computations do exhibit such limits.

The second kind of bounds are for computations limited by input/output. The assumption is that you can do no better than to get data into or out of a chip over the entire surface of the chip. We know that areally rather than peripherally distributed input/output pads are used by some companies. More interesting cases of areal input/output are chips such as Carver Mead's retina circuits [27], which perform optical input over the entire area of a chip. So, if you could perform input/output over the whole surface of the chip, you could think about taking a snapshot of the chip at later

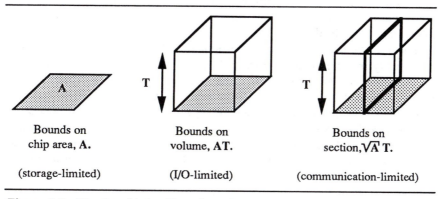

Figure 1.2   The three kinds of lower bounds.

and later times. Thus, the volume AT represents a limit on the amount of information that you can take into the chip over the course of the computation.

The most interesting kinds of computations are those in which, again, we have to look at the evolution of the state of information within the chip at successive times. Clark Thompson's model of VLSI computation is a particularly simple one, saying that "For each unit of time, one bit of information may be conveyed across each wire on the chip." Yaser Abu-Mostafa could tell you that this model is not quite realistic, because there are many ways of encoding information on wires. For example, we might encode information on a wire by the time of a signal transition or by analog signals. Nevertheless, Thompson's VLSI complexity model is reasonably realistic for conventional digital design styles.

The wires that we focus on in producing any of the AT^2 proofs are those that cross bisections. Imagine the chip as being able to exchange information between its left and right halves across a limited number of wires. Perhaps you could also show that for a particular computation at least a certain amount of information has to be exchanged between the circuitry on the left and the circuitry on the right on successive time units, and that there is no way to organize the computation such that less information would be passed across the bisection. I am by no means proving that you can do this for a particular problem; the discussion is just to give you an intuitive notion—courtesy of Jeffrey Ullman—as to why AT^2 bounds come about. But when you can do this, the total amount of information exchanged between the two halves is represented by a cross-sectional area. This area is T high and, for roughly square chips, $\sqrt{A}$ across. The bounds on this section are bounds on $\sqrt{A} \cdot T$. Because this expression would be ponderous to write over and over again, the theoreticians put these bounds in terms of the square of that measure, namely, bounds on AT^2. These are computations limited by their internal communication requirements.

A subject that will come up again later in my remarks and also in Bill Dally's is the significance of the bisection. Multicomputer communication networks are typically bisection-limited, just as chips are.

Because the two plan-view dimensions on a VLSI chip allow much more complexity in interconnection than does the vertical dimension, chips can be regarded as two-dimensional. I have heard people conjecture that the brain is efficient because it is wired together in three dimensions, but that is not so. The brain has a much more complex structure in two of its dimensions than in the third, just like a chip. In fact, it can be described as a folded sheet measuring a meter or so on a side. However, suppose we could build computers that were wired together in three dimensions. Instead of the bisection varying as the one-half power of the area, it would vary as the two-thirds power of the volume. Presumably you would sell these computers by the cubic meter, or some volume unit, so the cost per operation would be limited by the volume-time product, which varies as $1/\sqrt{T}$. So, we would do a little better, as we might expect, if we could wire computers together in three dimensions; however, there still is a dependence such that as we go to larger times, the costs per operation go asymptotically to zero.

Now that we understand how to interpret these AT^2 results in terms of the cost of an operation as a function of its latency, I want to turn to some energetics arguments.

I went through the interpretation of the AT^2 results first not because it seemed to me to be the easiest or strongest argument in favor of the cost of an operation being a decreasing function of latency, but because the effort of trying to understand the meaning of these AT^2 results back in 1980–82 first started me thinking along these lines. I should mention that I had many interesting discussions about these ideas with Professor Richard P. Feynman of Caltech between 1982 and 1985. As many of you know, Dick Feynman died recently. We miss him.

We shall look at two examples of relationships between the energy and time required for a computing operation, but for you to understand how these results differ from the conventional digital-engineering wisdom, you need to know a little bit about the usual notion of the power-delay product in circuits. Of course, this product of power and time has units of energy. In CMOS circuits this energy is equivalent to the *switching energy,* and it is in this ideal, negligible-static-power VLSI technology that I will discuss circuit energetics. This material may be familiar to many of you, but I am going to inflict it on you anyway.

Here is a favorite qualifying-exam question (Figure 1.3) that you can ask electrical engineering students when you become professors. This is a very simple circuit with a battery (voltage source), an ideal switch, a resistance, and a capacitance. The capacitance is initially discharged when

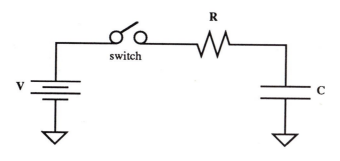

Figure 1.3   Simple energetics example.

we close the switch. (A variant on this problem has the battery replaced by a capacitance of the same value as the one shown, but initially charged to a certain voltage.) You ask the student this question about the energetics of this circuit: After the circuit reaches (approaches) its final state, how much energy has flowed out of the battery, how much has been dissipated in the resistance, and how much is stored in the capacitance? My rule of thumb is that if the student starts writing a differential equation, you should flunk the student on the spot.

A good student will note immediately that the final voltage on the capacitance is V, so the final charge on the capacitance is CV. We have to provide that charge from the battery at potential V, so the energy provided by the battery is CV^2. The student also knows that the stored energy in the capacitance is $\frac{1}{2}CV^2$ because the stored energy is the integral of vdQ or of $(Cv)dv$. Thus, exactly half of the battery energy ends up stored in the capacitance; the other half must then be dissipated in the resistance. This result is independent of the value of the resistance, even if the resistance is nonlinear or varies with time. It is important that you notice that this answer needed to appeal only to fundamental physical laws: conservation of charge, the electrostatic definition of energy, and conservation of energy.

At this point you might ask the student what happens if the value of the resistance is zero. The final stored energy in the capacitance will still be $\frac{1}{2}CV^2$. What happened to the other half of the energy? It is not dissipated in the wires, because these are superconducting wires; R really is zero. The question does not appear to admit an answer, but it is important that the student recognize that conservation of charge and energy are fundamental, so that there is no way to avoid losing this $\frac{1}{2}CV^2$ energy. What might happen physically in the case where R is zero is that the inductance of the connection between the battery and the capacitance could not be zero, so the charge would slosh back and forth for a while until the other half of the energy managed to radiate itself away.

With this exam question as background, normal digital-circuit energetics are nicely illustrated by a CMOS inverter (Figure 1.4). When we want to switch the output from 0 to 1, the power supply provides charge CV at potential V, so that $\frac{1}{2}CV^2$ is dissipated in the p-channel transistor (which you can think of as a time-varying resistor), and the other half of the energy ends up being stored in the capacitance of whatever the output is connected to. In switching the output back from 1 to 0, the $\frac{1}{2}CV^2$ that was stored in C simply gets dissipated in the n-channel transistor. Every switching event has inevitably associated with it a dissipation of $\frac{1}{2}CV^2$. Do not let anybody deceive you that superconductors provide a way to avoid this energy cost; it is an inevitable result of switching the potential difference across the transistors. Of course, if you had inductors you could do better.

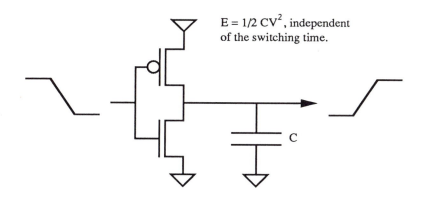

Figure 1.4   Normal digital-circuit energetics.

From the dissipation associated with this normal digital switching comes the notion of the switching energy, $\frac{1}{2}CV^2$, where C is generally taken to be the input capacitance of the circuit. This much energy is associated with an input switching event, which you can think of as a very primitive constituent of a computation. This energy is independent of how fast you operate the circuit. When you look at any catalog of circuits or projections of high-speed computers, the power-delay product is the figure of merit that is used. The switching energy is the same measure as the power-delay product, which is defined as the power dissipated by the circuit when switching at maximum rate times the delay of the device. You can see that this is the same as the energy dissipated through one switching event; thus, this view does not result in any dependence between energy and how fast the circuit operates.

There is a sneaky way to revise the way in which circuits operate. I call this revision *hot-clock* circuits [33]. A great majority of the signals on VLSI chips make transitions in synchrony with a clock. Instead of driving a capacitive load from the power-supply rails, we can drive it directly from an externally provided clock signal. This is a practical technique that works very well for driving the long select lines and word lines in memories; for driving the literal, implicant, and output lines in PLAs; and for driving the long buses in stored-program computers. You can drive all of these signals in synchrony with some externally provided clock.

We will use a model of a CMOS switch as an ideal switch in series with a resistance. In a CMOS hot-clock circuit (Figure 1.5), the switch is placed between a clock signal and a line that is to be driven. The clock signal has a certain rise time and fall time. If the enable signal is a 1 during the

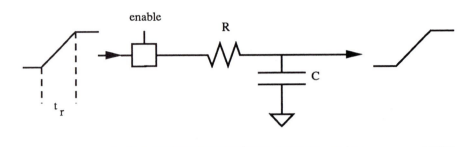

Figure 1.5   CMOS hot-clock circuit energetics.

interval shown, the clock is reproduced at the output, slightly delayed by just RC. On the other hand, if the enable signal is 0, you get an all-zero signal for that period. Now, by perfectly ordinary circuit analysis you can discover that for the interesting case in which RC is much less than the rise time t_r, or the fall time t_f, the energy associated with a single transition of this circuit is RC/t_r times the usual CV^2 energy term. If $RC = t_r$, it looks like this coefficient becomes 1, but the expression is not valid in that range. This formula $E \approx (RC/t_r)CV^2$ is valid only for $RC << t_r$.

RC is a built-in constant of this circuit. If I decide that I want to operate the circuit more slowly, I can extend the clock period; but when I do, I am also entitled to extend the rise times and fall times. Thus, I can operate this circuit in a fashion in which the energy per switching event varies inversely with the time! This very simple circuit seems to defy the usual notion that the switching energy is independent of the speed of operation.

Let me stress that the energy that is decreasing with increasing clock period is the energy per switching event. Because there are also proportionately fewer switching events per unit time, the power dissipated is decreasing quadratically with the clock period. Here we have another case of the $1/T$ dependence of cost (energy) with operation time for the very simple operation of driving a wire. For conventional circuits, the energy associated with driving the wire is independent of how quickly the wire is driven. But, if you are a little sly about how you drive the wire, you can get the dependence down to $1/T$.

When we were regularly designing hot-clock nMOS circuits, the typical values of RC that we would use (in 3-micron nMOS) would be 0.5ns. Making $RC = 0.5$ns was not at all difficult. That is roughly a 10-square transistor driving a select line in a RAM. This is a perfectly practical technique.

For high-performance chips in this technology, typical values of t_r and t_f were 5ns, with a clock period of about 50ns. The ratio is thus about

1/10. In fact, having designed the circuit in that way, if we operate it progressively more slowly, the power that is used by the chip decreases because the number of clock transitions per unit time decreases, but it also decreases because the energy associated with each transition decreases. In other words, it is a power-time-squared-bound design style. The energy associated with each switching event could sensibly be made a hundredth or thousandth of what people were used to in that technology. This high-performance technique would also make sense for watches and other devices that have no serious speed requirement and are powered by batteries.

Before I get found out, I had better show you where the dissipation is hiding. I do not want to be accused of selling perpetual-motion machines.

Do you know how to eliminate pollution from a given area? You out-law the industries that are producing the pollution. Of course, you do not eliminate the demand for the goods that the industries produce, and so those industries go offshore, as we say in the United States. We export them to Canada or other places. Hot-clock logic exports the task of switching the clock potential. We can deal with this problem in a variety of ways, but we have very few ways to deal with it on a high-complexity MOS chip. So there is no magic—I am not trying to sell you a perpetual-motion machine, I am just going to factor away any of the dissipative parts.

When you do factor the clock switching off the chip, you can attach the clock generators to resonant circuits that use inductances, or even to something with flywheels or motors if you prefer the mechanical analogy. You can operate the external clock driver as a dissipationless circuit, except for small losses. In actual practice, because we can export the driving of these high-capacitance loads external to the chip—where we can use technologies that are made for driving lots of power (as opposed to MOS technologies, which are made for high complexity)—our experience is that the chips that result are somewhat faster than those designed using conventional design styles. We have exploited an opportunity to specialize the technology to the function. We can use resonant drivers externally; we cannot fit flywheels or inductances very well inside the chip.

I have to inflict just one more such example on you. This is one of my favorites, a mutual-exclusion circuit (Figure 1.6). I shall go through this example very quickly and not worry too much about the analysis. In building self-timed circuits, we like to have an element in which you can have concurrent requests but in which acknowledges are mutually exclusive. Also, the paired requests and acknowledges (such as R_1 to A_1) have to follow a particular protocol, such as a request on R_1 being followed by an acknowledge on A_1 to grant the use of some resource to requester 1.

The implementation of a mutual-exclusion element employs a bistable circuit that is composed of cross-coupled NOR gates. When both R_1 and R_2 are 0, both of the acknowledge output signals are 0. The condition

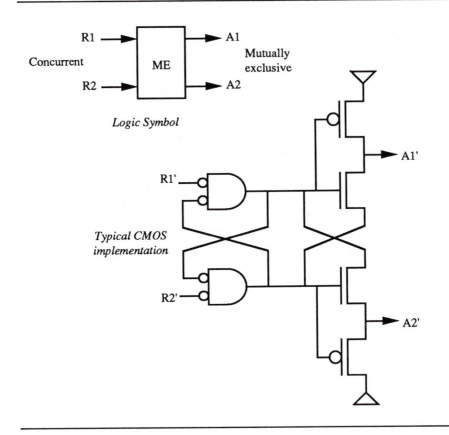

Figure 1.6   Mutual-exclusion circuit.

to produce A_1 is that R_1 must be 1 and A_2 must be 0. Because of the symmetry of this circuit, you would not be surprised, then, that if R_1 and R_2 were to switch from 0 to 1 at the same time, the voltages at the two NOR outputs would also start to switch to 1 at the same time. In fact, in MOS circuits they both switch to a half-voltage or metastable point, and it takes a certain amount of time for the circuit to exit from this metastable point. Neither acknowledge output should be produced during the metastable period, so an additional circuit is used to filter out the metastable states. This second circuit detects when a solid decision has been made by producing one of the two acknowledge outputs only after the two voltages differ by a sufficient amount.

If you look at what is going on in these circuits when R_1 and R_2 switch to 1 "simultaneously," conventional circuit analysis predicts that the circuit simply remains in the middle state forever. Of course this is physically unrealistic because the space of zero time difference is vanishingly small, and real circuits have some noise. It is more useful to examine the probability of the response as a function of the difference in time between R_1 and R_2 switching to 1. In this case, an initial state is established from which the circuit will evolve exponentially away from the metastable point until the two voltages are sufficiently different that you can reasonably say, "It cannot come back into this forbidden region, so it is safe to say that a decision has been made."

Let me state the results of this probability analysis in much the same form as it is presented in section 7.5 of Mead and Conway [26]. For uniform distributions of differences in time between R_1 and R_2 switching to 1, there is an exponential distribution of decision times; that is, the probability that a decision has *not* been made by time t for $t \geq 0$ is $e^{-\rho t}$, where ρ is the rate of a Poisson process. Equivalently, the probability can be written as $e^{-t/\tau}$, where τ is the time at which the probability is reduced to $1/e$. This τ can be related easily to circuit parameters. Then you can show for the example electrical circuit (as you have probably anticipated) that the energy associated with making one of these decisions varies as $1/\tau$. In other words, if you want to be able to make this decision faster, you must design a circuit in which τ is smaller, but to do so you need to expend more energy *for each decision*. For example, if you expend twice as much energy for each decision, you can drive ρ up or τ down by a factor of two (within technology limits).

You can analyze classes of mutual-exclusion circuits, and, in fact, classes of physical mechanisms for mutual exclusion. A few years ago, my friend Dick Feynman became intrigued by my analysis for electrical circuits (or perhaps by an amusing mechanical mutual-exclusion device in my office) and produced a far more general result that applies even to quantum-mechanical systems.

You end up with the same result once again. Here it is cropping up in the form that the energy associated with each mutual-exclusion operation varies as $1/T$. If you are not in a hurry for a decision, you can use very little energy. If you are in a big, hot, busy hurry to get that decision made, you are going to expend more energy doing it. This is a wonderful result for those of us who like to procrastinate in making decisions. However, there certainly are good reasons for people to be in a hurry. Our human metabolism paces our existence at certain rates, so indefinite procrastination is not feasible.

Let me summarize. I regard the AT^2 results as rather well established in the VLSI complexity model (at least if you judge by the number of papers published). It is purely conjecture that an invariant exists in the form that the energy per operation times the operation time exceeds some constant. Of course one wonders whether the AT^2 bounds are one case of a more general result, although machines wired in three dimensions (and nature has given us three dimensions in which to build machines) reduce the dependence to one that is not quite so dramatic. However, in approaches such as hot-clock logic or mutual-exclusion circuits, in which we have strong reasons to believe that we have examined the limiting case, we end up seeing this $1/T$ dependence again. These circuits illuminate the limiting case in which we might try to save energy by performing operations more slowly, and these results differ from the conventional power-delay product results. Finally, we notice empirically that the costs of doing all sorts of things in computing do tend to decrease with increasing T.

By the way, the observation that the cost per operation decreases as the time per operation increases is exactly the reverse of Grosch's Law, a purely empirical observation that remained more or less valid until about 1975, or roughly until the advent of the single-chip processor. This "law" stated that the cost per operation was the smallest on the biggest, fastest machines. In fact, the reverse is true in today's technologies. The lowest cost per operation is associated with rather small and slow computing engines.

Finally, it is time to point out that if the cost per operation varies as $1/T$, you can also make the cost of a computation vary as $1/N$ for problems that distribute with N-fold concurrency. This is probably already evident to many of you, since I have consistently been dealing with the *cost per operation*. If we fix the time allowed to perform a computation, and if the computation can be formulated to be performed in N concurrent threads, then the cost per operation—and hence the total cost of the computation— can be made to be $1/N$th of what it would be if the operations had to be performed sequentially, each N times faster than if they were performed concurrently. Similarly, if you fix the cost for such a computation, you can perform it N times faster if you exploit concurrency.

We have finally crept up on the loophole, which is that if we have a problem that distributes, there is a way to compute it at a lower cost or faster by distributing it—but I can see that many of you are still skeptical. Let me try a more detailed cost-accounting exposition for those of you for whom the cost-per-operation point may be too direct to see.

Here is the down-to-earth explanation of AT^2 complexity results for hard-core engineers: You first take as a given that, for example, to multiply two n-bit binary numbers, any multiplier design with chip area A and latency T is constrained such that $AT^2 \geq kn^2$. As it happens, binary multiplication has $AT^2 \geq O(n^2)$. Furthermore, these are tight bounds over a very interesting range of A and T, namely, any multiplier that deserves to be called "fast." Preparata [30] shows one family of multipliers that meet these bounds, and others have constructed other families, so you can find in the literature multiplier designs that are right on the AT^2 limit (sometimes within a log factor or two or three).

Now you are given a problem: Let us say that you have an FFT that requires 10^9 multiplications, and you must complete it in one second, perhaps because otherwise the target would get away. If you dislike this reason, you might say, "I have this FFT to perform, and I have to do it within four hours; otherwise, I will get hungry before it finishes." At any rate, you have to pick *some* time limit. I get to pick the numbers here, and I picked one second. When you are talking to hard-core engineers, you have to use numbers. You cannot just get away with symbols. I will put the symbols in parentheses for the rest of you, but the hard-core engineers will look only at the numbers.

An arrogant engineer might say, "Terrific, we know how to do that. We'll build a 1ns multiplier and just do it straight out." The engineer starts looking in the catalogs—arrogant engineers do not design their own multipliers; they buy them—but he has difficulty finding *any* 1ns multipliers in the catalog.

A little later a clever engineer comes along, and she notices that the problem can be done in 100 approximately equal parts with relatively sparse interaction. (For 100 here, the rest of you can read N. The 100 is just for the benefit of the engineers.) She comes up with a design in which she uses 100 100ns multipliers. You notice that the cost argument here is brutally direct: Each one of these multiplier circuits costs 10^{-4} as much (N^{-2} as much for the rest of you) as a 1ns multiplier, because of the AT^2 dependence, so a system with 100 multipliers costs 10^{-2} as much. (In symbols, you could use N multipliers that cost N^{-2} as much, so the system costs $1/N^{th}$ as much.)

There is nothing terribly new about approaches that use many parallel multiplier circuits in algorithmically specialized computers, such as FFT engines. People who have been doing signal and image processing have been building special-purpose computers for a long time in which they have employed, if not recognized the generality of, these engineering tradeoffs.

Let me explain now why this interpretation of the AT^2 results might be a bit controversial. Having said that these AT^2 bounds apply to all sorts of problems, then why not also to the FFT? In fact, the same AT^2 bounds apply to binary multiplication and to the FFT, which happens to be structurally the same problem. However, all I need to appeal to for this example is multiplication as a constituent operation of this computation, and that the multiplier circuits are AT^2-optimal. The other costs of wiring the multiplier circuits together are not limiting in this example, because the FFT is not AT^2-bound when N is only 100 and the computation is so large. To understand this point intuitively, you can check that each of the 100 multipliers has so many operations (10^7) to perform that the *latency* of the FFT computation is limited by the multiplier circuits rather than by its own internal communication requirements. Let me assure you that even when you include the appropriate coefficients of the technologies involved, the communication between the multiplication operations in a large FFT with only 100 multiplier circuits is quite sparse in comparison with the communication within the multiplier circuits themselves. It is sparse enough to be handled in interchip wiring technology, rather than the intrachip wiring that the multiplier will use.

You would also find that to do good engineering, you must pay attention to the coefficients and the range over which the AT^2 results apply. Let us look at the real-world constraints on these observations. If we were considering a large FFT, these results could not be applied indefinitely. For an n-point FFT, there are $n/2$ times $\log_2 n$ butterfly operations to perform, and up to $n/2$ of them can be performed concurrently, or in parallel. If $n = 10^6$ points, $\log_2 10^6$ is about 20, so that whole computation can be done in $10^6/2 \times 20$, or 10^7, operations. To employ the most parallel form, you do $n/2$ butterfly operations in parallel, move the data as required (no mean feat), and repeat this $\log_2 n$ or 20 times. So if we could do $n/2$ operations in parallel, why not use 500,000 elements capable of doing a butterfly operation (rather than the 100 multipliers in the example above)?

There are several different reasons why this ceases to work. For one, you also have bounds on AT and on A. The AT^2 results apply only over certain ranges of areas and times. For example, you could certainly build a $1\mu s$ multiplier circuit in $2\mu m$ CMOS technology in 10 mm^2. Then, according to an unrestricted interpretation of the AT^2 results, you should be able to build a 1ms multiplier in $10\mu m^2$, about the area of one transistor. At this point you realize there must be limits on how far the AT^2 results apply to the constituent operations.

Another reason that we have already discussed is that the multiplier circuit is not the whole show—there will be other circuitry to support the communication between the multiplier circuits. Obviously, anybody with any sense of greed is going to try to push N as high as possible until the constituent operations cannot be made smaller and still be AT^2-optimal, or until the amount of communication required to support the computation becomes limiting. We avoided this problem by using small N on a large problem, but if you reduce the problem size you will run out of concurrency sooner or later, and in the case of highly branched AI computations, perhaps sooner. Besides limited concurrency in the problem domain, you may also be limited by the target architecture; for example, the bisection of the communication network in a concurrent computer is limited by economic constraints, and may limit the amount of concurrent communication that may take place in a given algorithm.

You can, of course, do this computation as if the results of each multiplication were needed for all of the results that follow, but then you would be doing it in a way that is physically difficult. There are pathological computations, such as computing 2^{2^n}, in which there is surprisingly little possible concurrency. But for a computation such as an FFT, each result is not needed for the next step, and so you do not have to perform them as if they were. To do these operations sequentially requires us to rush each one of them, and each operation ends up costing more to perform.

Let me give you a punch line of sorts, which you can argue about to your heart's content: *Nature loves concurrency.* If you have this notion that there is a fixed cost associated with doing a given computation, it should not matter to you if you do 10^9 multiplications one after another 1ns apart, or whether you do them in chunks of 100 multiplications of 100ns each. But, in general, if you have a computation to perform that can be done in N approximately equal parts, with sufficiently sparse interaction (where the wires and circuitry that support the interaction are much less than those of the constituent operations), then these AT^2 results (not to mention the E varying as $1/T$ results) show that you can do it concurrently either N times cheaper at the same performance, or you can apply exactly the same technique to do it N times faster at the same cost. It is not that the computation has to be done one way or another; it is just that we have apparently been doing computations in a style that is physically the hard way. Nature loves concurrency, and does not reward you for treating all computations as being in that case in which each operation must be done serially with every other.

If these observations about the relationship between the cost per operation and the speed of an operation are a part of some future theory of the physical design of computers, we might ask whether such a theory sheds any light on what is happening in the development of computers beyond the microscopic examples that can be analyzed quantitatively.

Is the performance of supercomputers—the fastest conventional sequential computers—advancing? Only slight multiplicative gains occur per year in sequential speed. Gordon Bell has studied supercomputers carefully over the past several years and noted at a recent meeting that the designers are pushing clock rate by about 1.17/year. Machines such as the Cray Y-MP manage to get to three times the X-MP performance by increasing the instruction rate per processor by only about 1.5 and doubling the number of processors. For that matter, if you look at the X-MP versus the Cray 1S, you find that the peak performance of a single processor increased only slightly, from 160 to 210 megaflops. Most of the peak speed increase came about through parallelism—four processors rather than one. In other words, supercomputer designers are not doing very well by pushing clock speed, because that approach requires decreasing the latency of individual circuits against diminishing returns. Of course, speed-of-light wire delays are also an issue in these machines, although not a significant limitation inside chips. Overall, even supercomputer designers have been getting most of their performance gains by pushing N rather than clock speed.

Within the realm of parallel machines is an enormous range of possible systems. I brought this picture (Figure 1.7) of the Digital Orrery [1] mostly for the purpose of showing that I am not a MIMD message-passing bigot.

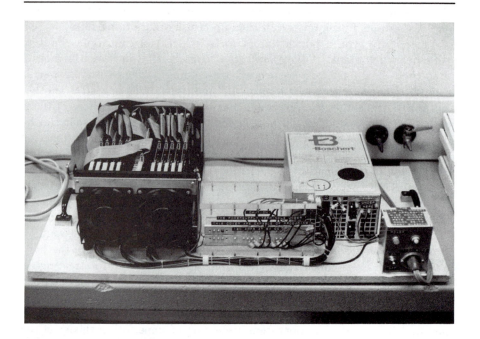

Figure 1.7 The Digital Orrery.

This is a SIMD message-passing machine built at Caltech the year that Gerry Sussman was there on sabbatical. This little box contains ten small computers that do floating-point operations. They are connected in a ring, which you can see in the wiring on the top of the card frame. Each of the computers is called a "planet computer," and there are ten of them because our solar system has a sun and nine planets. This machine was designed for calculating planetary orbits (more seriously, for investigating problems of the long-term stability of orbits). Gerry liked to call this machine "60 VAXes in a cubic foot."

For this particular machine, N is ten, the operation is SIMD, and the communication structure is a ring, because nothing more was required by the problem for which this machine was designed and built. Unless we had more bodies to keep track of, it would not help to make N larger. What makes the problem difficult is that one must compute orbits over very long periods—billions of years. Parallel machine design need not always be an issue of whether you are doing it shared-memory, message-passing, MIMD, SIMD, or with neural networks. The technical problems that we keep finding ourselves up against in making real systems—and my students and I typically like to make programmable computers because we like to write programs—are the difficult and ill-defined tradeoffs between cost, performance, application span, and programmability.

Engineering tradeoffs will be driven by the greed for speed, rather than a greed for massive concurrency *per se.* The tendency to push N is only a consequence of the greed for speed, and generally creates difficulty for application span and programmability. This graphic attempt (Figure 1.8) to portray the range of message-passing systems (adapted from [31]) is a log-log plot of the number of nodes versus the node complexity. The log-log form is a nice representation for this design space, because the hyperbolas of constant cost (constant total area) are straight lines. Another nice thing about log-log plots is that you can wedge a very wide range into a small space and do not need to be very precise.

If you look at the various kinds of VLSI chips that we know how to make, you find memory on the left side and conventional computers on the right side. Memory tends to have a very large replication of small elements;

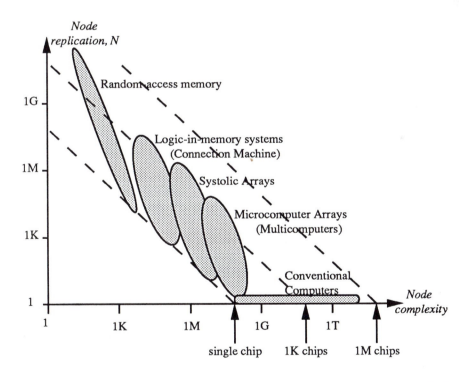

Figure 1.8 Taxonomy of message-passing systems.

conventional computers have a very small replication of large elements. People have been exploring a great variety of different architectures that are in the middle ground. The connection machine at least started out being a logic-enhanced memory. Along the line of constant total system cost, the systolic or computational arrays and the microcomputer arrays of similar cost spread down and to the right.

The one thing that you notice about these alternatives is that peak performance increases principally with replication, but this is performance on some problem that the machine is able to do at all. Generality tends to increase principally with node complexity. Some designers prefer maximal performance on a limited set of big problems. Others seek generality. The way in which you devote silicon area can be biased in either of these directions.

If you want to write a program that can be run reasonably efficiently on any machine in sight, you might design the program to exploit relatively fine-grain concurrency. If you have lots of concurrent processes—many more concurrent processes than you have computing engines to run them on—you can just scatter the processes at random to the computing engines you have, so that load balancing is neatly assured by the weak law of large numbers. Load balancing is relevant to generality, because you must pay some attention to utilization of the concurrent resources. It does not help you to have ten times as many computing nodes if increasing the distribution of your computing problem decreases the utilization in proportion.

Also, in practice, the application span is inevitably related to the programmability of a parallel machine. Many people believed that parallel machines were no good for irregular problems, I suspect because some of the older literature in parallel computing gave the impression that the usual grist for parallel mills was inverting matrices and other similar, highly regular computations. But the notion that irregular computations are not within the application span of parallel machines is erroneous. It is the amount of concurrency, not the regularity, in the formulation of a computation that matters.

In highly regular computations you may do *a priori* balanced partitioning of the work from an analysis of the problem, and this is the sort of problem you usually find described in papers. In a highly irregular but highly concurrent computation, you may need to "use up" some of the concurrency by mapping several processes per computing node to satisfy requirements of load balance and consequent "efficiency" in the utilization of the machine. Bill Athas has shown you some of the statistics of running irregular problems in which the structure of the problem is unknown *a priori* because it is the answer, such as: "What is the tree of all solutions to this particular problem?"

1.2 Multicomputers

In this section I will give you a status report on message-passing multi-computers, a category of distributed-memory, MIMD (multiple-instruction, multiple-data) concurrent computers. What these words will mean to some of you is that I want to tell you about the new "hypercubes." I am going to concentrate in this lecture on multicomputer architecture, hardware, and system software. My next two lectures will be concerned with multicomputer programming methods and applications.

This would have been a good talk to give in an evening session, partly because you could get a better look at the photographic slides if the room were darker, and partly because—unlike subjects such as area-time complexity or programming—this is lightweight, recreational material. If I slip into what sounds like a salesperson's jargon, it is because I am quite proud of what my Caltech research group and the multicomputer manufacturers have been able to accomplish with the introduction of the second-generation multicomputers.

Let me begin with definitions, terminology, and history. Both of the common MIMD architectures, the shared-memory *multiprocessor* and the message-passing *multicomputer*, are obvious extrapolations (Figure 1.9) of the conventional single-processor, or von Neumann, computer. Even the computers of the 1960s often had a memory bus to which multiple memory units could be attached, either for modular memory expansion or for increased performance. Devices such as disk controllers, which behave like processors in generating their own addresses to read and write memory, could also be attached to the memory bus. From this structure it was only a small step to machines that had multiple instruction-interpreting processors. So, one MIMD architecture that appeared early in the evolution of concurrent computers was the shared-memory multiprocessor, so called because its multiple processors share a memory that spans a single address space. To provide sufficient aggregate memory bandwidth, most multiprocessors require as many memory units as processors. Multiprogramming and time-sharing systems provided an immediate way to benefit from multiple processors, but the application of multiple processors to a single task is still seldom used outside of the research community.

There are as many ways to design a multiprocessor as there are ways to design the switch that allows any processor to access any memory. Bus contention causes a processor to idle until the memory-access cycle completes; hence, the performance of such a multiprocessor eventually becomes limited by the throughput of the bus. A high-performance bus will suffice for up to a modest number of processors, say, $N = 10$ or so, before the bus is no longer able to support the volume of memory-access traffic that originates from the processors. The volume of memory-access traffic can be reduced by using a local cache memory with each processor, but either

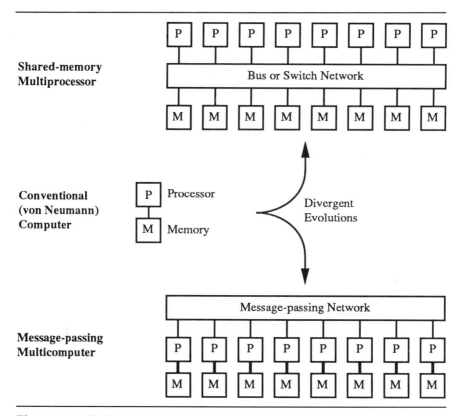

Figure 1.9   Evolution of the multiprocessor and multicomputer.

caching shared variables must be disallowed or some scheme for cache coherence (such as the cache "snooping" on the bus transactions) must be employed.

When N is large enough that the bus becomes saturated by the volume of cache-miss and shared-variable traffic, the bus can be replaced with a switch network that provides concurrent paths between the processors and memories. The $N \times N$ crossbar switch connecting N processors to N memories is one such organization, but is out of favor with the VLSI community because its cost scales as N^2. Various indirect binary n-cube networks that are N wide by $n = \log_2 N$ deep, such as Banyan or Omega networks, are preferred for large-N multiprocessors. However, the latency of the switch eventually limits high-performance multiprocessors to $N \leq 10^3$ or so. The latency and throughput of processor-to-memory communication—the so-called von Neumann bottleneck—is widely recognized as the most severely limiting factor in computer performance. It certainly does not help to interpose the longest wires or largest component of a concurrent computer between the processors and the memories. (More fundamentally, the flaw is the absurd concept that all memory references can be made equivalent by making them all equally expensive.)

The message-passing multicomputer explores a different evolutionary path. As its name suggests, it is a multiple-computer structure in which the interprocess communication takes place by message passing. In common with computers on a network, the individual computers in a multicomputer are called *nodes.* The physical advantages of the multicomputer architecture derive principally from the localization of a processor and memory, for example, depending on the technology and "grain size" of a node, to a single circuit board, multichip carrier, or even a single chip. Because multicomputers use different mechanisms for processor-storage and interprocess communication, they scale successfully to large N; it is certainly feasible to build high-performance multicomputers with $N = 10^6$. Even if you count the world population of machines in which $N \geq 64$, there are substantially more multicomputers than multiprocessors; in fact, there are more multicomputers than all other types of highly concurrent machines combined.

Of course, people started connecting computers together with networks or data channels even before they started building multiprocessors. I do not know precisely when this first occurred, but my guess is that it was shortly after the first occasion in which two computers were installed in the same room. The design space of multicomputers is a large one: It includes alternative organizations for both the message-passing network and the node computers. For example, the nodes may be either single-processor computers or multiprocessors. If all of the nodes are the same, we refer to the machine as *homogeneous.* Because the processors in a multiprocessor may share the same code, it is relatively more important for multiprocessors to be homogeneous than it is for multicomputers. The mixed types of computing nodes in a *heterogeneous* multicomputer need only employ compatible data representations and message-passing protocols.

The low-level programming model for multicomputers is based on processes and messages. The nodes run a multiprogramming operating system that interleaves the execution of multiple processes that may reside in the same node. All interprocess communication takes place by messages. There are numerous precedents for message-based interprocess communication in multiprogramming operating systems. Because multicomputer processes are distributed across the nodes, their operating system support is best described as *distributed multiprogramming.* There are also many useful algorithms that can be borrowed from theory and experience in distributed and parallel computing. Thus, it was not long after the introduction of the first multicomputers that programmers had learned to use them reasonably effectively.

In order to give you a status report on multicomputers, I have to tell

Figure 1.10   The 64-node Cosmic Cube.

you a little bit of their history. The early commercial multicomputers descended from the Cosmic Cube [32] (Figure 1.10) in their organization, system software, and programming method. My students and I designed the Cosmic Cube in 1981, and built several of these machines in 1982–83. The design was based on research on message-based programming methods and architectures that started at Caltech in about 1978. Some of the early development of these ideas can be found in the doctoral theses of Sally

Browning [6], Bart Locanthi [23], and Dick Lang [22]. The name "Cosmic Cube" comes from the Greek *cosmos,* meaning beautifully structured and open-ended, and *cube,* for its direct binary n-cube or "hypercube" message-passing communication network. Machines of this design are essentially open-ended in the number of nodes.

The Cosmic Cube nodes are identical, or nearly so. One node with one extra channel is required to connect the binary n-cube network to a host computer. The complexity of the nodes is similar to what we thought in 1981 would fit on a single chip in 1991. That projection still appears to be accurate. Each node has a megabit (128KB) of memory, a processor, a floating-point coprocessor, and a group of communication channels. The operating system in each node takes care of routing messages from any process to any other process. That the underlying graph is a binary n-cube is meant to be at least partly hidden. However, because the message performance was not spectacular in these early machines, the programmers did well to localize processes that had to communicate often.

Although the Cosmic Cubes were experimental systems built to test our architecture and programming ideas, these machines were very soon thereafter commercialized. In 1985, three different commercial multicomputers were announced: the Intel iPSC/1, the Ametek S/14, and the N-CUBE/ten, all of which were various evolutions of the original Cosmic Cube. If you study an Intel iPSC/1 node (Figure 1.11), you will observe that it is about the same physical size and has the same types of components as a Cosmic Cube node. However, instead of an Intel 8086 processor, it uses an Intel 80286 processor, which is about four times faster. Instead of 128KB of memory, it has 512KB. It also has circuitry to accept extensions on each node; for example, floating-point accelerators, extra memory, or I/O devices. Because ethernet chips were available "off the shelf," Intel used seven of them to implement the communication channels necessary to build machines up to a binary 7-cube (128 nodes). An eighth ethernet chip provided a global connection from each node back to a host. There are, I believe, well over one hundred of these iPSC/1 machines currently in use. Often when I visit other research groups, someone will tell me, "Look, we have one of your cubes," and will open a closet to show me an iPSC/1 on which they have been doing programming experiments or heavy-duty computing.

You will notice that what is happening with these commercial *medium-grain multicomputers* is that the architecture is scaling successfully with advancing technology by keeping the nodes about the same physical size and cost. However, the capability of single-circuit-board nodes is scaling rapidly with advances in VLSI technology. Just as the performance and memory size scaled by about a factor of four from the 1982 Cosmic Cubes to the 1985 iPSC/1, we have come to expect continued advances along this scaling track. There is, of course, another interesting scaling track,

Figure 1.11   An Intel iPSC/1 node.

more in line with the idea behind the Cosmic Cube experiment. The nodes might instead retain about the same complexity, but use VLSI technology to scale to smaller size and lower cost, so that a multicomputer system might employ more nodes. I will come back to this other scaling track later.

Between 1983 and 1986, in addition to transferring the Cosmic Cube technology and software to industry, doing a lot of programming, and working on other projects, my students, staff, and I were developing and refining a set of ideas for a second generation of medium-grain multicomputers. We expected that the processor performance and memory capacity could be increased by a factor of about ten just by exploiting the advances in single-chip processor and RAM technology, but we had other goals as well. Some of our ideas were so grandiose and radical that we knew we had little chance of persuading our industrial partners to incorporate them into their next products. So, we selected and integrated into a coherent design a few key techniques to enhance multicomputer performance, application span, and programmability. The Cosmic Cube experiment yielded timely results because its limited objectives and the leverage gained by using off-the-shelf chip and software technology allowed us to do the experiment in a short time. Evidently, an important part of the art of devising architecture experiments is to control your ambitions. Fred Brooks's advice about the "second-system effect" (Chapter 5 in [5]) was taken very much to heart in devising plans for the second-generation multicomputers.

The design assumption we made for the second-generation medium-grain multicomputers was that the nodes would exploit the same commodity chip technology that is used in personal computers and workstations. There is no terribly pressing reason to design another unique processor just to be used in a multicomputer. A Motorola 68020 works just as well in a multicomputer node as it works in a Macintosh or a Sun-3 workstation, and saves you a decade or so of chip designing and compiler writing. The production programming systems are based on writing explicitly concurrent programs with a concurrent process model, in which each process is an instance of a program. Instead of devising unique programming notations for process code, we use existing ones, such as C, FORTRAN, Lisp, and Pascal, extended with a set of functions for creating processes and for sending and receiving messages. The nodes are able to host the same variety of programming environments as contemporary workstations. The one thing we regulate fairly closely is the message semantics, which are portable across all of the existing multicomputers. These design assumptions are exactly the same as the design decisions we made for the Cosmic Cube experiment.

Our principal goal for the second-generation multicomputers was to improve the message-passing performance. Instead of trying merely to keep pace with the improvement in the processor performance, we set an aggressive goal for ourselves of trying to improve message-passing latency by up to 1000 times for the critical case of short, non-local messages. Relative to the expected gain of 10 times in processor performance, the message-passing latency for this critical case would be improved by 100 times.

Why did we adopt this large reduction in the latency of short, non-local messages as our central goal for these second-generation machines? To be honest, it was partly because we believed we could reach this goal with technology and with new organizational techniques that would appeal to any engineer. But this is not the intent of the question. Why would users care about improving the relationship between communication and node performance by such a large factor?

Even the relatively low-flux computations that were the mainstay of the first-generation machines tended to be communication-limited. We and our industrial partners could justify taking aggressive measures in this area because message performance was a consistent priority of users whose applications were encountering the limits of the capabilities of the message system. Multicomputers have proven to be reasonably "general-purpose" concurrent machines, so our choice was to try to "push the envelope" toward increased generality. Our own programming experiments indicated that we could extend the application span from easily distributed problems (such as matrix computations, partial-differential-equation solvers, local- and distant-field many-body problems, and distributed simulations of systems of many loosely coupled physical processes) into a range that included searching and sorting, concurrent data structures [13], various AI computations, graph problems, signal processing, and distributed simulation of systems of many tightly coupled physical processes [38]. These high-flux [18] computations tend to generate large volumes of short, non-local messages.

Our other consideration was that making the latency essentially insensitive to the process placement would simplify the programming. We had been asking people to solve two problems simultaneously: One is to place processes so as to balance the load; the other is to place processes so as to minimize the communication. In fact, these two objectives are often at odds with each other. As I discussed in my previous lecture, we can do much better with load balancing if we can disperse the computation rapidly and widely; for this, good performance in non-local communication is essential.

For reasons that will later become apparent, we needed to select a centerline design point for the number of nodes, N. Machines with thousands of nodes can encounter mechanical packaging problems that are unique to their very large size. Also, for the types of message networks we were considering, the point at which the network might become congested was a function of the network size. We selected $N = 256$ as the centerline design point.

Another goal was that medium-grain multicomputers should have open interfaces. They should have a standard hardware interface that can be used for floating-point accelerators or I/O controllers. They also need to provide standard software interfaces. For example, an explicit part of the programming system that we developed for the second-generation multicomputers is referred to as the "UNIX annex." The manufacturers are able to implement any subset of UNIX together with the standard programming system. It is perfectly normal in the programs written for these machines for people to include familiar packages such as "standard IO."

We learned a lesson about software interfaces from the first-generation machines. For the Cosmic Cube, we had provided some fairly elaborate but universal communication primitives. All messages had types and always included the identity of the process that was the source of the message. Each manufacturer used a slightly different variant of these primitives. For the second-generation systems, we have changed to a streamlined form of low-level message primitives that allows very simple and efficient layering of other message functions. One of the most useful things we can do in a university, where we do not have the commercial interests that drive companies, is to produce software that can be licensed on a nonexclusive basis. For example, Berkeley produced a form of UNIX to which no company could claim exclusive rights, and this operating system was largely responsible for creating a market of products that are highly competitive precisely because the systems made by different manufacturers are program-compatible. Following this example, we developed a programming system that can run on any multicomputer. This system software allows the user to emulate the system calls used by other systems, so that nobody gets locked into a single manufacturer because of system primitives and the expense of "porting" software from one manufacturer's system to another.

In summary, our "wish list" for the second-generation medium-grain multicomputers—tempered by what we thought was feasible, and backed up where necessary by working prototype hardware and system software— included ≈10-times-faster processors, ≈10-times-larger node memories, ≈1000-times-smaller message latency, open interfaces, and a more modular and streamlined low-level programming system. How did we persuade two multicomputer manufacturers to incorporate these ideas into their new products? The recent phenomenon of computer manufacturers basing product designs so directly on university research (Berkeley UNIX, RISC processors, multicomputers, systolic arrays, the connection machine, and many other examples) is an interesting technical, social, and economic development, and deserves some comment.

Both Intel Scientific Computers and Ametek Computer Research Division are industrial partners in these efforts, and that they are competitors makes life quite interesting. Of course, at Caltech we know what both companies are doing, but each company naturally does not want the other to know its plans. So, we are very careful that the information that we get from the companies is appropriately compartmentalized. We are also careful to report new research developments to these companies at the same time, and often well before the work is written up in a form suitable for external publication. I wrote the design document for the second-generation medium-grain multicomputers over the summer of 1986, but the descriptions of the message-network and programming techniques employed in these machines appeared in print [34, 35, 3] only after the techniques had been demonstrated in announced products. The companies do a very good job of incorporating new ideas once we have developed them to the point of a credible demonstration, simulation, or prototype, or have otherwise taken the risk out of including them in products.

Last September (1987), Intel announced their iPSC/2, which is a one-and-one-half generation machine that provides an upgrade path from the iPSC/1. Intel replaced the Intel 80286 in the iPSC/1 node with the Intel

80386, which is nearly ten times faster; increased the memory to up to 8MB; and boosted the performance of the communication network, particularly for non-localized message traffic. However, instead of increasing message performance by a factor that was equivalent to the improvement in the processor performance, the bandwidth of the channels was increased only by a factor of about four. However, the latency for non-local messages was improved by as much as a factor of twenty by using "wormhole routing," a Caltech development that is replacing the store-and-forward routing that was used in all of the first-generation machines.

Another amusing thing about academics working so closely with companies is that the companies always give their own name to technical developments. I can understand why Intel was reluctant to advertise that they were using something as flaky sounding as "wormhole routing," but I suspect that they would have picked their own name even if we had selected a more attractive name. Although wormhole routing was devised at Caltech, and Caltech has patents on it, I believe that Intel calls it "direct-connect" routing.

Now I want to show you the very latest multicomputer, the Ametek Series 2010, and use it as a vehicle for describing in detail the architecture and design of a second-generation multicomputer. Whereas the development of the iPSC/2 had started prior to the summer 1986 design document, the Ametek Series 2010 development started in October 1986, and was based very closely on the programming system, message-passing network, node architecture, and packaging scheme described in this design document. The development required 16 months to the first prototype, and the product was announced at the Hypercube Conference in January 1988. Our goals for second-generation medium-grain multicomputers have largely been achieved in the Ametek Series 2010. The node computing performance and memory size are increased by a factor of about ten. This machine approaches and in certain conditions exceeds the goal of increasing the communication performance by a factor of one thousand for non-localized message traffic. Also, it has all of the open hardware and software interfaces we might hope to find.

The way you should visualize one of these second-generation "hypercubes" (Figure 1.12) is, first of all, to forget the hypercubes. They do not use a hypercube network; instead, they use a routing-mesh communication network that forms a modular backplane. The nodes, which are single-board computers with a message interface on one end and a standard VME interface on the other end, plug into the backplane. The modular division into network and nodes is not only physical, but logical. A packet injected into the network leaves the network only at the destination; the nodes do not participate in message routing. The number of nodes is not constrained to be a power of two. The entire edge of the routing mesh is available to hook up other devices, such as interfaces to your satellite dish,

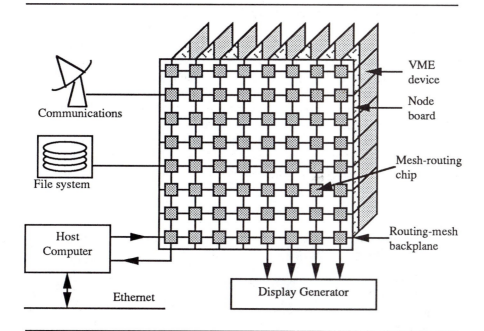

Figure 1.12   Designer's-eye view of a second-generation multicomputer.

displays, disks, and network hosts. Such devices can also be attached to the node VME interfaces.

The raw speed of the communication channels is about 225Mb/s, or 28MB/s. The routing mesh is composed of *mesh-routing chips* (MRCs), one for each node. For modularity and to minimize the number of signals that have to go through connectors, the MRCs are physically located on the backplane itself. Each node connects to the routing mesh via just two channels, one input channel and one output channel, each 28MB/s. This bandwidth is fairly close to that of the memory of the node, so there was little incentive in the Ametek Series 2010 to make the channels operate still faster. However, we know right now how to push the routing-chip technology up to a gigabit per second at modest cost. At 225Mb/s, we are just at the beginnings of the capability of this communication technology; we have at least another factor of five available to us.

In each node there is a specialized message interface that takes care of some of the send- and receive-queue maintenance tasks that were performed in earlier machines by the operating system. The node also contains up to 8MB of memory and a 25MHz Motorola 68020/68882 processor. There is just this one instruction processor, but there is nothing that precludes a multiprocessor node. As it is, the instruction processor, message-interface processor, and VME-connected accelerators or I/O controllers all access the memory.

Let us switch now from the designer's-eye picture of the machine to the salesperson's picture (Figure 1.13). The small cabinet is for machines up to 32 nodes; the pair of large cabinets will house a 128-node system. What is behind each of the panels is a 4×4 mesh unit on the back side of the machine, and 16 node boards that plug into the front side. Inevitably, LEDs are associated with each node to give the machine a Hollywood-set appearance. The cantilever design is quite functional in providing airflow from underneath. The blowers are on top, as you can tell from the lack of lights for that panel. The power dissipation is about 30 watts per node.

Figure 1.13   Ametek Series 2010 cabinets.

A 32-node machine can be put in the small cabinet, two of these cabinets side by side make a 64-node machine, and four cabinets side by side make a 128-node machine. The 4×4 mesh units stack vertically and horizontally. For example, the best configuration for 64 nodes (to maximize the bisection) is a 2×2 array of 4×4 units. For 128 nodes you can put two 64-node machines side by side in four small cabinets or stacked in two large cabinets. For 256 nodes, you stack 4×4 mesh units four wide and four high in four large cabinets. (The larger configurations of these cabinets start approaching the Star Trek set in appearance.) If you want to make a 1024-node system as a 32×32 mesh, you should not start looking for an eight-unit-high cabinet. If anyone were to order such a machine, Ametek would design a 2×8 mesh unit.

From the salesperson's-eye picture of the machine, let us switch to the programmer's view of the system. The set of processes that participate in a computation is called a *process group* and includes not only *node processes* in the multicomputer nodes, but also *host processes* that run in UNIX computers that are on the same network as the multicomputer. There are usually only a few host processes. For example, there may be a host process that allocates the multicomputer, creates the initial process structure in the multicomputer, and reads keyboard input on the user's workstation; and another process to update a display on the screen of the workstation. Node processes are created (spawned) dynamically, and identified by an ordered pair, (n,p), where n is in the range $0, 1, \ldots, N-1$ and specifies the node number and p, the process identifier within the node. Host processes are simply UNIX processes that enter the process group by calling a function whose arguments specify the (n,p) identifier for the process.

When the process code is written in C, host and node programs are linked to a library that includes the following low-level functions:

spawn("name",n,p,m) creates an instance of the compiled program "name" as a node process with the identifier (n,p) and mode m. (Process placement is determined by the programmer.)

aspawn("name",&n,&p,m) creates an instance of the compiled program "name" as a node process with mode m and returns its identifier (n,p). (Process placement is determined by the system.)

ckill(n,p,m) kills the node process (n,p) and removes any messages that may be queued for it.

exit() kills the calling process and removes any messages that may be queued for it.

xmalloc(k) allocates a message buffer of k bytes and returns a pointer to the message buffer.

xfree(b) frees the message buffer pointed to by b.

xrecvb() returns a pointer to a message buffer, but only after a message arrives for the calling process. This function is semantically similar to **xmalloc**, except that the size and initial contents of the message buffer are determined by the earliest arriving message.

xrecv() is a nonblocking variant of **xrecvb**. It may return a NULL pointer if a message has not arrived for the calling process.

xsend(b,n,p) sends the contents of the message buffer pointed to by **b** as a message to process **(n,p)** and also frees the message buffer. This function is semantically similar to **xfree**, except for the side effect of sending the message.

xlength(b) returns the length of a message buffer pointed to by **b**.

nnodes() returns N.

mynode() returns the node part of the identifier of the calling process.

mypid() returns the pid part of the identifier of the calling process.

All of these functions are available both to node and to host processes. The interface between the host runtime system and the node operating system makes the physical boundaries between the systems "seamless."

Most programmers dislike verbose programming manuals; hence, the "C Programmer's Abbreviated Guide to Multicomputer Programming" [35] is only 23 pages long, plus some appendices. This programming guide seems to include all that someone familiar with C and UNIX needs to know to start writing multicomputer programs. It shows, for example, how higher-level message functions can be expressed in terms of the **x** primitives. The rationale for the choice of these functions is based largely on their layering and portability properties [3].

Having given you an overview of *what* one of these machines is, let me go back to the *why* and *how* story, with particular emphasis on the message-passing performance.

The advance in message-passing performance was achieved by two organizational improvements and by integrating the routing and flow control into the custom-VLSI mesh-routing chips (MRCs) that I showed you in the designer's view of the Ametek Series 2010. The first organizational improvement was wormhole routing, a blocking form of *cut-through routing* [19]. The second organizational improvement was the network topology, a two-dimensional routing mesh of byte-wide channels rather than a binary n-cube of bit-wide channels. It is difficult to achieve a factor of 1,000 all in one way; the factor of 1,000 was achieved as the product of these three partially interdependent improvements.

An organizational technique that was preserved essentially intact from the Cosmic Cube message system is the flow-control hierarchy (Figure 1.14).

A **Message** is the logical unit of information exchange:

A message is composed of a sequence of **packets,** each with its own routing header:

A packet is composed of a sequence of flow-control units (**flits**):

Figure 1.14   The message flow-control hierarchy.

Unlike geographically distributed networks, a multicomputer network is physically compact. The small transmission delay from node to node allows flow control over small bundles of data called flow-control units, or *flits.* The channels are queue-connected; each time the sending circuit places a flit on a channel, the receiving circuit must acknowledge that it now stores the flit before another flit will be placed on the channel. A flit can be blocked due to contention; the queue discipline then blocks the following flits. The flits do not themselves carry routing information, so they cannot be interleaved on a channel except by techniques such as virtual channels [13]. (See Bill Dally's lectures in Chapter 3.) A *packet* is a sequence of flits in which one or more initial flits form a *header* that specifies the destination node, and the last flit is tagged as the *tail.* A maximum number of flits in a packet is adopted as a system convention to increase network fairness and simplify storage management in the nodes. Messages that are longer than the maximum packet length are fragmented into a sequence of packets at the sending node, and these packets are reassembled into a message at the receiving node. Thus, with fair arbitration for channels, a long message can block a short message requiring the same channel for no more than the maximum packet transmission time. Of course, the mechanisms below the level of messages are invisible to the programmer.

Geographically distributed networks generally use the packet as the unit of flow control, routing, and error control. An incoming packet is stored within the node; then, if the check sum is correct, the packet is forwarded to the next node and an acknowledgment packet is returned to the sending node. (Most systems request retransmission for erroneous packets, so the sending node must retain a packet until the acknowledgment is received.) The first-generation multicomputers used this same *store-and-forward* routing method. However, at least for the Cosmic Cube, the acknowledgment packet was neither needed nor used, because the flow control was provided at the flit level. Nor was there any need for error control, because the channel error rate was negligible. A multicomputer network operates in a physically protected environment and employs high-energy signals to drive the wires that make up a channel. Thus, these channel signals are even less subject to transmission errors than the signals internal to the node computers; indeed, the channel error rate is so small that it cannot be measured.

Wormhole routing is an utterly simple idea that takes advantage of the very low error rates on multicomputer channels. The essential idea of wormhole routing is to advance a packet directly from incoming to outgoing channel, without storing it, as soon as enough information has been received in the packet header to select the outgoing channel (Figure 1.15). Kermani and Kleinrock [19] reported a similar idea in 1979, but their *virtual cut-through routing* reverts to store-and-forward operation when packets are blocked. If we regard cut-through routing as the generic technique, then virtual cut-through routing, wormhole routing, and adaptive cut-through routing [29] are the specific methods that deal with blocking by diverting the packet to local storage, by blocking the packet in place by a queueing discipline, and by misrouting, respectively. Even the Cosmic Cubes and the first-generation commercial machines use cut-through routing at the packet level, but in the Ametek Series 2010, the routing chips use the blocking form of cut-through routing at the flit level. It is wormhole routing that makes the message latency largely insensitive to message distance, but that was not the first motivation for considering it.

In 1984, while I was worrying about the impact of store-and-forward routing on the node storage bandwidth, I did some simulations of cut-through routing. I was sure that blocking cut-through routing would be

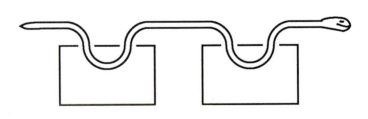

Figure 1.15   The idea of wormhole routing.

a complete disaster, because congestion in the network would consume resources that would lead to more congestion. Much to my surprise, the simulation showed that the blocking form of cut-through routing did very well. Network performance is often depicted by plotting throughput against applied load (Figure 1.16). Because these are simulation results, you will have to take my word for it, but the throughput under random message traffic achieves a *stable* level at nearly 50 percent of the maximum determined by the network bisection. We did not invest all the effort of designing routing chips until we had completed extensive simulations, which required over 1000 hours on a 128-node Intel iPSC/1.

This approach also offered the possibility of a routing chip that would not have to buffer entire packets, but only a few flits. So, I got one of the project groups in my VLSI class that year to design a prototype cut-through routing chip for binary n-cube networks. It was the students who named it wormhole routing, perhaps because of the possibility suggested by theoretical astrophysicists that "cosmic wormholes" could provide shortcuts between distant points, or perhaps because of an opening presentation in which I illustrated this routing technique in terms of flit-segmented worms whose heads could be leaving the network while the tails had not yet entered.

The initial motivation for wormhole routing was to conserve storage bandwidth in the nodes. Store-and-forward routing is a terrible waste of

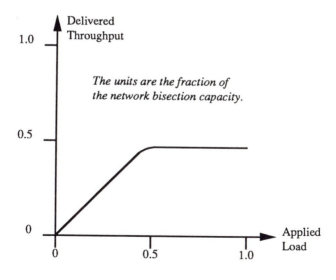

Figure 1.16 Characteristic performance of wormhole routing.

precious storage bandwidth; the packet consumes bandwidth both coming and going. Wormhole routing also allows us to segregate routing entirely into the network, and to use very little storage in the routing chips. In this respect, wormhole routing is a pipelined version of circuit switching: The head forms the path, and the tail breaks it. So many papers and reports have concentrated entirely on the latency-reduction advantages of cut-through routing that I wanted to emphasize that it has other equally important advantages.

What about the latency? If the cycle time of a channel is T_c, and the width of the data path is W, then the bandwidth of the channel is W/T_c bits/s. A packet L bits long will require time $T_c(L/W)$ to traverse a channel. Thus, the time required for the packet to traverse D channels in store-and-forward routing would be $T_c(L/W)D$, even if the packet were forwarded as soon as it was completely received. Notice that this expression for the latency of store-and-forward routing,

$$T_{\text{S\&F}} = T_c(L/W)D,$$

depends on the *product* of L and D. However, if the head of a packet can be advanced to the next channel in p cycles by using cut-through routing, the latency becomes

$$T_{\text{CT}} = T_c(pD + L/W),$$

which depends on a *weighted sum* of L and D. $T_c(pD)$ is the *path-formation* component, and $T_c(L/W)$ is the time required to spool the message through. If the routing information is contained entirely in the first flit, and if one cycle is allowed for the communication across the channel and one cycle to make the routing decision, a straightforward routing-chip design has $p = 2$.

Let us spend a moment being engineers by looking at the numbers. The actual parameters for the routing chips in the Ametek Series 2010 are $T_c = 0.035\mu s$, $p = 2.4$, and $W = 8$. The parameter p is not an integer because the chips are not synchronous, but are self-timed [26]. So, for example, a 64-byte message (eight double-precision floating-point numbers), which is a short message by present multicomputer standards, would require $0.084\mu s$ per hop to form the path, but $2.24\mu s$ to spool the message through the channels. The average distance in a binary 8-cube ($N = 256$) is 4, so the average path-formation component of the message latency would be only $0.336\mu s$. That the term dependent on message distance is so much smaller than the term dependent on message length provides an important clue that we can afford to make the average D larger with little impact on message latency, particularly if we get something back in return that would make $T_c(L/W)$ smaller.

In 1985 Bill Dally pursued this clue, and provided a convincing argument that the optimal networks for these machines would be of lower dimension and larger average D than the binary n-cube, which is the practical network of highest dimension and smallest average D. Because lower dimension networks are more wireable, we can afford to make W larger, which provides exactly the return we were seeking. If we can keep all of the wires short, we can also reduce T_c. The way in which the D *versus* W tradeoff is formalized is to assert that the cost and throughput of a network are closely related to wire bisection. (You will recall that we discussed wire bisection yesterday in the context of chips.) If the wire bisection is fixed, we should be free to select whatever topology minimizes the latency. Bill Dally will undoubtedly be showing you his analysis [12] of latency *versus* network dimension for the class of k-ary n-cube (hypertorus) graphs (see section III-A in [31]). This analysis shows that the optimal number of dimensions for machines in the range $N = 256$ is two!

Instead of working out the mesh version of the analysis, let us continue to be engineers and look at the numbers comparing a 256-node binary 8-cube and a 16×16 two-dimensional mesh. One useful preliminary is to know that the average distance on a square 2-D mesh is $\frac{2}{3}(N^{\frac{1}{2}} - N^{-\frac{1}{2}})$, which for large N is approximately $\frac{2}{3}\sqrt{N}$. Networks with the same wire bisection B would have the following parameters:

Parameter	Binary n-cube	2-D mesh
Number of nodes	$N = 256$	$N = 256$
Radix	$k = 2$	$k = 16$
Dimension	$n = 8$	$n = 2$
Channel width	$W = 1$	$W = 8$
Wire bisection	$B = 256$	$B = 256$
Average distance	$D = 4$	$D = 11.6$
Pipeline depth	$p = 2$	$p = 2$

Now we can tabulate the number of network cycles, $pD + L/W$, required to route packets of length L. With $T_c = 0.035\mu s$, the time is shown in parentheses:

L(bits)	binary n-cube	2-D mesh
0	8 ($0.3\mu s$)	23 ($0.8\mu s$)
8	16 ($0.6\mu s$)	24 ($0.8\mu s$)
16	24 ($0.8\mu s$)	25 ($0.9\mu s$)
32	40 ($1.4\mu s$)	27 ($0.9\mu s$)
64	72 ($2.5\mu s$)	31 ($1.1\mu s$)
128	136 ($4.8\mu s$)	39 ($1.4\mu s$)
256	264 ($9.2\mu s$)	55 ($1.9\mu s$)
...	...	...
2048	2056 ($72\mu s$)	279 ($10\mu s$)

What we observe, of course, is that the path formation ($L = 0$) is such a small component of the latency in either case that the mesh's wider path gives it a slight advantage for messages even as short as 32 bits, and an asymptotic advantage of a factor of eight. The maximum packet length in the Ametek Series 2010 is 256 bytes, which corresponds to the last line in this table. One of these packets requires only $10\mu s$ to traverse a channel. Of course, people could make binary n-cubes with 8-bit-wide channels, but the packet latency in a lightly loaded network would be little better than the 2-D mesh, and I would hate to think about the cost and the size of the cables between cabinets.

Bill's analysis for torus networks was published in his doctoral thesis [12] in June 1986. It was amusing to hear rumors during 1987 that the iPSC/2 and Ametek Series 2010 would use a 2-D torus network, while I knew that the iPSC/2 used a binary n-cube, and the Ametek Series 2010 a 2-D mesh. Although the 2-D mesh is an aperiodic variant of the k-ary 2-cube, it is not a torus (Figure 1.17). Let me try to list our reasons for using a mesh rather than a torus, in approximate order of importance:

(1) Because the mesh is aperiodic in each dimension, the deadlock-free routing is simpler than it is on a torus. The simplicity translates into simpler and faster mesh-routing chips.

(2) We considered the communication capability around the edge of the mesh network to be an asset. It is difficult to figure out where to attach to a torus.

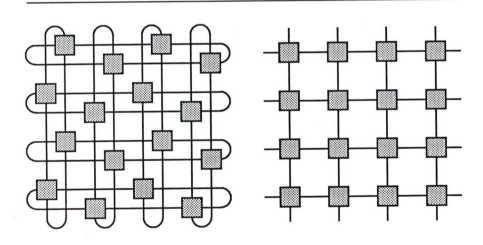

Figure 1.17   Two-dimensional torus (left) and mesh (right) networks.

(3) These machines are normally *space shared* rather than time shared. For example, for binary n-cube machines we allocate subcubes to users, and the subcubes of binary n-cubes are still binary n-cubes. It may be important to a user whose message communication has been carefully localized to provide the same communication plan as a smaller machine. The submeshes of meshes are still meshes. However, there is no general way to partition a torus so that the parts have a torus topology.

(4) Under the usual constant-wire-bisection assumption, the mesh slightly outperforms the torus. When projected onto a plane, the "end-around" channels of a torus are part of its bisection. For equal N and wire bisection, the mesh channels can be twice as wide. However, the average distance is not twice as large, but only about four-thirds as large.

(5) In order to avoid long wires, a 2-D torus would be folded (as shown in Figure 1.17), which complicates the packaging and results in channels that are twice as long as those of a 2-D mesh.

There are a few more reasons for preferring one network over the other, but these were our reasons for choosing the mesh rather than the torus.

The users are not encouraged to pay attention to the network topology. The programming guide [35] does not even say what the topology is. From the user's standpoint, it is a complete graph, an all-points-to-all-points network. Also, unless otherwise specified by a user, the system reserves the right in space sharing to assign the specified number of nodes from any nodes in the physical machine, even if they are not contiguous, let alone in the shape of a mesh.

I have omitted talking about the properties of the network and routing algorithm that make it deadlock free; however, it is the same dimension-order routing that we used on the Cosmic Cube, but here there are only two dimensions.

The scaling properties of mesh networks are quite different from those of binary n-cubes, which are scaled to increasing N in powers of two by scaling the network dimension, $n = \log_2 N$. The bisection increases in proportion to N, but the nodes are of increasing degree. The mesh is a fixed-degree network in which a submesh is composed with another by abutment (using only short wires). The trouble is that the bisection of the mesh network is growing only as $\sqrt{N}$. So, for a sufficiently large machine, you are going to get yourself into some mischief with this kind of scaling. I mentioned that our centerline design point for the network was 256 nodes. You can figure out for yourself that with randomly selected packet destinations, the nodes of a 256-node machine would have to be emitting packets a quarter of the time to saturate the bisection. The corresponding number for a 64-node machine is half of the time, which can just barely be achieved by system programs, let alone user programs, so a 64-node machine has a surplus of communication capability. However, it would be possible with intense, non-localized message traffic to drive the network of a 512-node machine into a state of moderate congestion. It is for these large-N machines and programs that message locality becomes important.

The last critical step in advancing the message-passing performance was just pure technology: casting the message handling and routing functions into custom VLSI chips. It was quite exciting for me to see the backplane of the Ametek Series 2010 (Figure 1.18) for the first time, because I had been sketching pictures of this package over a period of a couple of years. When I saw the real thing put together, it was amazing how simple it was. You can observe in this photograph the routing chips, the channels between them, the vertical connections between adjacent 4×4 mesh units, the connectors for the node boards, and the 300Amp power buses. The routing chips are arrayed physically as a mesh, but the node boards are arranged in a row; the wiring between them unravels this slight discontinuity. The signals from chip to chip are of sufficiently high energy that they have no problem driving the short ribbon cables that carry the channels from cabinet to cabinet. Because the routing chips are self-timed, the longer the wiring run, the more delay is added to T_c. You may notice that the routing chips are biased toward the edges to try to minimize the length of the cables between mesh units. The vertical cables are very short, but the horizontal cables are about six inches long.

The real complexity of the message network is hidden inside the mesh-routing chips (Figure 1.19). The MRC is a communication chip, so it should be no surprise that it is severely pad limited. The core of the chip is composed of an x router and a y router, which are, of course, identical. Because of the dimension-order routing, the x and y routers are simply cascaded. What is fun about the individual routers is that each of them is, in turn, composed of still more elementary routing circuits [17].

To complete this reverse-engineering tour of the Ametek Series 2010, let us take a look at the node, first in block-diagram form (Figure 1.20). The node is organized around the memory bus, which is the type that hardware designers call a RAS/CAS (row-address-select/column-address-select) bus. This bus supports random access at the usual ≈200ns cycle time of dynamic memory chips, but also supports a "column-access" mode that is more than twice as fast as the random-access mode. For example, the message interface reads and writes the payload part of a packet from sequential memory locations in a single column, and so is able to perform one memory reference for each two 40ns clock ticks to achieve an access bandwidth of a 32-bit word every 80ns, or 50MB/s. The message interface also reads and writes control blocks to maintain the send and receive queues in memory. The VME interface is also able to use the RAS/CAS bus in

Figure 1.18   Backplane of the Ametek Series 2010 multicomputer.

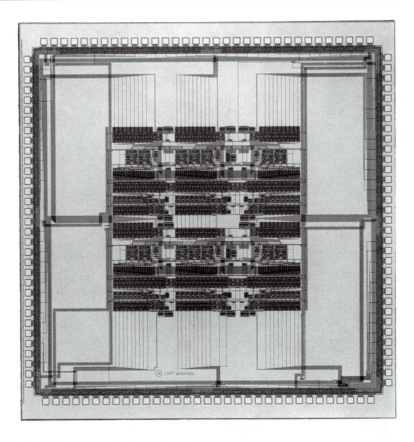

Figure 1.19 Photomicrograph of a mesh-routing chip.

either mode. The 25MHz Motorola 68020 processor is off in its own world: It has its own bus with a 68882 coprocessor, timer, and bootstrap ROM. It accesses the RAS/CAS bus through an address-translation unit that also provides what we call a page pseudocache. The pseudocache keeps track of whether the page in the particular memory bank that the 68020 is accessing has already gone through the first half of the cycle. If so, it only bothers to do the second half of the cycle. Thus, the pseudocache has much the same effect as a cache. In the usual escalation of how much memory is included with the node, this generation of node is 1–8MB.

Because the message network is so much faster than earlier machines, it was necessary to deal more expeditiously with packets. The message interface includes 256 40-bit words of microcode and FIFOs that hold full

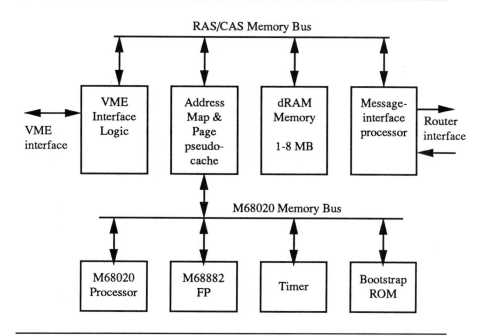

Figure 1.20   Block diagram of the Ametek Series 2010 node.

packets for static-column-mode transfer to and from memory. To give you some indication of the performance of the message system, it is faster to send a message from one node to another—a cross-node memory-to-memory copy operation—than for the 25MHz Motorola 68020 to perform a copy in its own memory. The message system has an advantage here because it uses two memories concurrently, both in a sequential-access mode, whereas the 68020 copies in a single memory in a random-access mode. You can fairly say that the message system is limited by the bandwidth of the node memory rather than by the network itself.

This node design also exhibits some of my obsession with conserving storage bandwidth. The custom memory-management unit maintains two kinds of pages. The usual 8KB pages are used for code, data, and stack. The 256B pages are dynamically allocated, and can be accessed both by the message system and by user programs. The packet also has a maximum payload of 256B (together with control information that is invisible to the user). Dynamically allocated buffers for messages are thus automatically aligned to these 256B pages, so fragmentation and reassembly of long messages is accomplished without copying.

The single-circuit-board node (Figure 1.21) is physically laid out (right

Figure 1.21 An Ametek Series 2010 node.

to left) with the message interface processor near the mesh-backplane connector, then the memory, then the main processor, and finally the VME-interface logic near the VME connectors. From the standpoint of a company, you might look at this node board and say, "What is to prevent me from replacing the Motorola 68020 with a fast RISC processor or a floating-point processor?" Obviously, nothing at all. The essentials of the node are the message interface and memory; the other parts might even be mounted piggyback to facilitate changing them as technology evolves.

You can tell from this description of the node that the design of the Ametek Series 2010 followed the "integrated hardware/software" approach. The design and implementation (on the Cosmic Cubes) of our new, streamlined operating system, the Reactive Kernel [36], preceded and guided the node design. The Reactive Kernel is structured (Figure 1.22) as an *inner kernel* and a set of reactive processes called *handlers.* Each message includes a *tag* that determines which handler to run when a packet is taken off of the receive queue. For example, the *spawn handler* creates new handlers, and links their calls for system services into the inner kernel. The *assembly handler* keeps track of packets that are fragments of larger messages, and

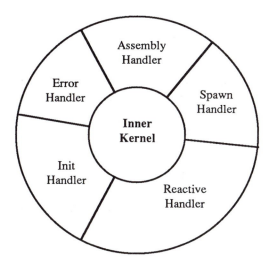

Figure 1.22 Structure of the Reactive Kernel.

when the packet received allows a complete message to be delivered, dispatches to another handler. The *reactive handler* runs user processes, and supports a set of system calls that are patterned after the services that the inner kernel provides for handlers.

Global operations between cohorts can be advantageously performed at the handler level, where the message latency is not dominated by context switching between protected user processes. The Reactive Kernel is designed to allow global operations, concurrent data structures, or programming-language runtime systems to run at either the system or user level. Its modular structure also makes it highly portable; for example, 90 percent of the Ametek Series 2010 version of the Reactive Kernel is the same code as the generic Caltech version.

The host runtime system, the Cosmic Environment, is a set of dæmon processes, utility programs, and libraries that implement this same programming environment on UNIX hosts or across a collection of network-connected UNIX hosts. The Cosmic Environment also supports simulated multicomputers, which are known as "ghost cubes." Thus, you do not need a multicomputer to be able to run message-passing programs. The Cosmic Environment system is used at about 150 different sites in addition to its use with multicomputers.

Where do we go from here with the architecture, hardware, and low-level programming systems? Multicomputers have a simple structure. Each node includes one or more processors, memory, read-only memory for boot-strap and initialization, a message interface, and a router. There is nothing that says that it takes a large circuit board for these functions. Our research group at Caltech is still looking toward performing all of these functions on a single chip.

Our first attempt at an integrated node was the Mosaic A (Figure 1.23). One of the attractive features of a single-chip node is performance. The memory bus runs right down the backbone of the chip, so the "von Neumann bottleneck" is very well localized. Although these Mosaic A chips worked, and the instruction rate was quite fast for the 3μm nMOS technology employed, they did not have the external communication and routing performance required to build a large multicomputer. We are currently working on the Mosaic C, a complete node integrated onto a single CMOS chip. The routing section is logically very much like the routers in the medium-grain machines; indeed, it was from studying how to build routing networks for the difficult case of large-N, fine-grain multicomputers that we developed the techniques that we used in the second-generation medium-grain multicomputers.

If the fabrication-process yield allows, each Mosaic C node will have 16KB of memory. The Mosaic C processor runs at about 14 MIPS. DARPA is sponsoring the construction of a 16,384-node Mosaic C system, which will be structured as a $32\times32\times16$ 3-D mesh. With a total of 256MB of primary memory, a peak performance of 200,000 MIPS, and the lowest-latency message handling yet built into a multicomputer, the Mosaic C will be quite an interesting experimental machine.

Whereas the medium-grain machines now have hundreds of nodes with megabytes per node, the fine-grain multicomputers will have tens of thousands of nodes, initially with tens of kilobytes per node. This is a two-orders-of-magnitude shift, which is more than enough to make an interesting experiment. In addition, I hope and expect that the Mosaic C will be the origin of a new scaling track for fine-grain multicomputers based on single-chip nodes.

1.3   Concurrent Programming

Let me start with a few comments on some of the other lectures. After all, if we fail to take advantage of the opportunity that workshops provide for low-latency interaction, we might as well just read each other's papers.

In a tutorial article on "Concurrent VLSI Architectures" [31] published several years ago in the issue of the *IEEE Transactions on Computers* that celebrated the centennial of the IEEE, I tried to convey my hope that architecture and design tradeoffs could and should be treated in a less ad hoc way today than they were in the early years of computing.

Figure 1.23 Photomicrograph of an early single-chip node.

The great variety of early digital technologies—with different technologies for performing logic operations, driving wires, and storing information—made it difficult to quantify architectural tradeoffs. However, VLSI has made the engineering situation for digital systems sufficiently uniform that many design problems are tractable for analysis. There is a new school of computer architecture developing in which we try (not always successfully, and sometimes too hard) to be analytical about design decisions. Look, for example, at Bill Dally's thesis, "A VLSI Architecture for Concurrent Data Structures" [12], or compare today's quantitative evaluations of instruction sets with the literary-criticism-style discussions of instruction-set architectures of a decade ago.

While we may have heard too much about message-passing networks, I do not think we have heard enough about programming. I tried to suggest in my first lecture that VLSI has given us not only the *means* for building highly concurrent computers, but also the *reasons*. Carver Mead has very appropriately described what has been taking place with VLSI and concurrent computing as "the first revolution in content in computing science." Other fields have gone through such revolutions in content when their premises or laws have proved to be false, for example, the "ultraviolet catastrophe" in physics. The algorithms and complexity theory that developed in the early years of computing account for the operations and storage utilization required for a computation, but neglect the communication. The complexity theory developing for VLSI and concurrent computing is starting to reveal the advantages of organizing computations in a way that admits concurrency and respects communication. Clearly, the revolution in the use of concurrent computing will depend most critically on programming—the formulation and expression of computing problems for concurrent execution. Since I did not lecture yesterday, I spent some time revising the emphasis of today's and tomorrow's lectures, with the following plan in mind:

First, I want to review with you the common misunderstanding caused by the differences in goals and strategies between programming small-N and large-N concurrent systems. Second, building on Ernst Mayr's lectures, I want to show you some of the *similarity* between the shared-memory and message-passing approaches from the viewpoint of the fundamental algorithms that are used at the lowest levels of many application programs. Third, I want to give you some examples of the design of concurrent application programs. These three segments are all intended to emphasize the advantages of designing concurrent programs from first principles, rather than trying to rewrite "dusty-deck" sequential programs.

To discuss "small N" versus "large N," I will use some slightly loaded language: The *conservative approach* aims at programming small-N concurrent systems, and is immediately attractive commercially. The *radical approach* aims at programming large-N concurrent systems, and is appealing to researchers because it promises major gains in the longer term and exposes fundamental computer-science issues.

"Conservative" is meant in both a positive and a negative sense. Commercial enterprises need to conserve the millions of person-years invested in the development of sequential application programs. They must be able to run "dusty decks." There is an established technique that allows sequential programs to be run on concurrent computers (Figure 1.24), but it is useful only for small N. The idea, of course, is for a compiler to detect the concurrency in a sequential source program, and to produce a concurrent object program. The source program might be written in C, FORTRAN, Pascal, Lisp, *etc*, but FORTRAN is generally preferred because its lack of pointers simplifies the data-dependency analysis on which much of the concurrency detection depends. The compiler may also be able to determine how to assign different program modules to the resources of the target concurrent computer. However, because some of the resource requirements cannot be determined in advance of program execution, the concurrent program also requires a suitable runtime system to assign resources during execution.

It has long been understood that the concurrency that can be detected even in large scientific programs, which tend to have more implicit concurrency than programs in other application areas, is typically only

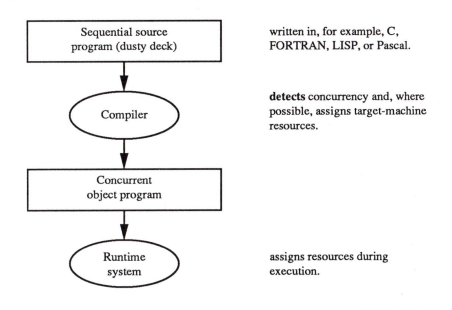

Figure 1.24   Conservative approach to concurrent programs.

about 10 to 30. For example, a milestone 1974 paper by David Kuck et al., titled "Measurements of Parallelism in Ordinary FORTRAN Programs" [20], reported an average theoretical speedup of about 10 for an ensemble of 86 FORTRAN programs. Actually achieving this speedup would require 30 or so processors operating at a utilization of about 30 percent and instantaneous interprocess communication. (The programs were not actually compiled and run, but were subjected to a data-dependency analysis to determine their potential concurrency and theoretical speedup.) If you were to insist on somewhat higher processor utilization and allow for the cost of interprocess communication, I believe that you would decide that compiled sequential programs can often exploit a concurrent system with $N \approx 10$ processors, but that diminishing returns would become increasingly evident for larger N.

Compilation techniques have advanced considerably since Kuck's paper, but the concurrency that compilers can discover is still found in the *innermost* expression evaluations and iterations of the source program. You can conclude from the nature of these concurrencies the necessary properties of the concurrent systems that execute these computations. They must have fast mechanisms for interprocess communication and process creation (or for assigning processors to execution "threads"). For example, in order to spread an iteration profitably across multiple processors, the system must be able to initiate multiple execution threads in much less time than it takes to execute the iteration, and to assign individual iterations in much less time than it takes to execute the body of the iteration. These are "fine-grain" control mechanisms that must work with little delay on relatively small execution units. So, if you can arrange to have small N and fast interprocess communication, which is the niche of the shared-memory multiprocessor, this compilation scheme is quite workable. There are a number of examples of successful commercial applications of this approach, including the Alliant, Encore, and Sequent mainframes, and the Apollo, Ardent, and Stellar workstations.

Instead of starting with a sequential program, the radical approach starts with the computing problem (Figure 1.25), and depends on the programmer to write an *explicitly* concurrent program. Why can a programmer find concurrency that a compiler cannot find in a sequential program for the same problem? Very simply, it is because a compiler is limited to semantics-preserving transformations that merely reorganize the sequencing of operations in the innermost parts of the sequential source program, whereas the programmer can work with the whole problem.

For example, a sequential program to compute the shortest (least-cost, with nonnegative edge weights) path in a graph by Dijkstra's algorithm [14] repeatedly expands the least-cost vertex from a list of vertices that have been assigned a cost but have not yet been expanded. A vertex is expanded by assigning a cost to the successor on each edge incident out

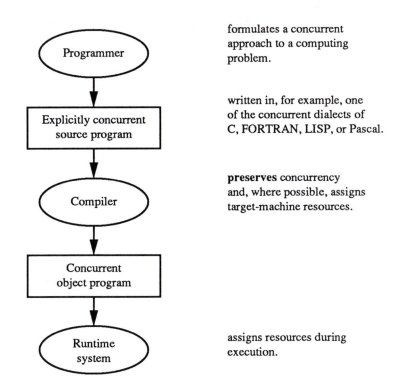

formulates a concurrent
approach to a computing
problem.

written in, for example, one
of the concurrent dialects of
C, FORTRAN, LISP, or Pascal.

preserves concurrency
and, where possible, assigns
target-machine resources.

assigns resources during
execution.

Figure 1.25   Radical, large-N approach to concurrent programs.

from this vertex. Because the vertices are expanded in order of increasing
cost, each vertex is expanded only once; however, the program requires
that the vertices be stored in a data structure that allows the least-cost
vertex to be removed and new costs to be assigned. This data structure
requires that all steps in the computation be sequential. Any program
transformation that may involve expanding multiple vertices concurrently
is disallowed, since the expansion of the least-cost vertex may assign new
costs that change the list. However, computing the shortest path in a
graph can be performed by an algorithm in which the vertex costs are
not kept ordered, and the vertices can be expanded concurrently. This
Chandy-Misra concurrent shortest path algorithm [10] does not assure that
a vertex is expanded only once (some redundant work may be performed)
and involves some care with termination detection, but exploits nearly all
of the concurrency inherent in the problem. The compiler that is able to

perform a translation as general as that from the Dijkstra to the Chandy-Misra algorithm has not yet been written, and probably never well be.

As a homework problem, you may wish to consider why the translation from the Chandy-Misra to the Dijkstra algorithm is not so forbidding. This translation is a program refinement that involves adding sequencing constraints.

The principal difference between the implicit concurrency that a compiler can discover in a sequential program and the explicit concurrency that a programmer can find in a computing problem is the difference between the innermost and outermost concurrencies. The analyses required to extract concurrency from sequential programs are tractable only within the narrow context of inner program blocks. The programmer, however, has the "big picture," and can work with the largest data structures within the program. This approach has the wonderful advantage that the available concurrency typically increases with the problem size. My favorite example of this phenomenon is computing ray-traced pictures. If you need to compute a million picture elements, they can all be computed independently from the same database of objects. The efficiency of ray-tracing computations is increased by operating on small regions determined by a quad-tree decomposition, but there is still much more concurrency in the problem than there are computing elements in today's most highly concurrent machines. Now, make the problem larger by increasing either the number of picture elements or the number of objects in the database. The amount of concurrency increases nearly in proportion.

The radical approach offers very large gains for some problems, but depends on the creativity of the programmer to come up with a concurrent formulation of a computing problem starting with an understanding or analysis of the problem. A good knowledge of the fundamental algorithms and an ability to analyze the problem are critical to this stage of the program design. However, the mechanics of writing the code are not very different from the mechanics of writing sequential programs. There are now many programming notations from which to choose. Unfortunately, most of them are specific to a single architecture. There are message-passing extensions to C (Cosmic C), FORTRAN, Lisp (Concurrent Common Lisp), and Pascal; notations specifically developed for MIMD message-passing programming, such as CSP, Occam, and Cantor; shared-variable extensions to all of the common sequential languages; and data-parallel variants of C (C*) and Lisp (*Lisp) for the SIMD Connection Machine. Let me mention also Chandy and Misra's Unity notation [11], not because it is a programming notation for which there are compilers for concurrent computers, but because it is a very good way to start the design of concurrent programs. These programming notations allow the construction of highly efficient programs, because the terms of expression are close to the ways in which the machines operate. Where we would like to see improvement is in convenience and portability, namely, higher-level concurrent programming notations that allow programs to be mapped to a variety of architectures.

Unlike the "parallelizing" compiler, which attempts to *detect* the concurrency hidden in a sequential source program, the compilers for these concurrent programming notations need only *preserve* the concurrency that is expressed in the source program. It is perfectly legitimate for such a compiler not to exploit all of the concurrency available; for example, two concurrent processes might be combined if the compiler can detect that there is no benefit in running them concurrently. However, except for inner-block concurrencies such as vectorization, it is not necessary that these compilers perform any reorganization of a program. Thus, the structure of the object program in execution very closely reflects the structure of the source program. The compiler and programming system may assign target-machine resources in advance of execution, but assignments that cannot be deduced from the source program end up being deferred to the runtime system.

In summary, the radical approach appears to be necessary for large-N concurrent systems such as multicomputers and the Connection Machine. Unfortunately, we cannot generally run dusty decks efficiently on these highly concurrent systems; however, the authors of concurrent programs often recycle procedures (such as code for physical phenomena) from existing sequential programs. There have been some thousands of concurrent programs written using this radical approach, so it is probably no longer regarded as radical. (It requires an entire file cabinet for me to store all of the papers that people have sent me about the interesting programs that they have written for multicomputers.)

This picture of the two distinct approaches (Figures 1.24 and 1.25), compilation for small-N systems and concurrent programs for large-N systems, was seen clearly by many researchers in concurrent computing as long as a decade ago. Even though my research group has concentrated entirely on the architecture and programming of large-N, or *scalable,* systems, many of the visitors to our laboratory end up asking, "Why aren't you doing anything about automatic translation of sequential programs into highly concurrent programs?" I wish that the dichotomy in the feasible approaches for small- and large-N systems were more widely understood.

Ernst Mayr (Chapter 2) will show you fundamental concurrent algorithms expressed in terms of programming with shared variables (shared memory). Because of the interest of the combinatorial-algorithm researchers in asymptotic results, these algorithms are, of course, aimed at large-N systems and are expressed as explicitly concurrent programs. I want to build on Ernst's excellent lectures by showing you that these fundamental algorithms are equally applicable to message-passing systems. They differ in expression because message-passing and shared-memory systems support somewhat different abstractions for the programmer. Operationally speaking, the difference is that shared-memory systems share data efficiently, whereas message-passing systems copy or move data efficiently. The features of message passing that may seem peculiar to the programmer accustomed to shared memory are that the copy operation occurs across processes (between disjoint address spaces) and is initiated by the process that holds the data rather than the process that requires the copy.

Let us compare these abstractions according to their function. The notion of a *process* applies to both shared-memory and message-passing systems. However, a shared-memory process can access both private and global variables, whereas a message-passing process can access only private variables, but can also send and receive messages. The global address space in shared-memory systems provides a global name space. The same function is provided in message-passing systems by the process identifier, which is typically represented as an ordered pair, (node, pid). The unique numbering of the nodes, together with the unique numbering of the processes within the nodes, establishes unique process identifiers. Distributed mechanisms for mapping names into other types of references can be constructed readily within either type of system.

Shared-memory and message-passing systems differ significantly in their low-level synchronization primitives. Shared-memory systems typically employ P and V operations to provide periods of exclusive access by one process to "critical sections." For simple binary semaphores, P and V correspond respectively to operations that *lock* and *unlock* a critical section. Implementations of P and V require only an atomic (hardware) test-and-set operation to a shared variable. The original reference is Dijkstra's 1968 paper on "Co-operating Sequential Processes" [15], which also introduces the concept of general semaphores that can take on integer values. Alain J. Martin's 1981 paper, "An Axiomatic Definition of Synchronization Primitives" [24], is an excellent exposition that shows the correspondence between the synchronization properties of $P–V$ operations and send–receive primitives.

The corresponding low-level message-passing synchronization mechanisms are the message-sending and message-receiving operations. Messages are rather elegant, because they are both data-carrying and synchronization mechanisms. There are two common types of send and receive operations:

Synchronous communication is defined such that the number of completed receive operations on one end of a channel is identical to the number of completed send operations on the other end. This type of communication, which is used by CSP and Occam, is also characterized as *zero-slack* and *unbuffered,* and can be thought of as a synchronization between processes. Because this form of send–receive communication induces the strongest possible synchronization, it has somewhat awkward compositional properties. When a buffered channel is required, a buffer process can be "spliced" into a channel; however, communication through any intermediate process is no longer synchronous. It is difficult to maintain such a strong property in composition.

Asynchronous communication   is defined such that the number of completed receive operations in a process is not greater than the number of completed send operations directed to that process. (It could hardly be the other way around!) Because the number of completed send operations can be arbitrarily greater than the number of completed receives, this communication is also described as *unbounded-slack, buffered,* and *arbitrary-delay.* The communication primitives used in the Cosmic Cube and its commercial descendants are of this type; in addition, these systems assure that message order is maintained between pairs of communicating processes.   Systems that employ asynchronous communication are able to execute processes and transport messages concurrently.   Asynchronous communication has relatively weak synchronization properties that are defined between processes and the messages that they produce and consume; however, stronger forms of synchronization, such as synchronous send and receive operations, can be implemented very easily and efficiently in terms of these asynchronous primitives.

In the examples that follow, we will be using the asynchronous form of send–receive communication.

In seeking common ground by using higher-level synchronization operations, *barrier synchronization* is a useful abstraction for both shared-memory and message-passing systems.   The concept is that a set of processes in execution crosses a "barrier" as an atomic action; that is, after all processes have reached the barrier, all traverse it at once. Barrier synchronization is useful for separating different phases of a concurrent computation.

In spite of the differences between shared memory and message passing, it is trivial to show that either can simulate the other reasonably efficiently. It is more common to simulate message-passing systems on shared-memory architectures than the other way around, both because it is somewhat more practical with today's machines, and also because people are so fond of the message-passing paradigm that they will often write explicitly concurrent programs in terms of message passing even if they are going to run the programs on shared-memory machines.

While compiler writers tend to prefer shared memory (for reasons that may have been illuminated by the small- versus large-N discussion), I believe that people writing explicitly concurrent programs tend to prefer message passing. If the users of the multicomputers designed at Caltech have had less trouble using concurrent machines than skeptics had predicted, it may be because the message-passing paradigm is intuitive and contains few pitfalls. For example, if a message arrives for a process that does not exist, the user gets an error message that says "message for nonexistent process," followed by information intended to identify the origin of the errant message. On the other hand, if a shared-memory program writes to some random place in global memory, it may be a long time before the consequences of this programming error are discovered. At a more formal level, the preference for message passing can be justified from composition properties. It is not generally possible to reason about a shared-memory process without knowing precisely how the other processes that access the shared memory behave. The protocols for accessing memory are unrestricted. However, the isolation of a message-passing process within its own address space and the interface discipline provided by the message functions make it possible to reason about the behavior and interfaces of a single process.

Programming systems that involve explicit message-passing communication encourage a division of effort between processes and messages. Processes are stationary, but active; whereas messages are inert, but mobile. Processes react to messages by changing their internal state, sending messages, and (in some systems, but not CSP) creating new processes. Messages are inert with respect to their internal data; you would not want the contents of messages to change. People seem to relate very easily to this process–message programming paradigm. In fact, I have noticed that Ernst Mayr describes even shared-memory computations with expressions such as "this processor wants that" or "asks that." Of course, "wanting" or "asking" is not really what is happening in a shared-memory machine; what a processor is doing is accessing in the common memory a variable that might also have been accessed by another processor. However, our short, operational ways of thinking about what is going on in programs are often conveniently expressed in terms of interactions of information-carrying entities.

Before we get hung up on preferences for shared memory versus message passing, let us look at an example that illustrates how the fundamental algorithms that Ernst Mayr was showing you in his first lecture are perfectly applicable to either scheme (Figure 1.26). This is an example of *doubling* applied to a message-passing program. When I first learned of such algorithms from a 1968 paper by Ken Batcher [4], I found it easiest to understand them in terms of a sequence of operations that exchange information on successive dimensions of a direct binary n-cube. One of the reasons why our early multicomputers used a binary n-cube network was these log-time algorithms for computing *global* functions, such as the sum, minimum, maximum, prefix sum, or rank over data items provided from each node.

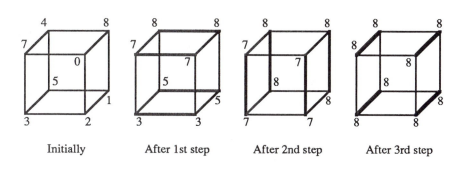

Initially	After 1st step	After 2nd step	After 3rd step

Figure 1.26   A message-passing version of a doubling algorithm.

Although Ernst developed these doubling algorithms in a tree form, I shall use a binary n-cube form and later show you the relationship to the "doubling tree." Just to keep the example simple, the number of processes and the number of multicomputer nodes are both equal to 2^n. Our problem is to find the maximum element over a set of numbers that are distributed across these 2^n processes. Whether or not the underlying communication structure in the machine is a binary n-cube, we will set up the *process structure*—the set of processes and their references to other processes—as a binary n-cube. For the binary 3-cube pictured (Figure 1.26), each vertex has reference to its three neighbors along each of the three dimensions.

The maximal element in each process can be computed locally, and what remains is to compute the maximum across this entire set. The scheme of exchanging on *successive* dimensions starts with processes sending messages to their neighbors across dimension 0 (the neighbors whose node numbers differ by 2^0), and receiving the corresponding message from the neighbor. Each process will make a comparison, and retain the larger of the two numbers as the current maximum. You might say: "Horrors! I'm doing redundant work. I only needed to do the comparison once." However, when the time required to perform the comparison is much smaller than the message latency, it is advantageous to perform the comparisons twice to allow the messages to be sent concurrently. If the comparison were expensive, you could use a protocol that starts with the even processes sending their local maximum to the odd processes, which would compute the maximum and communicate it back to the even processes. The processes can also keep track of the origin of the current maximum; this extra information is often useful in application programs.

Each subsequent step then exchanges the current maximum found on the previous step until the last dimension of exchange has been completed. What is the progress condition? The monotonicity argument is very easy for this example: Initially, the maximum element exists in one process. On each step, the set of processes holding the maximum element doubles.

Although there is some redundant computation, this variant on doubling not only determines the maximum (or any other associative function), but also distributes the result everywhere. It is just as if you computed the maximum using a tree communication structure, and then broadcast the result back along the same tree, except that we folded these two operations together. This communication pattern not only distributes the results, but can also indicate the identity of the process that holds the maximum element. This pretty algorithm illustrates how devious thinking (or an understanding of the beautiful papers that the theory people write) often pays off in writing these low-level concurrent programs.

If you trace the paths over which the maximum element propagates to all of the processes, you will see the relationship between the tree and binary n-cube forms of this algorithm. This *doubling tree* (Figure 1.27) (also known as a "time-on-target" tree) is frequently used in multicomputer computations to distribute or broadcast data items from a given node or to combine data items into a result at a given node. For example, to compute the maximum and leave the result only in process 0, the program would resemble the binary n-cube form, except that only the odd processes on the current dimension would send messages on each step and would "drop out of the game" after sending this message. Recursive divide-and-conquer computations use the doubling tree to split a problem recursively in a "keep half, give away half" pattern and the same communication structure to combine results. (See [3] for an example of mergesort.)

Let us move on now to the message-passing program (Figure 1.28) that implements this computation. This program fragment is not pseudocode; it was extracted from a real program in C with the usual multicomputer functions. The Cosmic-C functions that are used in this program are:

tmalloc(k) allocates a buffer of **k** bytes for a message that will carry a *message type* and returns a pointer to the message buffer.

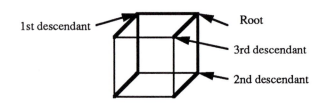

Figure 1.27   Doubling tree embedded in a binary n-cube.

```
int   N, node, pid;                     /* reference context   */
N = nnodes(); node = mynode(); pid = mypid();
...
typedef struct { float  max;            /* Each msg contains max */
                 int origin; } msg;     /* and origin elements. */
float max;                              /* Local max and origin. */
int   origin, D;                        /* D = 2**dimension.   */
msg   *p;

max = maxelement(my_subset);            /* Initialize maximum and */
origin = node;                          /* its origin from this   */
D = 1;                                  /* node, and D=2**0.      */

p = (msg*) tmalloc(sizeof(msg));        /* Get working msg buffer */
p->max    = max;                        /* and initialize its max */
p->origin = origin;                     /* and node elements.     */

while (D < N)                           /* Iterate logN times.    */
{
    tsend(p, node^D, pid, D);           /* Exchange msgs across   */
    p = (msg*) trecvb(D);               /* dimension log(D).      */
    if (max > p->max ||                 /* If my max > max in msg */
       (max == p->max &&                /* or they are equal and  */
        origin > p->origin))            /* my origin is larger,   */
    {
        p->max    = max;                /* assign my values to    */
        p->origin = origin;             /* working msg buffer;    */
    }
    else                                /* else                   */
    {
        max    = p->max;                /* assign msg buffer max  */
        origin = p->origin;             /* and node to my vars.   */
    }
    D = D * 2;                          /* Next dimension.        */
}
tfree(p);                               /* Free working buffer.   */
...   /* max now contains the maximum element, origin            */
      /* specifies the node from which it originated, and a      */
      /* barrier synchronization has been accomplished.          */
```

Figure 1.28 Cosmic-C program fragment for computing the maximum element across a set of cohorts.

tfree(b) frees the message buffer pointed to by **b**.

trecvb(t) receives a message of type **t** and returns a pointer to the message buffer. This function is semantically similar to **tmalloc**, except that the size and initial contents of the message buffer are determined by the earliest-arriving message of the specified type.

tsend(b,n,p,t) sends the contents of the message buffer pointed to by **b** as a message of type **t** to process **(n,p)** and also frees the message buffer. This function is semantically similar to **tfree** except for the side effect of sending the message.

nnodes() returns N.

mynode() returns the node part of the identifier of the calling process.

mypid() returns the pid part of the identifier of the calling process.

Let us walk through this program just to see what the inside of a multicomputer program may look like. The integer variables **N**, **node**, and **pid**, ordinarily declared in the outermost program block and initialized as shown, establish the reference context for the process. The $N = 2^n$ processes have identifiers **(0,pid)**, **(1,pid)**, ..., **(N-1,pid)**. In this common process structure, called a *set of cohorts,* the cohorts are spawned from the same program, or, more poetically, they are all cast from the same mold. Each cohort has implicit reference through the node numbers to all of its cohorts. Of course, multicomputers are not constrained to running the same code everywhere, so this example does not show off what MIMD machines can do that SIMD machines cannot do. On the other hand, this example illustrates a fairly general technique for running SIMD computations on multicomputers.

The **typedef** declaration defines a **msg** structure composed of a floating-point member, **max**, and an integer member, **origin**. The declarations and initialization also include **max**, which is initialized to the maximum within the subset stored in this process; **origin**, which is initialized to the node number of this cohort; and **D**, the iteration variable that will specify the successive *dimensions.* A message buffer for the first message is allocated by calling the **tmalloc** function; the pointer **p** that is returned is used to initialize the **max** and **origin** members to the values for this cohort.

Within the limits of this one-page program fragment, I am trying to sneak in quite a few facts and concepts. One thing I do not want you to miss just because it is hidden in the single statement **max = maxelement(my_subset)** is that a process generally does not hold a single element of a distributed set, but a particular subset of the whole set. I am assuming for this example that the data has already been distributed. This assumption differs from Ernst's example, in which it is possible to pass a reference to the set S as an address in global memory. You would not try to fit the entire working set of a large computation into a single node of a distributed-memory machine such as a multicomputer or Connection Machine. Instead, programs such as matrix packages ordinarily employ a fixed convention that determines how the matrices are stored, hence, how they are loaded, unloaded, and computed upon. This distribution by subsets is typical of many application programs. For example, a 64-node medium-grain multicomputer is typically not used for matrix computations in which the matrices are 8×8, or even 64×64, but more like 1000×1000. The matrix will be distributed across a set of cohorts such that each cohort stores its own share of the matrix, such as 16 columns. (You will recognize that I am applying the same technique that I used for the FFT example in my first lecture.) To locate the maximum diagonal element of the matrix to serve as a pivot, the program first finds the local maximum, and then enters the distributed part of the computation to find the maximum diagonal element of the entire matrix.

All we have left to discuss in this example program is the iteration. From the **while** condition and the last statement in the body of the iteration, D = D * 2, we see that the body of the iteration is not executed for $N = 1$. This is correct; a single process passes directly through the barrier. For $N > 1$, the iteration will be executed with values of D in the sequence $1, 2, 4, \dots, N/2$. Each step starts by sending a message of type D across dimension $\log_2 D$ to the cohort in **node**$\oplus$D, where $\oplus$ is the bitwise exclusive-OR operation. The send is immediately followed by the receipt of the corresponding type-D message.

It is worth pausing here for a moment to think about the synchronization implications of this **tsend–trecvb** pair. In an MIMD system, we cannot actually force all of these processes to execute the same instruction at once. However, it is sufficient for a synchronization to assure that no process returns from the **trecvb** call until its synchronization partner has called the matching **tsend** function. Assuming that there are no other type D messages directed to this process, this is exactly what the message exchange assures: Neither process can return from the **trecvb** function until the other has entered the corresponding **tsend** function. Thus, this simple message exchange is a zero-slack synchronization (or a two-way barrier synchronization).

Following the message exchange, the program compares the local **max** with the maximum received in the message, **p->max**, together with a special condition in the event that these two values are equal, and then assigns the (**max,origin**) pair to the local variables or **msg** members such that the pair with the larger **max** prevails. You might notice how the message buffer that was initially allocated is "given away" by the **tsend** function, but another message buffer is then acquired by the **trecvb** function.

When the iteration terminates, the message buffer acquired in the last iteration can be freed, and the local variables **max** and **origin** specify the maximum element and the cohort from which it came. A barrier synchronization has also been accomplished in the usual distributed MIMD sense: No process executes beyond the iteration until all processes have at least entered the iteration. A formal demonstration starts with the property of the pairwise synchronizations mentioned above: The ith iteration in process p will not terminate until the ith iteration in process $p \oplus 2^{i-1}$ has been entered. Because these two-way synchronizations occur across dimensions *in sequence,* we can appeal to induction on i to show that the termination of iteration i in process p assures that the set of cohorts, $p \oplus q, 0 \leq q < 2^i$, has entered the iteration. Thus, any cohort is assured when the $\log_2 N$th message is received that all of its cohorts have at least entered the iteration, and an N-way barrier synchronization has been accomplished.

The use of a message type is an example of *message discretion*: It allows a process to *exercise discretion* as to which message it receives next. Messages are normally queued in the order in which they arrive; but if the messages are typed, they are also sorted into different queues according to type. Because the maximum function is commutative and associative, it may appear that the program would still get the right answer even if two messages were to arrive out of order. However, it is easy to construct an example in which the maximum element arrives out of sequence and late, and thus would not be distributed properly to all of the cohorts. So, the dimension-order sequencing really is important for this problem, and we employed typed messages to enforce the ordering.

The messages in the original Cosmic Cube always had types, and the message primitives always exercised discretion by type. In the second-generation multicomputers, we use lower-level primitives without types (the functions **xmalloc**, **xfree**, **xsend**, and **xrecvb**), and provide libraries that implement families of specialized message functions in terms of the **x** primitives. The **t** (typed) library functions simply put the type in a hidden place in the message, and maintain separate queues for different types of received messages. For other kinds of message discretion—such as by priority, sending node, sending pid, or length—there are other libraries.

The type field for the **t** functions is a 32-bit integer. I already mentioned that the message types used in the example must not conflict with message types used in the rest of the program. It is good insurance against bugs to assign a range of types for each phase of a program and for each barrier between these phases. This is easily accomplished by adding a defined constant to the type **D** in the example. A liberal use of message types is some insurance against mistakes. I have been familiar with the properties of this class of algorithms, and with the general transformation from SIMD to MIMD forms, for so many years that I could write this example by rote. More generally, however, we discover the necessity for the synchronization properties of programs by understanding why we cannot prove that a more weakly synchronized version is correct. The proofs need not be very formal, and it is easier to reason about the correctness than it is to live with bugs.

There is enough nondeterminacy in the process scheduling and message system that you have to be careful. Nondeterminacy is the essence of concurrency; it is what allows a concurrent system to make opportunistic choices of which process to run or which message to assign to a communication channel. Nondeterminacy makes these programs portable: They run on multicomputers, multiprocessors, and "ghost cubes" (network-connected UNIX processes emulating multicomputer nodes). Process scheduling is highly nonuniform across this set of execution engines. Perhaps the program-correctness situation is not so bad because the users who are writing relatively more sophisticated programs are also relatively more sophisticated programmers. There is certainly a "Peter Principle" being applied here: People write concurrent programs up to the limit of their own competence in writing concurrent programs.

Perhaps it would surprise you that we advocate a style of writing such programs that minimizes the sequencing constraints and thus maximizes the nondeterminancy. An alternative to using message types in this example would have been to place a barrier synchronization in the body of the iteration. The program would be correct, but a barrier within the iteration would enforce a stronger synchronization than is necessary. This example illustrates the style of using the weakest synchronization we can manage in a given computation. Weakly synchronized execution sequences allow the largest range of opportunistic scheduling choices in execution. By using a sequence of pairwise synchronizations, this computation allows the iterations to get more than one full step out of synchrony. For example, the computation of the maximum might be completed in some subset of the cohorts, allowing them to proceed, before it has been computed in the remaining cohorts.

As a homework problem, can you explain the equality part of the `if` condition in the example program (Figure 1.28)? I glossed over it in my earlier explanation. What if `max` and `p->max` are equal? Will the program get the right answer anyway?

1.4   Application Programming

Inevitably society will judge our research community's efforts in theory, programming, and engineering by its applications. Thus, the subject of applications may be quite appropriate for this last lecture of the workshop. You have heard from several people who have visions of concurrent systems that will have truly spectacular performance. Few of us doubt that this is where the *technology* is headed, but the real *opportunity* (and fun) in advancing the practice of highly concurrent computing is that we will be making possible computations that today are infeasible, or even beyond imagining.

This short section is devoted to examples of the formulation and design of application programs written for multicomputers. Most of the techniques illustrated by these examples apply to the design of explicitly concurrent application programs for other architectures. Multicomputers have been available long enough and in large enough numbers to have stimulated an interesting variety of programming projects. In place of a long list of application programs, and before individual examples, a short list will give you a sense of typical application areas.

Computational-linear-algebra problems   include matrix operations and the solution of systems of linear equations. The concurrency in these computations is typically at least as large as the rank of the matrix. Computations that can be expressed in terms of systems of linear equations or in terms of matrices appear in many areas of science and engineering. A later example will examine the concurrent formulation of these highly regular problems.

Interaction problems   involve computing the evolution over time of particles or bodies that influence each other. *Universal interaction* under such forces as gravity, in which all bodies influence all other bodies, is surprisingly easy in naive formulations for concurrent computers. *Local interaction* under such forces as the contact between particles in flowing granular materials involve additional complications, but present no serious difficulties in formulation. Interaction problems sometimes emerge in unexpected areas; for example, certain fluid-flow problems can be formulated as n-body problems in which the bodies are vortices.

Mesh or grid-point computations   are based on a systematic subdivision of space such that each process manages the computations for a region and exchanges state information with neighboring regions. Classical relaxation and finite-difference techniques are commonly used to solve field problems on a discrete grid; more generally, spatial or domain decompositions can be employed to expose the concurrency in many other types of problems, such as image processing and VLSI-layout analysis.

Monte Carlo methods   employ a random, representative sampling of a physical phenomenon or configuation space in combination with other techniques from which the concurrency is derived. For example, a Monte Carlo simulation of a jet plume may compute the interactions of only a representative sample of gas particles within a region, or a computation of a diffraction pattern may accumulate the effects of only a representative sample of photons. Probabilistic-iterative-improvement optimization computations, such as simulated annealing, are also classified as Monte Carlo methods. Simulated annealing has been shown to work well on concurrent machines, even in a form in which concurrent decisions of whether to apply a given move may be in conflict. The convergence rate is tolerant of a substantial latency between the move decision and the application of the move.

Signal-processing, FFT, encryption and decryption, and error-correcting coding and decoding computations are typically patterned after the pipelined or systolic algorithms for regular computational arrays [21].

Computer graphics is a wonderful application for high-speed concurrent computers. Graphics computations are demanding and exhibit large degrees of concurrency not only in transforming the object space to a particular viewpoint, but also in rendering or ray-tracing objects in the image space. Programs for nearly all of the common computer-graphics techniques have been written, and they perform very well on concurrent computers.

Circuit simulation, a transient analysis of the behavior of an electrical circuit containing nonlinear elements, is a specific application that is said to consume about one quarter of the world's high-speed computing cycles. This application is important also because it is generic to engineering analysis in other fields. For example, the CONCISE concurrent circuit simulator [25], which is functionally similar to the familiar SPICE circuit simulator, has been adapted to perform simulations of systems such as distillation columns [37]. CONCISE computes a piecewise-linear model to which linear-equation-solving techniques are applied. If the solution is outside of the linear range, it is used as the initial condition of another linearization and solution. Both the model evaluation and the equation solving contain large degrees of concurrency for large systems.

Distributed discrete-event simulation, also known as event-driven simulation, computes the behavior of systems composed of large numbers of elements whose behavior and influence on each other can be modeled by discrete events. Typical applications are the simulation of logic circuits, computer networks, or military engagements. In contrast to circuit simulators, which operate by solving simultaneous equations, discrete-event simulators determine the behavior of a system by computing from cause to effect. The physical processes can be represented by processes, and their influences on each other can be represented by messages. The concurrency in the simulation reflects that in the system being simulated.

Graph computations include concurrent versions of shortest-path, maximum-flow, partitioning, clique-finding, and other common graph algorithms. (Message-passing versions of several concurrent graph algorithms can be found in [12].) Graph computations are central to many applications in computer-aided design, program analysis, and discrete optimization.

Concurrent data structures   [12], which might also be called *distributed data structures,* are the same abstractions as are common data structures, such as dictionaries, heaps, sets, and ordered sets, which encapsulate data together with the operations defined on this data. When the data is distributed across a set of cohorts, operations defined on the data structure can be performed concurrently. An operation is invoked by sending a message to any cohort and will generally involve communication and synchronization between the cohorts. The hope is that by encapsulating the "tricky" synchronization and locking requirements of a concurrently accessed data structure, the construction of certain types of concurrent programs can be reduced to writing processes that operate on the concurrent data structure.

Game-playing programs   are an inevitable recreation. These programs generally employ the same search strategies as their sequential counterparts, but divide the search among a set of processes.

Simulated evolution,   also known as genetic algorithms, might be regarded as a form of probabilistic-iterative-improvement optimization. My favorite example is a program written by Reese Faucette, a Caltech student who, as a staff member in my research project, also wrote the Cosmic Kernel, the original node operating system for the Cosmic Cube. Reese decided one weekend to make the Cosmic Cube play the three-dimensional, 4×4×4 version of tick-tack-toe. Only after finishing the program did Reese realize that he had made a terrible conceptual mistake: The program could not be made concurrent in the way he had planned. However, he had a working sequential program, so he said, "Ha! I am going to resurrect this code to get some good out of it, anyway. I will write a host program that spawns these tick-tack-toe programs in the cube nodes, and plays them against each other." The decisions made by Reese's tick-tack-toe program were based on an objective function with 22 parameters. Processes that tended to win were allowed to reproduce, but with small perturbations in their parameters. Processes that tended to lose were killed off. An overall control kept about 200 such processes in play at once. The program started out playing such a pathetic game that anybody could beat it, but, after about twenty-four hours, it was invincible. This recreation was one of the early simulated-evolution programs for concurrent computers; many others have since been written. If there is a lesson in this story, it is that making these machines available to creative people often has unforeseen consequences and contributes to the fun and excitement.

Neural-network models   consume a lot of time on our multicomputers at Caltech. For example, one of Yaser Abu-Mostafa's students was running a program every night for months on our 128-node iPSC/1 to test the conjecture that Boolean functions of greater information-theoretic complexity are more difficult to learn. Although one can find a substantial amount of concurrency in neural-network simulations, this program worked simply by "dealing out" cases to processes in the nodes.

Numerical sieves   are listed here only because they are an old hobby of mine and can be depended on to provide an endless supply of jobs to fill idle cycles. Most of these programs employ a recursive divide-and-conquer search strategy to partition the search to a very large number of processes.

All of these programs were formulated and written using a process model in which the concurrency and distribution of processes is explicit. I hope you get some sense from these descriptions of how the design of an explicitly concurrent program depends on the programmer's overall understanding of a problem.

Although the source of the concurrency is evident in these examples, the design of an application program also involves considerations of load balance, communication, and program structure that may favor one formulation over another. A highly regular problem makes a good initial example.   LINPACK [16] and EISPACK are standard packages used in scientific-computing circles for performing matrix operations, inverting matrices, solving systems of linear equations, and finding eigenvalues or eigenvectors. Cleve Moler, one of the original authors of LINPACK, worked at Intel Scientific Computers during their formative period and wrote "cube" versions that he called LINCUBE and EISCUBE (pun intended). We shall examine only the basic operation of multiplication of full matrices.

Most matrix packages for concurrent computers have employed a submatrix approach in which, for example, a 1000×1000 matrix would be mapped onto a 64-node machine as an 8×8 matrix of processes, each of which can store a number of 125×125 submatrices. With matrix-element subscripts starting with 0, element a_{jk} of matrix A would be stored in process $(y, z) = (\lfloor \frac{j}{125} \rfloor, \lfloor \frac{k}{125} \rfloor)$. We refer to the submatrix in process (y, z) as A_{yz}. The 8×8 matrix of processes can be embedded systematically and optimally into either a hypercube- or mesh-connected multicomputer.

Multiplication of matrices $C = AB$ that are stored in this way involves

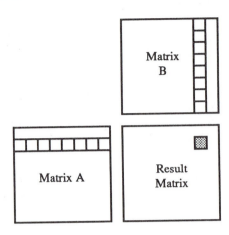

Figure 1.29   Matrix multiplication by submatrices.

the multiplication and accumulation of 8 submatrices in each of the 64 processes:

$$C_{jk} = \sum_{m=0}^{7} A_{jm} B_{mk},$$

where A_{jm}, B_{mk}, and C_{jk} are the submatrices. The possible communication patterns are the same as those for an 8×8 multiplication, except that submatrices rather than individual elements are conveyed in the messages (Figure 1.29). If the submatrix product C_{jk} is to be computed in process (j, k), in which it will be stored, it is necessary for the processes (j, m) and (m, k), $0 \leq m < 8$, to send copies of their submatrices to process (j, k). This communication can be accomplished using any one of a variety of coordinated patterns that control the size of the message queues and initiate a submatrix multiplication on each step. Perhaps the simplest approach conceptually is to cause submatrices A_{jm} to be broadcast in rows and submatrices B_{mk} to be broadcast concurrently in columns, for successive values of m. The broadcast can be accomplished using pipelining or the doubling algorithm discussed in my preceding lecture. When a process has received the pair of submatrices, it multiplies them; then, it prepares to participate in or originates the next broadcast. There are many other ways to traverse the range $0 \leq j, k, m < 8$ that have suitable properties. In any case, it is evident that each matrix element that is received in a message will participate in 125 multiply-accumulate operations; hence, the message bandwidth need be only a small fraction of the node's multiply-accumulate bandwidth.

Cleve Moler, with his deep understanding of LINPACK and the performance metrics of these machines, used an alternative representation that places matrix element a_{jk} in node k mod N. Thus, for our 1000×1000 matrix on a 64-node machine, columns $0, 64, 128, 196, \ldots$ will be stored in process 0, columns $1, 65, 129, 197, \ldots$ will be stored in process 1, and so on. (You may wish to compare this periodic assignment with the mappings that Lennart Johnsson shows in Chapter 4.) On the surface, this so-called "mod-N" or "wrap" mapping appears to be less efficient for matrix multiplication than the submatrix formulation. Visualize the matrix multiplication $C = AB$ as multiplying each row vector from A by the entire matrix B to obtain a row of C.

The beauty of this formulation is that matrix B may remain in place while matrix C is computed in place. Although the rows of A must be broadcast to all of the processes, they are distributed across the processes. What shall we do? It requires only $N-1$ messages from each of $N-1$ concurrent processes to transpose A. Because the time required to send these short ($\approx 1000/64$ data items) messages is dominated by the per-message software overhead, the time required to transpose the matrix depends (linearly) on the size of the machine rather than on the rank of the matrix. In addition, transpose operations can often be optimized out of a sequence of matrix operations. Once we have A^t, the set of columns stored in each process can be broadcast simply by pipelining them along the linear array of processes. To generate each element of the result matrix, the inner product is formed between each row vector of A that is received and each column vector of B that is stored in a process.

Moler's formulation is not as efficient in its use of the message system as is the submatrix formulation. Even discounting the transpose operation, each matrix element that is received in a row of A participates in only $1000/64 \approx 16$ multiplication operations. (If the rank of the matrix is r, the submatrix formulation produces $r/\sqrt{N}$ multiply-accumulate operations per element received, whereas the "mod-N" formulation produces r/N.) For sufficiently large problems, it would not seem to matter which formulation is used; however, other characteristics of Moler's formulation allowed his program to outperform programs based on the submatrix formulation on the Intel iPSC/1. Those iPSC/1 computers that were used for intensive floating-point computations were VX models that had floating-point-vector accelerators on each node. The mod-N formulation maximizes the floating-point execution rate by initiating the longest possible vector operation in a single instruction. A more important advantage of Moler's formulation is that algorithms for solving systems of linear equations, such as Gaussian elimination, cause the rank to be reduced on each step. The submatrix approach generally does not leave the load balanced as the rank is reduced, but the mod-N formulation keeps the load maximally balanced.

As an aside: For numerical stability, Gaussian elimination computes a pivot element on each step by finding a maximum across a set that is distributed across the matrix. The routines I showed you in my preceding lecture would be used in the find-pivot phase of this computation. When the location of the pivot is determined, all of the cohort processes can start working on the forward-elimination phase concurrently.

What lessons can we learn from this example? The most important lesson is that the decision as to how the data will be distributed to processes determines nearly all aspects of how the program will be written, and what its performance characteristics will be. Of course, this observation applies to many program-design problems, both sequential and concurrent. There was more than one way to formulate this problem, and the less obvious approach has some advantages. Finally, large computational-linear-algebra problems might be regarded as nearly a solved problem, at least for multicomputers. These programs run with essentially perfect utilization; there is so little that can be improved in the way of efficiency that we might well organize the programs to be easy to understand and maintain.

My second example, the n-body problem, is trickier but less tedious than matrix computations. I included this example not because I believe that any one of you is interested in calculating the orbits of planets, but because I suspect that the same universal-interaction technique is effective for simulating neural networks. Offhand, it would seem that universal-interaction problems should be difficult on machines in which locality is important. Should multicomputers stick to matrix, grid-point, image-processing, and local-interaction computations? Actually, universal interaction is one of the easiest and most efficient computations on all forms of concurrent computers, at least under a naive but unequivocally precise formulation [32].

For n bodies, n processes compute and accumulate the forces due to the $n - 1$ other bodies, followed by the time-step integration. Each of the n bodies interacts with the $n - 1$ other bodies, yielding $n(n - 1)$ forces to compute were the interaction not symmetrical. However, for a symmetrical force such as gravity, it is sufficient to calculate only half of these forces. Once the program has calculated the attraction of body A for body B, it does not need to compute the attraction of body B for body A. The n processes will compute all $n(n - 1)/2$ interactions in $(n - 1)/2$ steps, a time complexity for the interaction phase that is $O(n)$. (To simplify the discussion, we will assume that n is odd so that $(n - 1)/2$ is an integer.) The n processes then perform the time-step integrations concurrently for the n bodies, a time complexity for the integration phase that is $O(1)$.

A ring is an appropriate process structure for calculating the interactions of n bodies (Figure 1.30). Each process is *home* to one body. The computation starts with each process sending the position and mass of the body it houses to the next process around the ring. Initially, concurrent messages are sent $1 \to 2, 2 \to 3, \ldots, n \to 1$. Let us focus on the message that originated from process 1: When it reaches process 2, the interaction

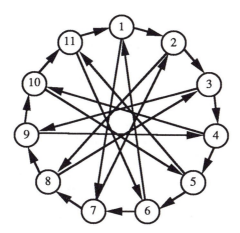

Figure 1.30   Process structure for the 11-body problem.

from body 1 to body 2 is computed; when it is passed on to process 3, the interaction from body 1 to body 3 is computed, and so on. If the force were not symmetrical, the processes could usefully pass the message describing body 1 all the way around the ring until it had arrived at its home process. Concurrently, the messages that originated in each of the other processes have visited all of the processes around the ring, so that all of the pairwise interactions have been computed by this simple structure. If you visualize the "rotation" of the messages against the ring, and have ever seen an old-fashioned DC generator, you will understand why I refer to this process structure as a *commutator.*

For neural-network simulations, each process simulates a neuron, or a set of neurons, and the states are reported in messages that revolve around the ring. The arrival of a state-carrying message triggers the multiplication of the states it contains by the appropriate weights. The accumulated $w_i x_i$ products are compared with the threshold when the neuron's own message returns, and the neuron may then change state. Indeed, as observed in the message flow at any process, the states of all of the neurons will appear to change in synchrony, even though the circulation on the ring is not actually synchronous. If I were asked to build a neural-network simulator, I believe that this is the way I would build it. Commutation is a way of serializing operations that would be performed with needless rapidity in a more direct implementation.

In my enthusiasm for this approach to neural-network simulation, I do not want you to forget the gravity of the situation. For the symmetrical-force n-body problem, it is sufficient to send the message directly home when it gets $(n-1)/2$ steps around the ring. The forces computed in these first $(n-1)/2$ interactions can be accumulated in a field in the message. Meanwhile, the home process has been visited by the messages containing the positions and masses of the other $(n-1)/2$ bodies. When the message returns home, the forces can be combined, and the process can integrate this force into its own position.

These first two examples are formulations that would apply equally to SIMD and MIMD machines. (Examples that can be grasped in the early stages of learning to "think concurrently" often have this unfortunate property.) However, the next two examples are strictly in the domain of MIMD programs and machines: The processes cannot act in lock step.

I mentioned computer graphics as a wonderful application area for concurrent machines. The standard graphics pipeline (Figure 1.31) transforms a collection of points, lines, or polygons to the viewpoint of an observer. The coordinates that describe these objects may be represented as a vector in the form $[x\ y\ z]$, the usual Cartesian coordinates in a three-dimensional space. The first step is to multiply these coordinates by a 3×3 rotation matrix, and to add a translation vector $[t_x\ t_y\ t_z]$. The resulting objects must then be "clipped" to the pyramid of vision. Clipping is an operation in which the endpoints of a line or the adjacent points of a polygon may both be inside the pyramid of vision (trivial accept), both outside the same plane (trivial reject), one inside and one outside (compute the intersection point), or both outside but with the possibility of passing through the pyramid of vision (compute both intersection points). Because clipping is so case laden, it is inefficient on SIMD types of computers. The clipping step also generally incorporates the perspective division and mapping of points to the coordinate system of the display. Finally, it is useful to sort the objects, for example, according to the z coordinate in order to reduce the

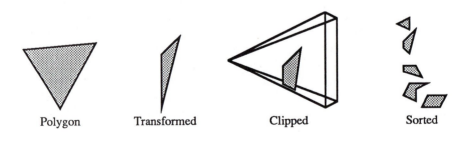

Polygon   Transformed   Clipped   Sorted

Figure 1.31   The graphics pipeline.

buffering requirements for subsequent z-buffer hidden-surface elimination, or according to the y coordinate in order to reduce the buffering requirements for scan-conversion.

Part of my youth was spent designing special-purpose digital engines that performed this computation by streaming the objects through a pipeline composed of successive matrix-multiplication, clipping, and display stages. This approach fit the technology of the time by minimizing the amount of storage in each pipeline stage. Although it would be possible to formulate this computation in the traditional pipeline form, the concurrency is limited and the ratio of computation to communication—the amount of computation associated with each graphic object received in a message—is small. The preferred formulation for multicomputers is the dual approach, in which the parameters of the transformation are sent to a stationary database of objects. In a pattern that should by now seem familiar, this graphics computation is made concurrent by distributing the largest information structure, the database of objects, across a set of N processes. The available concurrency is thus tied to the size of the database. The small message that carries the tranformation is applied to the data set in each process. The database of objects for demanding graphics computations is typically quite large, so that even $\approx 1/N$th of the database is large enough to assure a large ratio of computation to communication.

For each frame that will be computed, the rotation matrix and translation vector are broadcast to these N processes. The broadcast may advantageously be performed for this application using a binary-tree process structure (Figure 1.32) that will be reused to sort those objects that are to be displayed. Each of the N processes reacts to the message containing the rotation matrix and translation vector by transforming its own part of the database of objects, and then sorting the resultant list. These lists are merged into sorted lists by the tree processes. (If the concurrent computer includes a distributed frame buffer, then objects that are to be displayed in different areas could instead be dispatched directly to the appropriate processes.) It is possible for this same structure to be working on more than one display frame at once: New tranformations can be injected into the root of the tree before the previous frames have been produced. This formulation is quite simple and intuitive, and illustrates several issues that arise frequently in concurrent formulations for multicomputers.

Graphics databases are generally composed of objects that share points, such as connected lines or coincident polygon edges. It is most efficient if each point is transformed only once, even if it is part of more than one object. Thus, it may seem to be a good idea to distribute the database so that objects that share points are stored as much as possible in the same process. However, if you were to use this approach for distributing the object database for an application such as a visual flight simulation, you would immediately notice a problem. As the simulated airplane sweeps around

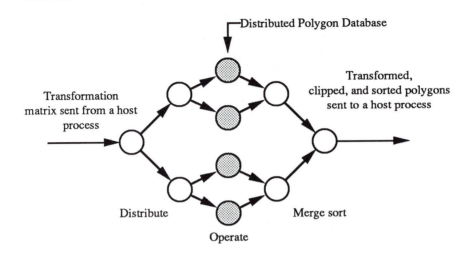

Figure 1.32   Process structure for the graphics program.

the scene, the great majority of objects would be outside of the pyramid of vision encompassing those objects that the pilot would see through the window. Visible objects, which take substantially more processing time than do objects that are outside the pyramid of vision, would tend to be clustered in a small subset of the processes. In order to avoid such "hot spots" in the load balancing, you must fragment the database differently. Rather than trying to localize computations that are logically connected, it is often necessary to disperse the problem by using techniques such as randomization. Depending on the size of the object database, it is advantageous to cluster objects that share points, but only to an extent that would leave a sufficient number of clusters to permit good load balancing in the dispersal of these clusters.

The program that is the template for the N processes that deal with the transformation and clipping of a set of objects is itself able to do the entire computation; thus, it is a sequential program for this same task. Programs for easily distributed problems often have this characteristic and are constructed and tested initially as sequential programs. For a problem as common as this graphics-pipeline computation, the programmer would likely adapt an existing graphics package rather than write the program from scratch. Fortunately, there are no language barriers to this approach, at least for medium-grain multicomputers, because the processes are written in the same programming languages that we use for sequential computers.

This graphics program is a coarse-grain and irregular instance of a category of formulations that are sometimes referred to as *broadcast-combine* problems. You could easily imagine using this same formulation for a problem such as pattern matching: Broadcast the pattern you are trying to match against a distributed set of representative patterns; then combine the responses to obtain only those with the highest scores. Although I am not aware of anyone assembling a systematic library of these kinds of distribution operators, such a library would certainly expedite the construction of these simpler kinds of concurrent programs.

As a final example, distributed discrete-event simulation is an elegant and important application. An effective method for distributing discrete-event simulations was worked out in the late 1970s by Mani Chandy and Jay Misra at the University of Texas at Austin, and independently by Randy Bryant at MIT. In the Chandy-Misra-Bryant distributed event-driven simulation algorithm [9, 28, 7], physical processes are represented by logical processes, and their interactions are represented by messages. The messages contain both state information and simulation time.

Some unnecessary complications are avoided if we assume that the system being simulated is closed and strongly connected, with the environment represented as a simulation process (Figure 1.33). A sufficient condition for progress is that the delay associated with every process be greater than zero. Consider the situation at the process whose output has advanced to the minimal simulation time $t = m$ (Figure 1.34). This process's inputs must be defined at least up to simulation time $t = m$. If this process has a characteristic delay of δ, then it is able to advance the simulation time at its output to at least $t = m + \delta$. Thus, the simulation time advances locally by virtue of the delays associated with the physical processes being simulated. In situations in which processes may have zero or negative delays, the progress argument is easily extended to the sufficient condition that the delay in every closed path be greater than zero.

Of course, the delay of a process might depend on its state; such

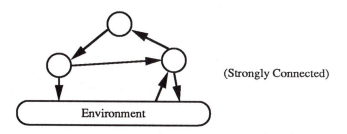

(Strongly Connected)

Figure 1.33 Process structure of a closed system.

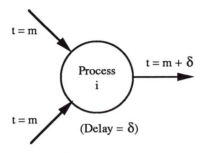

Figure 1.34   Progress of a simulation process.

temporal irregularities are part of its code. A simulation process might also be *nonstrict:* A change of state—an *event*—at an input may or may not cause an output event. For example, consider a two-input logic gate that computes $OR(a, b)$ with delay δ. If input a is defined to be 1 up to simulation time $t = 14$, the output is assured to be 1 up to simulation time $t = 14 + \delta$. If input b were to change $0 \rightarrow 1$ at simulation time $t = 11$, this input event would not initiate an output event. However, if a subsequent message extends the period during which input a is 1 up to simulation time $t = 17$, the output can be advanced to be 1 up to simulation time $t = 17 + \delta$. Messages such as these, which advance the local simulation time but do not contain any events, are called *null messages*.

According to the progress argument, it is necessary for the simulation processes to advance the local simulation time at every opportunity; otherwise, the simulation may deadlock. An alternative to this *eager* message sending is employed in practical simulators for such systems as networks of logic gates [38]. It would otherwise be very costly to simulate low-activity-level circuits that exhibit a small delay in a closed signal path, such as cross-coupled gates that form a latch. Whereas a sequential simulator organized around an ordered event list would need to simulate the gates only when an input changes, an eager-message-sending distributed simulator would produce a flood of null messages simply to advance the local simulation time around this circuit. In order to combine as much information into each message as possible, a simulator can employ a *lazy* message-sending strategy. For example, null messages can be combined with subsequent null messages and with event-containing messages. In order to detect locally the deadlocks that result from the failure to send messages eagerly, we can use the nonblocking **xrecv** function as follows:

```
while (1) {
    if (p = xrecv()) simulate(p);
    else send_queued_messages();
}
```

If the pointer returned by **xrecv** is non-NULL, it is passed to the **simulate** function, which performs the simulation and may either send or queue any resulting output messages. If the pointer is NULL, this process has run out of work to do, and will send at least one queued message (if it has any). A deadlock cannot occur without "starving" processes; when this occurs, the starved processes take actions to break the deadlock. The simulation functions may employ arbitrary or adaptive strategies for reducing the message traffic. It is most often profitable to combine null messages with subsequent event-containing messages because the null messages are less likely to be required for progress in the destination process.

The lessons that one can learn from discrete-event simulation are deeper and more subtle than those of the earlier examples. Because the simulation processes do not have access to a global event list, but instead must keep track of simulation time entirely locally, the distributed simulation necessarily involves more work than the sequential simulation. We can approach the efficiency of sequential simulation only by using guesses as to what information will be required by other processes. Much of our research in scheduling multiple-process computations is based on studying simulations; indeed, because of their reactive character, these processes are equivalent to the program objects of object-oriented programming. The origins of object-oriented programming are closely intertwined with simulation; for example, Simula was originally developed as a simulation language.

In place of a detailed summary of discrete-event simulation, let me try to interest you in a possibility. Logic simulation is an important tool for designing VLSI chips, and I can report to you that logic simulation runs extremely well on multicomputers. There are great opportunities for running other VLSI computer-aided-design and -analysis tools on concurrent and parallel machines. Part of what makes the present state of VLSI CAD and analysis unsatisfactory is that we are trying to design very complex chips using machines that provide 4 MIPS or so. Overnight or several-day runs are common. What could we do if we could put 1,000 MIPS behind some of these operations?

One of our Caltech undergraduates recently wrote a multicomputer program that reads CIF, flattens it, computes its cell bounding boxes, distributes the geometry into strips, and performs geometrical-design-rule and well checking. For a large chip on a 64-node second-generation multicomputer, it takes about 5 seconds to read and flatten the CIF, and then 0.2 seconds per rule for geometrical-design-rule checking. The whole chip is checked in a minute instead of an hour. Geometrical-design-rule checking is only a beginning. The tools that take the most computer time have generally been analysis tools such as design-rule checking, circuit simulation, and switch-level simulation. I mentioned the CONCISE circuit simulator earlier. Randy Bryant tells me that he is confident that his new switch-level simulator, Cosmos, will run very well on multicomputers [8]. What about synthesis tools? What could you do if you put 1,000 MIPS behind your silicon compiler?

If you get the impression from these examples that each concurrent program involves a piece of ingenuity that is idiosyncratic to the problem, you should attribute this impression to my obligation to select examples that exhibit some diversity. These concurrent formulations also exhibit many common features that are suggestive of a fairly small set of techniques or abstractions that cover the great majority of problems. However, there are many opportunities for new and unique approaches. Programming concurrent machines may involve putting on the hat of a manager and figuring out a scheme by which a collection of somewhat independent entities can work together. We humans have a lot of experience with cooperation and teamwork, and we should not hesitate to use this experience in developing paradigms for formulating concurrent computations.

Bibliography

[1] Applegate, J. H., M. R. Douglas, Y. Gursel, P. Hunter, C. L. Seitz, and G. J. Sussman. A Digital Orrery. *IEEE Trans. on Computers*, C-34(9):822–831, September 1985.

[2] Athas, W. C. Fine-grain concurrent computation. Caltech Computer Science Tech. Rep. (Ph. D. Thesis) 5242:TR:87, 1987.

[3] Athas, W. C. and C. L. Seitz. Multicomputers: Message-passing concurrent computers. *Computer*, 21(8):9–24, IEEE, August 1988.

[4] Batcher, K. E. Sorting networks and their applications. *1968 Spring Joint Computer Conf., AFIPS Proc.*, 32:307–314, Washington, DC: Thompson Book Company, 1968.

[5] Brooks, F. P. *The Mythical Man-Month.* Reading, Mass.: Addison-Wesley, 1975.

[6] Browning, S. A. The tree machine: A highly concurrent computing environment. Caltech Computer Science Tech. Rep. (Ph. D. Thesis) 3760:TR:80, 1980.

[7] Bryant, R. E. Simulation of packet communication architecture computer systems. Tech. Rep. MIT-LCS-TR-188, Massachusetts Institute of Technology, 1977.

[8] Bryant, R. E., D. Beatty, K. Brace, K. Cho, and T. Sheffler. COSMOS: A Compiled Simulator for MOS Circuits. 24th *Design Automation Conference*, pp. 9–16. ACM and IEEE, 1987.

[9] Chandy, K. M. and J. Misra. Asynchronous distributed simulation via a sequence of parallel computations. *CACM*, 24(4):198–205, April 1981.

[10] Chandy, K. M. and J. Misra. Distributed computations on graphs: Shortest path algorithms. *CACM*, 25(11):833–837, November 1982.

[11] Chandy, K. M. and J. Misra. *Parallel Program Design.* Reading, Mass.: Addison-Wesley, 1988.

[12] Dally, W. J. *A VLSI Architecture for Concurrent Data Structures.* Norwell, Mass.: Kluwer Academic Publishers, 1987.

[13] Dally, W. J. and C. L. Seitz. Deadlock-free message routing in multiprocessor interconnection networks. *IEEE Trans. on Computers,* C-36(5):547–553, May 1987.

[14] Dijkstra, E. W. A note on two problems in connexion with graphs. *Numerische Mathematik,* 1:269–271, 1959.

[15] Dijkstra, E. W. Co-operating sequential processes, in F. Genuys, ed., *Programming Languages,* 43–112, New York: Academic Press, 1968.

[16] Dongarra, J. J., J. R. Bunch, C. B. Moler, and G. W. Stewart. *LINPACK User's Guide.* Philadelphia, Pa.: SIAM Publications, 1979.

[17] Flaig, C. M. VLSI mesh routing systems. Caltech Computer Science Tech. Rep. (M. S. Thesis) 5241:TR:87, 1987.

[18] Hartmann, A. C. and J. D. Ullman. Model categories for theories of parallel systems. MCC Tech. Rep. PP-341-86.

[19] Kermani, P. and L. Kleinrock. Virtual cut-through: A new computer communication switching technique. *Computer Networks,* 3(4):267–286, September 1979.

[20] Kuck, D. J., et al. Measurements of parallelism in ordinary FORTRAN programs. *Computer,* 7(1):37–46, IEEE, January 1974.

[21] Kung, H. T. The structure of parallel algorithms, in *Advances in Computers,* 19, New York: Academic Press, 1980.

[22] Lang, C. R. The extension of object-oriented languages to a homogeneous, concurrent architecture. Caltech Computer Science Tech. Rep. (Ph. D. Thesis) 5014:TR:82, 1982.

[23] Locanthi, B. N. The homogeneous machine. Caltech Computer Science Tech. Rep. (Ph. D. Thesis) 3759:TR:80, 1980.

[24] Martin, A. J. An axiomatic definition of synchronization primitives. *Acta Informatica,* 16:219–235, 1981.

[25] Mattisson, S. CONCISE, a concurrent circuit simulation program. Ph. D. Thesis, CODEN: LUNTEDX(TETE-1003), Dept. of Applied Electronics, Univ. of Lund, Sweden, 1986.

[26] Mead, C. A. and L. A. Conway. *Introduction to VLSI Systems.* Reading, Mass.: Addison-Wesley, 1980.

[27] Mead, C. A., *Analog VLSI and Neural Systems.* Reading, Mass.: Addison-Wesley, 1989.

[28] Misra, J. Distributed discrete-event simulation. *Computing Surveys* 18(1):39–65, March 1986.

[29] Ngai, J. Y. and C. L. Seitz. A framework for adaptive routing in multicomputer networks. *Proc. 1989 ACM Symposium on Parallel Algorithms and Architectures,* ACM, 1989.

[30] Preparata, F. P. A mesh-connected area-time optimal VLSI integer multiplier. *VLSI Systems and Computations.* Rockville, MD: Computer Science Press, 1981.

[31] Seitz, C. L. Concurrent VLSI architectures. *IEEE Trans. on Computers* C-33(12):1247–1265, December 1984.

[32] Seitz, C. L. The Cosmic Cube. *CACM,* 28(1):22–33, January 1985.

[33] Seitz, C. L., A. H. Frey, S. Mattisson, S. D. Rabin, D. A. Speck, and J. van de Snepscheut. Hot-clock nMOS. *1985 Chapel Hill Conf. on Very Large Scale Integration.* Rockville, MD: Computer Science Press, 1985.

[34] Seitz, C. L., W. C. Athas, C. M. Flaig, A. J. Martin, J. Seizovic, C. S. Steele, and W.-K. Su. The architecture and programming of the Ametek Series 2010 Multicomputer. *Proc. Third Conf. on Hypercube Concurrent Computers and Applications,* 1:33–36, ACM, 1988.

[35] Seitz, C. L., J. Seizovic, and W.-K. Su. The C programmer's abbreviated guide to multicomputer programming. Caltech Computer Science Tech. Rep. 88-1, 1988.

[36] Seizovic, J. The reactive kernel. Caltech Computer Science Tech. Rep. (M. S. Thesis) 88–10, 1988.

[37] Skjellum, A. and M. Morari. Waveform relaxation for concurrent dynamic simulation of distillation columns. *Proc. Third Conf. on Hypercube Concurrent Computers and Applications* 2:1062–1071, ACM, 1988.

[38] Su, W.-K. and C. L. Seitz. Variants of the Chandy-Misra-Bryant distributed discrete-event simulation algorithm. *Proc. 1989 Eastern Multiconference, Distributed Simulation Conference,* ACM/IEEE, 1989.

[39] Thompson, C. D. A complexity theory for VLSI. Carnegie-Mellon Univ. Computer Science Tech. Rep. CMU-CS-80-140, August 1980.

[40] Ullman, J. D. *Computational Aspects of VLSI.* Rockville, MD: Computer Science Press, 1984.

CHAPTER 2

ERNST W. MAYR
Stanford University and
J. W. Goethe-University Frankfurt

Theoretical Aspects of Parallel Computation

Introduction

In a sense, this chapter is about what the ideal world looks like—about how theoreticians look at the problems and possibilities of parallel computation. Here is a list of the topics:

(1) Parallel-Machine Models and Complexity Classes

(2) Fundamental Parallel Algorithms

(3) The Dynamic Tree Expression Problem

(4) $\mathcal{P}$-complete Algorithms

(5) Parallel Approximation Algorithms

First I will talk about *parallel machine models*—the underlying stuff with all the buzzwords that are used in theory. Then there will be a list of what I call *fundamental parallel algorithms.* The idea is to present some techniques that work in many cases, for algorithms on many parallel architectures. In the third section, something I call the *dynamic tree expression problem* will be the main topic; it provides a nice way to think about and write programs for many different problems in parallel computation, independently of the underlying architecture. The fourth section is about $\mathcal{P}$-complete algorithms. These algorithms and the corresponding problems provide a theoretical way of showing that, in some cases, it is very hard to transform good sequential algorithms into good parallel algorithms. This is something many people are trying to do, so it gives a complexity-theoretical background why such a thing may not work in all cases. Finally, we want to talk a bit about computational problems that are not really parallelizable. Sometimes we simply cannot get good solutions, so we have to deal with approximations. We shall talk about cases where good and efficient approximation algorithms exist and others where they don't.

2.1 Parallel-Machine Models and Complexity Classes

We are talking about a theoretician's view of parallel computation. The first thing a theoretician always does is create models to simplify the whole thing. There are many issues in parallel computation one can think about that are shown in Table 2.1:

Table 2.1   Issues in parallel computation

1. Loosely versus tightly coupled
2. Parallel versus distributed
3. Shared memory versus message passing
4. Synchronous versus asynchronous
5. SIMD versus MIMD
6. Special versus general purpose

You can have parallel systems that are loosely coupled as opposed to systems that are tightly coupled. Distributed systems tend to be loosely coupled systems, while parallel systems are usually more tightly coupled; but one can think of many variations and intermediate cases. Very often we may assume that the parallel computation (at least, say, on some reasonably abstract level) is synchronous, whereas distributed computation is completely asynchronous. Further, we may have machines or machine models that use shared memory (with all the technological complications) versus other models where processors communicate via message passing.

In message passing, if a processor wants to talk to another processor it has to send a message and deliver it to a mailbox belonging to the other processor.

Synchronous versus asynchronous computation is, of course, largely a theoretical issue because in practice most computation is asynchronous, and special measures are taken to make it look synchronous.

In an SIMD system we have one stream of instructions (i.e., one program) working on many sets of data at the same time, whereas in MIMD different processors may execute different programs. Systolic computation is also very akin to SIMD, stressing the execution of the same instruction on many data items in lock step.

Finally, there are special-purpose parallel machines—say to compute an FFT—that do certain types of image processing or, maybe, sorting. They have to be contrasted with general-purpose machines such as Hypercubes or the RP 3, a project by NYU and IBM. The goal for general-purpose machines is to be able to execute any parallel algorithm with reasonable efficiency.

This list of issues is by no means complete. We just listed some of the more important ones.

Table 2.2   Models of Parallel Computation

1. Data flow
2. Actors
3. Vector machines (Pratt/Stockmeyer)
4. Wide area networks
5. LANs
6. Ultracomputer
7. Connection machine
8. VLSI
9. Parallel Random-Access Machine (PRAM)
10. Boolean circuits
11. Alternating Turing machines
12. Synchronizing Turing machines

There are also many models of parallel computation. In Table 2.2 is a list that is bound to be incomplete.

In this table, we could have also listed the brain, beehives, or ant communities. Data flow is a concept that has been drifting around for a long time. There is also a very close relationship to the actor model. There are very successful vector machines out there in industry, but then there were a few theoreticians who defined a model of vector machines that can simply do operations on whole vectors in one step.

The Ultracomputer is Jack Schwartz's model of a shared-memory machine. It consists of an array of processors, an array of memory modules, and an interconnection network built of communication links and constant degree switches in between.

VLSI is thoroughly covered in other parts of this volume.

The PRAM is a theoretical model for parallel computation. It stands for *parallel random-access machine*, and we are going to define exactly what that is. One way or another, it can be viewed as a formalization of electronic circuits.

Another, purely theoretical, model for parallel computation is that of alternating Turing machines (ATMs). They represent a generalization of nondeterministic Turing machines. Yet another, quite recent, model is that of synchronizing Turing machines, beefed up versions of ATMs, which we are not going to discuss in any more detail.

Figures 2.1 through 2.4 give some of the more relevant examples of fixed interconnection networks.

Later on, we shall mostly concentrate on the PRAM model. But it is a theoretical model and idealizes a lot. Closer to reality are fixed interconnection networks. They have a set of processors, a set of memory cells or memory blocks, and connections between them via channels. The simplest example would be a complete crossbar, in which everybody is connected to everybody else. That becomes infeasible in practice if the

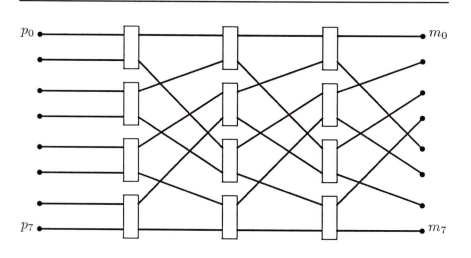

Figure 2.1   An 8 × 8 Omega network.

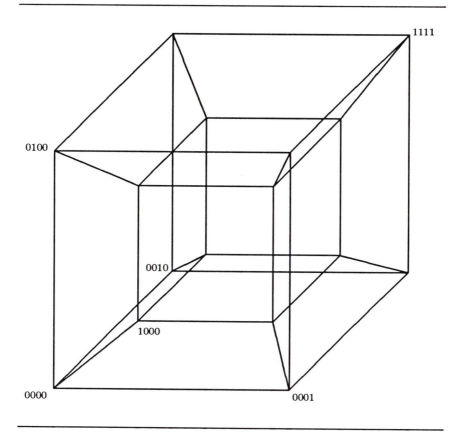

Figure 2.2   Four-dimensional hypercube.

numbers get larger. We then replace the crossbar switch with a network
consisting of bidirectional channels and switches, as shown in Figure 2.1. In
this example, the rectangles represent two-by-two switches that can either
route through horizontally so the first input goes to the first output and
the second input goes to the second output, or they can cross over with
the first input going to the second output and the second input to the first
output. You can set each of the switches to one of the two positions. You
can even look at switches that may have the broadcast capability—data
arriving at one input can be sent to both outputs at the same time. The
network shown in Figure 2.1 is called an Omega network. It is basically the
one that was proposed for the NYU ultracomputer. It is implemented in
the IBM RP 3 project where there are really 64 processors and 64 memory
modules. So it is a 512 × 512 computer. The actual switches are also a

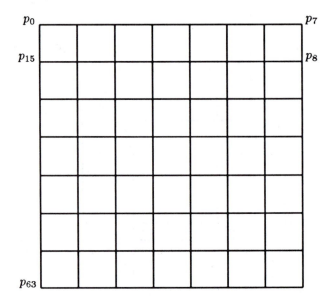

Figure 2.3   An 8 × 8 mesh.

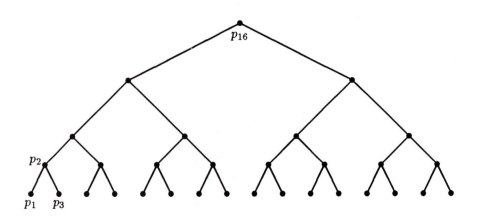

Figure 2.4   Five-level, complete binary tree.

little bit different from what we indicated here, but the figure gives the basic idea. If a processor wants to access a memory cell, it routes a request through the interconnection network to the memory cell. It turns out that from every processor there is exactly one path to a given memory cell. The path can easily be determined from the address of the memory module. Let's take 6 or 110 in binary as a simple example. We look at the bits of the destination address, starting from the low-order bit. When the bit is a 1, we take the lower output of the switch; if it is a 0, we take the upper output. This way we always get to the right destination. We leave it to the reader to verify this simple routing algorithm.

Figure 2.2 shows a four-dimensional hypercube. Obviously, you have seen zero-, one-, two-, and three-dimensional hypercubes before. A four-dimensional hypercube is simply two, three-dimensional hypercubes: one on the outside, one on the inside. Instead of arbitrarily naming the vertices, you attach a bit position to every dimension; the first dimension goes horizontally in the figure, the second goes into the page, the third vertically, and the fourth connects the outer cube to the inner. Thus, the figure shows what a four-dimensional hypercube looks like in two dimensions.

Another type of network—which is a little more realistic—is a mesh. It is planar, and you have processors arranged in rows and columns. Figure 2.3 shows such a two-dimensional mesh with eight processors on each side. The processors in the figure are numbered in a way that is called snakelike row major. You number the processors row by row, but you do it snakelike— one row from left to right, and the next from right to left. One of the earliest models of a parallel computer, the ILLIAC IV, had a connection pattern like the mesh, and in addition there was a wraparound from the processors on the right edge to the leftmost processor on the next row. Another variant is to wrap around the rightmost processor in every row to the leftmost in the same row, and the topmost processor in every column to the one at the bottom. This way we obtain the topology of a torus or doughnut.

Meshes and their variants are relatively simple to build in VLSI and to route, so they are quite practical.

Some of the theoreticians' most favorite models are trees. They are ubiquitous and present in many situations. When drawing them, we often turn them upside down, showing the root at the top. Figure 2.4 depicts a five-level, complete binary tree.

We now look at a few important parameters of networks like those shown in the figures. The parameters, and their values for the networks, are listed in Table 2.3. These properties are of interest if we want to look at their performance and their usability.

Remember that the nodes in the graphs correspond to processors and/or switches, and the edges correspond to channels or links. The distance between some pair of nodes in a network is the minimum number of links

Table 2.3   Important properties of networks

	Tree	Mesh	Shuffle-exchange	Hypercube
Size	$2n - 1 = 2^h - 1$	$\sqrt{n} \times \sqrt{n}$	$n = 2^k$	$n = 2^k$
Diameter	$2h - 2$	$2\sqrt{n} - 2$	$2k - 1$	k
Degree	3	4	3	k
Permutation	$\Theta(n)$	$\Theta(\sqrt{n})$	$\Theta(\log n)$	$\Theta(\log n)$

you have to traverse to get from one node to the other. The diameter of a network is then the maximum distance between any pair of nodes. The degree of a node is the number of edges incident on that node.

How easy is it to route a permutation on such a network? You have data in one arrangement and you just want to reshuffle it in some predetermined way—how easy or how hard is this?

We now look at the networks listed in the table in turn. First, consider a complete binary tree. If it has, say, n leaves, where n is a power of 2, then it has $2n - 1$ nodes altogether, and its height would be h. So the total number of nodes is $2^h - 1$.

A two-dimensional mesh with n nodes has side length $\sqrt{n}$, and we assume without loss of generality that in this case n is a square number.

So far, we didn't really define the shuffle-exchange graph. It basically is one rank of the Omega network, as shown in Figure 2.1. There are 2^k nodes connected like a perfect shuffle of cards. The first half of the nodes is connected to the even-numbered nodes in order, and the second half to the odd-numbered nodes, also in order. These connections are called the shuffle edges. The exchange edges simply connect an even-odd pair of nodes. They don't show up directly in Figure 2.1 but are hidden in the switches (rectangles).

The definition for the (binary) hypercube should be clear.

We'd like to remark here that there is quite a number of other networks that are very similar to the hypercube. Examples are the indirect binary n-cube or the butterfly, or FFT, graph. However, we are not going to discuss these any further.

The diameter is the maximum number of hops you have to make to get from one node to some other node. In the tree, the largest distance occurs if we want to go from a leaf in one subtree of the root to a leaf in the other subtree of the root. This distance is roughly twice the height, and the diameter is $2h - 2$. In the mesh, the largest distance is between diagonally opposite corners. This distance is twice the side length, counted in terms of edges.

For the shuffle exchange, the argument is a bit more difficult. Numbering the nodes of a shuffle-exchange graph with $n = 2^k$ nodes in the canonical way with k-bit integers, it turns out that the exchange edges connect nodes whose numbers differ in exactly the last bit, while traversing a shuffle edge corresponds to cyclically shifting the processor number by one position—left if we go forward, and right if we go backward. With this intuition, it is not hard to see that the diameter of the shuffle-exchange network is $2k - 1$, given for instance by the two nodes with numbers 0 and $n - 1$. Since k is the logarithm of n, the diameter is really logarithmic in the size of the network.

The same situation occurs for the hypercube, where, with the canonical numbering of the nodes, two nodes are adjacent if and only if their binary numbers differ in exactly one bit position. Thus, the distance between any two nodes is equal to the Hamming distance between their binary numbers.

Quickly, a thought about a lower bound for the diameter is that if the degree of each node is bounded by some constant D, then the diameter must grow asymptotically like $c \log n$ for some $c > 0$ since with i steps from a node we can reach at most D^i nodes.

The degree of a network is the maximum number of channels incident upon a node. For a binary tree, this number is three: one edge from the node to its parent, and one to each of the two children. For the mesh, the degree is four, corresponding to the four directions north, east, south, and west. The shuffle-exchange graph has degree three: one exchange edge, one outgoing-shuffle edge, and one incoming-shuffle edge. And for the hypercube, the degree is equal to the number of dimensions. If you are in a four-dimensional cube, you have four edges or channels going out, one for each dimension.

Suppose you have one element sitting at each node of a network, and you want to permute the elements in some way. How many parallel steps are needed if we assume that in one time step, every node in the network can send off and receive at most one element? The (asymptotic) time requirements for performing a permutation on the networks are also given in Table 2.3. They are worst-case bounds. We are not going to prove these bounds here, though some are quite easy to see. Assume, for instance, for the case of the binary tree that the permutation sends all elements in the left subtree of the root to the right subtree, and vice versa. Then $n - 1$ elements (all but the one in the root) have to pass through the root, clearly requiring time $\Omega(n)$. To get an upper bound of $O(n)$ for this problem on the tree is a nice little exercise. It is not too hard, but it's not trivial either. It is a wonderful example of divide and conquer. To do a permutation on the mesh is even more involved, and time $O(\sqrt{n})$ suffices. There is also a matching lower bound that follows directly from the size of the diameter of the mesh. For the shuffle-exchange graph and the hypercube, there is a very nice way to decompose any given permutation into a sequence of $O(\log n)$ simpler permutations, each of which can be executed on the shuffle-exchange network respectively by the hypercube in a single step, and whose composition results in the originally given permutation.

It turns out that the complexity of performing a permutation is closely related to the complexity of moving the elements in any subpart of the network (of, say, at most half the size of the whole network) out of this subpart. This latter notion is sometimes referred to as the flux of the network and can also be formally defined. One might be tempted to assume that the flux is characterized by the number of edges in the network, but this is not true. For instance, the number of edges in the tree and in the shuffle-exchange graph is roughly the same, namely, linear in the number of nodes. It is the topology that one really has to look at.

In summary, the tree has good diameter but poor flux since the root acts as a bottleneck. The mesh is in between in either case. The shuffle exchange is good in that it has constant degree, small diameter, and high flux at the same time. But it is not quite as uniform or "simple" as one might wish. The hypercube does not have constant degree; rather, the degree grows slowly, like the logarithm of the number of nodes. That may be bad in practice if the numbers get bigger. It has high flux, it can permute very fast, and it is regular. This is an advantage both in terms of hardware implementation and algorithm development.

When actually building interconnection networks and programming algorithms for them, there are many more aspects that come into play than those listed in Table 2.3. Many of these aspects depend on the available technology or on the particular values of parameters that have been chosen for a particular case. These issues are often extremely complex, but we are not going to consider them here in any way.

We shall now go on to a more theoretical model for parallel machines, the so-called PRAM model. It has the following features:

- An unbounded number of identical RAMs.

- An unbounded number of global memory cells.

- Each processor can access any memory cell in one step.

- Concurrent read or write access to a memory cell by more than one processor may or may not be allowed:

  — EREW: no read or write conflicts;

  — CREW: no write conflicts;

  — CRCW: read and write conflicts allowed. We further distinguish the *identical, arbitrary,* and *priority* model.

A RAM is a random-access machine, and we wouldn't go too wrong imagining something like a Motorola 68000 processor. A PRAM has available an unbounded supply of such RAMs that work synchronously, in lock step. There is an unbounded number of global memory cells. A memory cell again is an idealization of a storage cell because it really can store any binary number. It does not have a fixed number of bits. We fake the connections between the processors and the memory cells, and say that every processor can access any memory cell in one step. So we take away the cost of communication almost completely. This is where the idealization comes in.

If we allow several processors to access the same memory cell at the same time we must take care of conflicts that may arise. For simplicity, we shall assume that in any one time step, read accesses always occur before write accesses, and hence that there is no conflict between read operations on the one hand and write operations on the other. We are still left with conflicts among reads and conflicts among writes to the same memory cell. Depending on whether we allow or disallow these conflicts, we get different types of models. They are called exclusive-read, exclusive-write PRAMs in the case where any concurrent access is disallowed. So you cannot simultaneously read or write the same memory cell by more than one processor. The concurrent-read, exclusive-write PRAM allows you to read the same memory cell by more than one processor, but it disallows concurrent writes. This is logically okay since concurrent reads don't actually alter the data in the memory cell. The most powerful model—the concurrent-read concurrent-write, or CRCW-PRAM—permits concurrent reads and concurrent writes. Again, the concurrent reads cause no logical problems, but we have to deal with the concurrent writes.

Depending on how we resolve write conflicts, there are at least three types of submodels here. In the *identical* CRCW model, we assume that processors writing to the same memory cell in some step simply write the same data. Any algorithm for the identical CRCW model has to make sure that this condition is satisfied. If not, the algorithm is illegal, and the machine will explode. In the *arbitrary* model, we allow the processors to write different data to the same memory cell. We assume that exactly one of the processors will succeed in writing its data to the memory cell (i.e., no hybrid results occur), but we do not know, and make no assumptions about, which one. One way to implement this model is to have arbiters, but this can be costly in terms of time. In the *priority* model, if several processors want to write to the same memory cell, then the lowest numbered processor wins.

We are going to restrict ourselves mostly to the CREW PRAM model and refer to it simply as PRAM. Actually, there is not much difference between the models since all the various models can be simulated by one another without too much overhead. The overhead is really the cost of sorting. We can simulate the most powerful model, the priority CRCW, by an EREW in $O(\log n)$ time, using the same number n of processors. The trick behind it is sorting, and sorting can be done within those bounds; though presenting any of the presently known, optimal sorting algorithms would by far exceed the scope of this presentation. We simply assume that sorting n items on n processors of an EREW PRAM can be done in $O(\log n)$ parallel steps, and leave it as an exercise to design the simulation.

But let's try to keep our feet on the earth. Are we really up in the air with PRAMs? We write PRAM algorithms without worrying about communication. They can be simulated on more realistic models—such as the hypercube, the shuffle exchange, or the Omega network—quite efficiently. The loss in time is something like $\log^3 n$ or $\log^4 n$. Now, $\log^3 n$ is an interesting function. If you have $n = 1000$ processors, $\log^3 n$ is still 1000. So that's not too good. But, of course, if n gets much larger, like 10^6 or so, then $\log^3 n$ is still quite small. We call this the *law of small numbers*. It states that for small numbers, powers of log are really worse than square roots for instance.

How do we do this simulation? It is fairly involved, and there is no chance that we can discuss it here. There are basically two problems that need to be solved. The first is routing the requests for memory access from the processors to the memory modules. The other, and somewhat more difficult problem is to simulate the global shared memory by the memory modules. The difficulty here lies in mapping the data to memory modules in such a way that most of the data accessed in one parallel step by the processors is available in different memory modules. This is necessary since otherwise serializing the accesses at each memory module would slow down the overall computation.

One might argue that the PRAM model is too idealized, too theoretical. Of course, this criticism in itself is too general. In many situations the PRAM model works perfectly well, and solutions for it can readily be transferred to the real application. In other cases problems remain. There are many more models for parallel computation around, some of which might be more appropriate in such cases. But the PRAM model allows us to factor the problem of parallel algorithm design into two subproblems: First, find the parallelism available, without worrying about the cost of communication, and then map the PRAM algorithm onto the real-world architecture.

There are some basic facts that relate parallel computation on, say, the PRAM model, and ordinary sequential computation. One such fact is commonly called the *parallel computation thesis*. It is something like Church's thesis, which says that everything computable is computable by one of the formally defined machine models: Turing machines, Lambda calculus, or similar things. Of course, when somebody comes up with a new formal notion for *computability*, it has to be shown that this new notion is equivalent to the old notion—namely, Turing-machine computability. Similarly, the parallel computation thesis is something that has to be proved for every parallel model of computation under consideration. In essence, it says that parallel time is roughly equivalent to sequential space. This is in the asymptotic—not the engineering—sense. For instance, if a problem can be solved in polylogarithmic time on a PRAM, then it can be solved in polylogarithmic space on a Turing machine, and vice versa.

More precisely, we have

Theorem   1   *A PRAM running in time $T(n)$ can be simulated by a deterministic Turing machine using space $T^2(n)$.* ☐

We are not going to formally prove this theorem here, but we want at least to outline an idea for such a proof. In time $T(n)$, at most $2^{T(n)}$ processors of the PRAM can be active. The reason for this is that initially only the first processor is active, which then activates another processor. In the following step, each of the two processors can now activate a new processor, and so on. Also, the largest number in any memory cell has $O(T(n))$ bits, since per step the number of bits in a memory cell can only grow by a constant. The state of any memory cell or processor at time t depends on the state of the memory cells and processors at time $t - 1$. These dependencies can be checked in a depth-first manner, requiring only $O(T(n))$ storage per level. Since the depth of the depth-first search is $T(n)$, we obtain $O(T^2(n))$ space altogether. Note that a (multitape) Turing machine has a read-only input tape, a set of work tapes, and a write-only output tape. The space that counts is only the space that is used on the work tapes, i.e., the scratch space needed for the computation. We do not count the input tape, which has n cells where n is the length of the input.

It should be clear already from this very superficial explanation that the transformation from the parallel computation to a sequential computation is completely counterintuitive because it is depth first. This is completely different from the parallel computation, which goes in an orderly step-by-step way. That should warn you. Efficient parallel algorithms are different from efficient sequential algorithms.

Now the theorem for the other way around:

Theorem   2   *A deterministic Turing machine M requiring space $S(n)$ can be simulated by a CREW-PRAM in time $O(S(n))$. The PRAM uses $c^{S(n)}$ processors, where $c > 0$ is a constant depending on M.* ☐

Again, we only give the idea for a proof. In time $O(S(n))$, the PRAM computes all possible configurations of M, one per processor. Here, a configuration is a straightforward encoding of the state of the Turing machine together with the contents of its work tapes. Then, by simulating the Turing machine for one step, each processor computes the successor configuration for its configuration, i.e., the configuration reached by the Turing machine after running for one step. Then the PRAM uses *path doubling* to determine whether the initial configuration of the Turing machine leads to an accepting final configuration. Path doubling is a technique that will be explained later, but it basically consists of the following procedure. First, each processor finds the successor of its successor configuration by looking up the successor configuration computed by the processor that handles the successor configuration of the processor's own configuration. This way, the processor now knows the configuration two steps ahead. Repeating this step, it knows the configuration four steps ahead, and so on. $O(S(n))$ such doubling steps suffice in our case since an accepting computation of the Turing machine cannot have more than $2^{O(S(n))}$ steps.

Next, we present some complexity-theoretic background. Theoreticians like to use complexity classes to characterize the complexity of problems. A complexity class is a set of problems that can be solved, say, in time n^2, or in linear space, or something like that. In general, it is given by a bound on some resource or combination of resources, like space, time, or the number of parallel machines.

We have all heard about the class $\mathcal{P}$—sequential polynomial time. The class called $\mathcal{NC}$ is a parallel complexity class. By definition, it is the class of all those problems that a PRAM can solve in polylogarithmic running time. If the input is n bits long, the running time is $\log^c n$ for some fixed c, i.e. the cth power of $\log n$. For $\mathcal{NC}$, the number of processors that the machine employs must also be bounded by a polynomial in n. The idea is that the number of processors is small. In practice, we will want the polynomial to be linear or close to linear. Just consider a problem size of $n = 1000$. Then n^2 is already 1 million and a parallel processor with 1 million processing elements is still a little bit in the future.

The name $\mathcal{NC}$ means Nick's Class, after Nick Pippenger who defined this class in 1979. Actually, he didn't define it for PRAMs; he defined it for circuits, which are also models for parallel computation. But the definition carries over to PRAMs right away.

We also look at the class $\mathcal{NC}$ as capturing the idea of *efficiently parallelizable*. So, when we say "efficiently parallelizable," we really mean that the problem is in $\mathcal{NC}$.

Now, let's see whether we have learned our facts. Problems in $\mathcal{NC}$ can be solved by parallel algorithms using a polynomial number of processors and running in polylog time. It is easy to simulate a parallel algorithm by a sequential algorithm. We just simulate all processors for one step in turn. The simulation will take us only polynomial time, because there are only a polynomial number of processors. So $\mathcal{NC}$ is a subset of $\mathcal{P}$. But is it all of it? This is a difficult open problem. From the Parallel Computation Thesis we know that any algorithm in $\mathcal{NC}$ can be simulated by a polylog-space deterministic sequential algorithm. Now polylog is small compared to the length of the input, which is n. It is, for instance, much less than $\sqrt{n}$, and much less than $n^{.01}$, or things like that.

A $\mathcal{P}$-complete problem is a hardest problem in $\mathcal{P}$. Complexity theoreticians have shown that if such a hardest problem can be solved in polylog space, then every problem in $\mathcal{P}$ can be solved in polylog space. This situation is similar to what you know about $\mathcal{NP}$. When an $\mathcal{NP}$-complete problem can be solved in polynomial time on a deterministic Turing machine, then every $\mathcal{NP}$ problem can be solved in polynomial time on a deterministic Turing machine. We think that this is rather unlikely. But we do not have a proof, in the same way as we do not have a proof that $\mathcal{P}$ is actually different from $\mathcal{NC}$.

The above discussion justifies using the fact that a certain problem is $\mathcal{P}$-complete as evidence that there are no efficient parallel algorithms for this problem. Similarly, if you prove something $\mathcal{NP}$-complete, you don't try to solve such a problem efficiently in the sequential world.

By the way, being $\mathcal{P}$-complete does not mean that algorithms for such a problem have to have running times that are n^{10} or n^{100}. There are $\mathcal{P}$-complete problems for which we have linear-time algorithms. Rather, it is the combinatorics encoded in these algorithms that makes them hard to parallelize—everything depends on everything computed before.

Here are a few examples: The *circuit-value problem* arises in VLSI simulation. Given a description of a circuit and the values of the inputs, you want to know what the output is. The best way we know to compute the output in general is to simulate the circuit layer by layer, which takes roughly as much time as there are layers. Assume we have a size n circuit with width and height roughly $\sqrt{n}$. The simulation then takes $\sqrt{n}$ time, which is much more than $\log^c n$. *Maximum flow* is another well-known $\mathcal{P}$-complete problem. We know good sequential algorithms for maximum flow. *Linear programming* is again something that everybody knows. Due to Khachyan's work we now know that linear programming is in $\mathcal{P}$. But it is hard to parallelize, since it has been shown $\mathcal{P}$-complete. *Unification* is a problem that occurs, for instance, in logic programming, such as Prolog. When you have formulas, you have to unify terms. The problem is to find substitutions for variables that make two given formulas equal. Finding the most general such substitutions is a $\mathcal{P}$-complete problem. Regarding the potential of parallelism in logic programming, this is a negative result. It has to be said, however, that many unification problems arising in logic programming are of more special types for which efficient parallel solutions are in fact known.

One might suggest that circuit simulation could be done in something like constant time simply by building the circuit and running it. However, this approach uses a different model. We assumed here that the propagation time between gates of the circuit is unit time. So the time required is still proportional to the depth of the circuit.

The circuit-value problem is the generic problem used for $\mathcal{P}$-completeness proofs, just as satisfiability is the generic problem used for $\mathcal{NP}$-completeness proofs. There is a certain similarity, because satisfiability can be phrased as follows: You are given a circuit, and you want to know whether there is an input that creates a 1 as output. So this is an existential question: Does there exist an input? In the circuit-value problem, we are given the input, and we want to know what is the output. Is the output a 1?

When we study complexity classes we also look at reductions from one problem to another. For $\mathcal{NC}$, we use log-space–bounded, Turing-machine computations for the reductions. For $\mathcal{NP}$, you generally use polynomial-time reductions. Reductions using log-space are very efficient. For programs written in Pascal, C, or some other high-level language, log-space usually means that there are no arrays and only a constant number of variables. However, this is only an informal characterization, and for a real proof more detailed constructions are needed.

Depth-first search is a sequential programming technique used in many circumstances. Its parallel complexity is not known at present. There is a fast parallel algorithm for general depth-first search that runs on a polynomial number of processors and in polylogarithmic time. However, it uses random coin tosses. On the other hand, it can be shown that the problem of computing the very same depth-first search tree that is computed by the sequential algorithm on an input graph given by adjacency lists is $\mathcal{P}$-complete, giving some indication that it may be hard to parallelize depth-first search.

2.2   Fundamental Parallel Algorithms

In this section, we are going to present a few algorithms and procedures that crop up in many applications and give some ideas how one could implement them in an abstract sense, because we resorted to the PRAM model, which disregards many interesting practical details. Table 2.4 contains a list of some of these techniques.

Certainly, the list in the table is not intended to be complete. Doubling and pointer jumping or path doubling are very fundamental techniques, as is parallel prefix computation. They occur in many applications. List

Table 2.4 Fundamental parallel algorithms

1. Doubling
2. Pointer jumping or path doubling
3. Parallel prefix
4. List ranking
5. Sorting
6. Euler contour path
7. Tree evaluation

ranking is something that you may not have seen so far. It occurs more in graph theoretical applications, as opposed to numerical, matrix-based, parallel computation. The next item on the list is sorting, which I guess everybody can relate to. The Euler contour-path technique may also be something that you have not yet heard about. And the last item, tree evaluation, will be discussed in much detail in the next section, together with applications. We are now going to look in more detail at all of these items.

2.2.1 Doubling

Assume we want to compute the sum of n numbers a_1 through a_n. The sequential algorithm simply computes $s = a_1 + a_2$, then $s = s + a_3$, then $s = s + a_4$, and so on. Usually, the main part of this summation is performed in a loop $s = s + a_i$. Some people would say that the computation looks inherently sequential because s is used and updated in every iteration of the loop.

But let's be careful about that. Computing the sum of n numbers is by no means sequential because the other way to do it is to take pairs and sum them up. Then sum up pairs of the resulting sums, and so on. The scheme of this type of computation is depicted in Figure 2.5. Note that all operations on one level of the tree can be performed in parallel since they access disjoint sets of variables. What we double in every iteration is the number of inputs whose sum we have already collected in a single variable.

The computation as indicated in Figure 2.5 looks appropriate for a PRAM or a tree-structured network of processors, but it can also be easily performed on a hypercube or shuffle-exchange network. We leave it as an exercise to adapt the method as we just outlined it to these networks. Another quite interesting exercise is to find an efficient solution for the summation problem on mesh-like architectures where we cannot embed

$a_1 + a_2 +$ $\cdots$ $+ a_{15} + a_{16}$

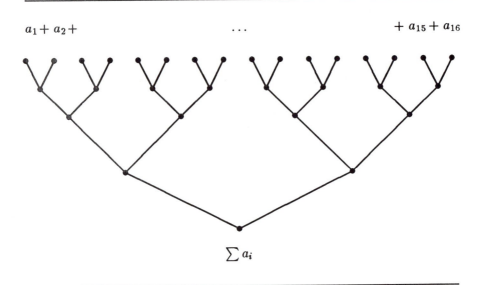

$$\sum a_i$$

Figure 2.5   Summation by doubling.

binary trees in such a way that all tree edges become short paths in the mesh.

In the above discussion of the doubling technique, we have used summation, such as addition on the reals. However, we can actually replace summation by any associative binary operator. Other examples for such operators would be the maximum, minimum, or most of the Boolean operations such as exclusive OR. A more outlandish example would be composition of finite functions, i.e., functions from a finite domain to a finite range. Imagine that each of the a_i in Figure 2.5 stands for such a finite function $\mathcal{F}_i$, and that the connector is a composition of functions. Then we are actually reading $\mathcal{F}_1$ after $\mathcal{F}_2$, after $\mathcal{F}_3$, and so on. Now composition is associative, and we can group the functions together any way you want, as long as we maintain the order. So we can also use the doubling trick for finite functions. Now, whether it is practical depends on how big the domains and ranges are. For instance, there is a simple way to derive a carry-lookahead adder in this way, with carry generate and carry propagate as the (essential) functions. We leave it to the reader to work out this example.

Another name used with regard to this doubling technique is census functions. In Figure 2.6, we give a detailed PRAM program to compute a census function. The program works as follows. Every processor uses

procedure *census_function*(n, s, res, o);
int n; **gmemptr** s, res; **binop** o;
co n is the number of elements in the input array starting at position s in global memory; *res* is the index of the global memory cell receiving the result; o is an associative binary operator **oc**
begin
    local *type_of_S*: *save*, **int**: *mask*, *myindex*;
    if PROC_NUM $< n$ **then**
       mask := 1; *myindex* := $s +$ PROC_NUM; *save* := $M_{myindex}$;
       while *mask* $< n$ **do**
          if (PROC_NUM AND $(2 * mask - 1) = 0$) and PROC_NUM $+ mask < n$
          then
             $M_{myindex} := M_{myindex} \circ M_{myindex+mask}$
          fi;
          mask := $2 * mask$
       od;
       if PROC_NUM $= 0$ **then** $M_{res} := M_s$**fi**;
       if *myindex* $\neq res$ **then** $M_{myindex} := save$ **fi**
    fi;
    return
end *census_function*.

Figure 2.6 PRAM algorithm for census functions.

a local variable *save*, which is needed in order to make the procedure free of side effects. All processors participating in the computation have indices less than n. The variable *mask* serves to select those processors that perform nontrivial computations in any given step. The algorithm starts with *mask* set to 1. The processor with index PROC_NUM takes care of the variable with index $s +$ PROC_NUM. We call this index *myindex*, and the processor initially saves away whatever there is in cell *myindex*. The variable *mask* is used to do the doubling. The expression (PROC_NUM AND $(2 * mask - 1) = 0$) is just a way of saying: What do the last few bits of PROC_NUM look like? In the first step, only those processors are active whose PROC_NUM is even, i.e., the last bit of their PROC_NUM is zero. Every such processor combines its value with that of the next processor. We also assume that all processors for which the **if** condition does not hold perform an appropriate number of no-op steps, such as to stay synchronized with the active processors. Then we double *mask*; remember the picture of the tree?

Thus, *mask* will be two in the next iteration of the loop, and we will collect the pairs into sums over four elements each. Then we'll double *mask* again, to collect pieces of four into pieces of eight, and so on. In the end, we just have to do some cleanup. The first processor takes care of storing the whole sum into the result position, and then every other processor restores the global memory cell that it initially saved away.

2.2.2   Pointer Jumping or Path Doubling

The next technique, pointer jumping or path doubling, has nothing to do with numbers. It is about digraphs, and applies to a special type of digraph called an in-forest. An in-forest is a set of trees with the edges all pointing toward the root. Every branch of such a tree is a directed path from a leaf toward the root of the tree. We assume that by some preceding computation the in-forest is stored in the global memory of the PRAM in such a way that for every node in the forest there is a pointer to its immediate ancestor in its tree. The pointer for the root just points to the root itself. We also assume that there is a unique processor associated with every node in the forest. The problem consists of finding, for every node in the in-forest, the root of the tree to which it belongs.

There is a simple algorithm to solve this problem. To simplify the following description, we shall identify each processor with the node it is associated with. In the first step, every node finds its grandparent by reading the location pointed at by its own parent pointer. It then replaces the parent pointer by a pointer to the grandparent. In the second step, every processor again reads the location given by its pointer, and substitutes the value found there. Thus, after two steps, every node knows its ancestor four generations away; after three steps, its ancestor eight generations away; and so on. If there are n nodes in the forest, then $\log n$ steps suffice, and every node will know the root of the tree to which it belongs. Figure 2.7 gives a simple example consisting of a single path with eight nodes.

We note that in the path-doubling algorithm no write conflicts will occur since only the processor associated with a node will update the pointer belonging to that node. However, in general there will be read conflicts because several pointers can point to the same node, as becomes immediately clear if we consider in-trees that are not just paths as in the example.

We would also like to emphasize that for the path-doubling algorithm, the pointers need not be stored in contiguous positions, and, of course, not in order as they are shown in Figure 2.7 for clarity only.

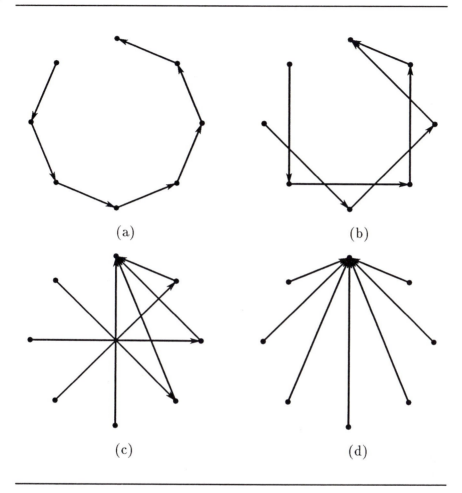

Figure 2.7   Example for path doubling.

2.2.3   Parallel Prefix

Parallel-prefix computation is a very essential technique, representing a generalization of the doubling technique considered earlier. Again, we are given an array of quantities $a_0, a_1 \ldots, a_{n-1}$. Instead of just computing the whole sum, we want to compute all partial sums starting at the first index. So, for $i = 0, \ldots, n - 1$, we want to compute $\sum_{j=0}^{i} a_j$. And again, the summation sign doesn't necessarily stand for the standard sum; it can be any binary associative operator. Figure 2.8 shows a program that does the job.

```
procedure  parallel_prefix(
      int n; co length of the input list oc
      global memory pointer start, result; co start indicates the place in memory
      where the input list begins and result where the list of results a₀ o ··· o aᵢ is
      to be put oc);
begin
      local type_of_S: save,save2; int: span, myindex;
      if PROC_NUM < n then
            span := 1;
            myindex := start + PROC_NUM;
            save := M_myindex co save global cell M_myindex since it may get changed
            during the computation oc;
            while  span ≤ PROC_NUM do
                  M_myindex := M_myindex−span o M_myindex;
                  span := 2 * span
            od;
            save2 := M_myindex;
            M_myindex := save co restore the original input values oc;
            myindex := result + PROC_NUM;
            M_myindex := save2
      fi
end.
```

Figure 2.8   Parallel-prefix algorithm.

The inputs for the procedure are again the length of the array and the starting point for where the result is supposed to go. Also, o stands for the binary operator. We need a few local variables, *save* and *save2*, to save away values. The variable *span* serves the same purpose as *mask* in the previous program, i.e., it denotes the distance spanned in a given step.

The computation is very similar to what we had before in the algorithm for census functions, only this time every processor—if we look at it after i iterations—will have the sum of the 2^i quantities immediately preceding it, including itself. In the first step, it combines the value in the preceding position with its own value, then these two with the two preceding them, and these four with the four preceding them, and so on until it bumps against the end. There are, of course, these boundary effects, but the algorithm just turns out to do the right thing. Then, there is some postprocessing that simply consists of storing away the results and restoring the global variables.

Again, we have omitted some nitty-gritty details in the description of the algorithm, in particular with respect to the synchronization of the processors. This is justified because we make the same assumptions as before.

2.2.4   List Ranking

The list-ranking problem is this: Given an array of n pointers that form a simply-linked list, determine, for each element in the array, its distance from the end of the list.

Typically, the list-ranking problem occurs as a subproblem in other algorithms. We want to point out that generally, of course, the list is not stored in a monotone fashion in contiguous memory cells. Obviously, the problem becomes trivial in this case. We may, however, assume that all list pointers are stored in a contiguous array. If this is not the case initially, we can use the parallel prefix routine to "compact" the representation of the list in memory. We leave the details of this operation to the reader.

As a possible application, think of the following (we shall see something similar when we discuss the contour-path technique below): Some parallel computation produces a simply-linked list of numbers, and we wish to compute the sum of these numbers for all initial segments of the list. This looks like a parallel-prefix problem. To apply our algorithm presented above, however, we first have to arrange the numbers in list order in a contiguous array in memory. It should be clear that this is a straightforward task once we have solved the list-ranking problem because the rank of an element in the list can be used to easily determine its position in the array.

We would like to mention that for sequential computation, list ranking is a rather trivial problem. We simply go through the list and push every element onto a stack; and, after arriving at the end of the list, we pop the elements from the stack and simply count.

For parallel computation, there is also no problem if we are given n processors—remember that n is the number of elements in the list. Why? Well, we just apply what we saw a moment ago, namely, the path-doubling routine. Each processor takes care of one element in the list and does the pointer jumping. After $\log n$ iterations, each processor will know where the end is. While it is doing the pointer jumping, it could actually count how many nodes are jumped over each time. It is straightforward to store this number together with the pointers.

Why should we be unhappy with this simple parallel solution? Well, let's consider the total work performed by the parallel computation, and let's compare this work to the work done in the sequential routine sketched above. In the parallel computation, we are using n processors running for $O(\log n)$ time. The total number of operations performed is therefore $O(n \log n)$, whereas sequentially we only used a linear number of operations. In general, we call a parallel algorithm *optimal* if the product of its running time with the number of processors it uses has the same growth rate as a corresponding optimal, sequential algorithm.

Maybe not too surprisingly, we have the following:

Theorem 3 *The list-ranking problem can be solved in $O(\log n)$ time on an EREW-PRAM using $n/\log n$ processors.* □

To actually prove this theorem would exceed the limits of this presentation. For two different solutions, both optimal, we refer the reader to [4] and [9]. Here, we give a brief sketch of one of the algorithms.

Sketch of optimal list-ranking algorithm:

1. Break array into stacks of height $\lceil \log n \rceil$.
2. Select top remaining element in each stack.
3. Determine chains formed by selected elements.
4. Splice out distinguished elements in singleton chains.
5. Have processor at tail of non-singleton chain splice out all elements in chain, one per step.
6. If there are elements left, go to step 2; otherwise, stop.

First, we break the array into roughly $n/\log n$ segments. The list could be completely disintegrated at this point because the list pointers are completely independent of the segments. There are $n/\log n$ stacks, and the basic idea is to assign one processor to each stack. Each processor looks at the top element. The nice thing is if this element points somewhere that is not at the top of a stack. The bad thing is if some of the selected top elements form a subchain of the list. Why? What we want to do is take the elements out one at a time and jump over them. That's what we did in the path-doubling routine. We removed each element and jumped over it. But we cannot take out two adjacent elements here because if we take the first one out we have to update the pointer pointing to it to point to the next element. If we take that one out also in parallel, that update becomes corrupted. So we have to break these long chains down into shorter chains. We select the top elements in each stack. They may form chains. Now each of these chains is handled by the processor at the tail of the chain. It will do that sequentially, one element per step. The other processors will go on and select the next element in their stack. The problem with this method is that the first processor that gets a long chain is busy for a long time dealing with its chain. Since, in the worst case, the length of such a chain could be $\Omega(n/\log n)$—which is far beyond $O(\log n)$—we have to take care of this situation. We do this by making sure that the chains are not too long. The technique used for breaking up the chains into shorter chains is called *deterministic coin tossing*, and is described in detail in [9]. It is of independent interest.

2.2.5   Parallel Sorting

Sorting is certainly one of the fundamental problems in computing, both in theory and in practice. A simple formulation of the problem is as follows. Given n keys $a_1, \ldots, a_n$ from a totally ordered universe, rearrange them into a sequence $a_{i_1}, a_{i_2}, \ldots, a_{i_n}$, such that

$$a_{i_1} \leq a_{i_2} \leq \ldots \leq a_{i_n}.$$

There are many special cases of this problem, and many algorithms making use of special conditions. Here, we are only interested in a very general type of algorithm for sorting, namely, those algorithms that access elements only in order to compare them with other elements.

On the theoretical side, the complexity of parallel sorting can be characterized as follows. It is well known that any sequential comparison-based algorithm requires $\Omega(n \log n)$ comparisons (and hence time) to sort n keys. Thus, a parallel algorithm requiring $O(\log n)$ time on a PRAM with n processors is (asymptotically) optimal. The following results are known:

(1) In 1983, Ajtai, Komlos and Szemeredi exhibited the construction of a sorting network of width n and depth $O(\log n)$. This network translates into a PRAM algorithm running in $O(\log n)$ time on n processors. However, the constant factor for the depth (resp., time), was (and still is, in spite of substantial improvements [25]) prohibitively large. For details of the original construction, see [1].

(2) In 1986, R. Cole designed a PRAM algorithm for sorting based on a "parallel merge" routine. This algorithm uses n processors and runs in time $c \cdot \log n$, for a "small" constant c. Even though the involved constants are much smaller than for the AKS network, they are still too large for practical considerations.

(3) The "Bitonic Sort" algorithm due to Batcher can be implemented on a number of networks, like the hypercube or shuffle exchange, as well as the PRAM model, and requires time $O(\log^2 n)$ on n processors. For many networks, the constant hidden by the big-O notation is $\frac{1}{2}$, and the real running time is very competitive for almost all practical values of n.

As far as realistic models of parallel computation are concerned, i.e., models based on fixed interconnection networks and not the PRAM model, Batcher's bitonic sort is still the most efficient method known. A lot of research is going on in this area, and recently some progress has been made for special cases—though Batcher's algorithm still remains to be beaten for the general case.

We'd like to make a remark about a problem sometimes mentioned in this context. This is the problem of topological sorting. Given a *dag*, a directed acyclic graph, we are asked to assign numbers to the vertices such that, whenever a vertex v is a successor of a vertex u in the graph, v's number is greater than u's. For parallel computation, the problem of topological sorting so far is closely related to the computation of transitive closure, which, in turn, is closely related to matrix multiplication. It is an interesting (and seemingly quite difficult) open question whether topological sorting can be achieved avoiding the "transitive closure" bottleneck of using an excessive number of parallel processors.

2.2.6   Euler Contour-Path Technique

We shall use quite a simple problem to motivate the contour-path technique. Assume we are given a tree in global memory, say in the form of an array of parent pointers or as an array of child pointers—it really doesn't matter. Now let's assume that we want to compute the depth for every vertex in the tree, i.e., its distance from the root. Note that this problem can be easily solved in linear time sequentially. If, for parallel computation, we allow read conflicts and n processors, we can use path doubling as was discussed earlier, counting the number of vertices we jump over.

Again, our goal is to find an optimal parallel algorithm, i.e., an algorithm running in $O(\log n)$ time on $n/\log n$ processors. We can characterize the contour-path technique as follows: Each edge in the tree is replaced by two antiparallel arcs, i.e., directed edges. Then, for each node, the arc pointing toward the root is chained with the arc pointing to the next sibling of the node. And, for each leaf node, the arc pointing to it is chained to the arc pointing from it.

On a tree with n nodes and, hence, $n - 1$ edges, stored as a length n array of pointers, all these operations can be performed in $O(\log n)$ steps by $n/\log n$ processors simply by splitting the array into $n/\log n$ roughly equal-sized segments and assigning one processor to each of the segments.

To solve the problem we were posed originally—namely, to compute the depth of every vertex in the tree—we assign 1 to every arc pointing from the root, and -1 to every arc pointing toward the root. Then we perform list ranking on the linear list given by the contour path, store the elements into an array in order of their list rank, and do a prefix computation on this array. It is easy to see that the prefix sum obtained for every node (as an endpoint of an arc) is exactly its depth.

To compute the depth of every node, we might just think of the following approach. We start from the root, broadcasting a counter to every child. A node processor, when it receives such a message containing a counter, increases the counter and further broadcasts it to its children. There are several problems connected with this approach. First, a node can have many children, and it could then take a long time until it broadcasts the new counter value to all of them. Second, the tree may be heavily unbalanced and contain long branches. Again, broadcasting along a long branch requires time proportional to its length, which may be prohibitively large.

This concludes our discussion of some of the most fundamental parallel-programming techniques and tricks.

2.3   The Dynamic Tree Expression Problem

In [30], Alternating Turing Machines using logarithmic space and a polynomial-size computation tree are studied. These machines can be thought of as solving a recognition problem by guessing a proof tree and recursively verifying it. Each internal node in the proof tree is replaced, in a universal step, by its children in the tree while leaf nodes correspond to axioms that can be verified directly. In every step, the machine is allowed to use only logarithmic storage (at every node) to record the intermediate derivation step.

We, in effect, turn the top-down computation of ATM's (as discussed in [30]) around into computations proceeding basically bottom-up, similar to the approach taken in [22], [23], and [33]. We present a uniform method containing and extending the latter results.

2.3.1   The Generic Problem

For the general discussion of the Dynamic Tree Expression Problem (DTEP), we assume that we are given

(1) A set P of N Boolean *variables*, $p_1, \ldots, p_N$.

(2) A set I of *inference rules* of the form

$$p_i \; :- p_j p_k \quad \text{or} \quad p_i \; :- p_j.$$

Here, juxtaposition of Boolean variables is to be interpreted as logical AND, and " $:-$ " is to be read as "if." In fact, the two types above are Horn clauses with one or two hypotheses, written in a Prolog style notation: $(p_j \wedge p_k) \Rightarrow p_i$ and $p_j \Rightarrow p_i$, respectively. We note that the total length of I is polynomial in N.

(3) A distinguished subset $Z \subseteq P$ of *axioms*.

Definition   1   Let (P, I, Z) be a system as above. The *minimal model* for (P, I, Z) is the *minimal* subset $M \subseteq P$ satisfying the following properties:

(1) The set of axioms, Z, is contained in M;

(2) Whenever the righthand side of an inference rule is satisfied by variables in M, then the variable on the lefthand side is an element of M:

$$\begin{aligned} p_j, p_k \in M, \ p_i \ :- p_j p_k \in I \\ p_j \in M, \ p_i \ :- p_j \in I \end{aligned} \quad \Rightarrow \ p_i \in M. \ \blacksquare$$

We say that (P, I, Z) *implies* some fact p if the Boolean variable p is in the minimal model M for (P, I, Z). For each such p, there is a *derivation* or *proof tree*: This is a (rooted) tree whose internal vertices have one or two children, and whose vertices are labeled with elements in P such that these four properties are satisfied:

(1) The labels of the leaves are in Z,

(2) If a vertex labeled p_i has one child, with label p_j, then $p_i \ :- p_j$ is a rule in I.

(3) If a vertex labeled p_i has two children, with labels p_j and p_k, then $p_i \ :- p_j p_k$ is a rule in I.

(4) The label of the root vertex is p.

The *Dynamic Tree Expression Problem* consists in computing the minimal model for a given system (P, I, Z); or, formulated as a decision problem, in deciding—given (P, I, Z) and some $p \in P$—whether p is in the minimal model for (P, I, Z). The algorithm in Figure 2.9 can be used to solve DTEP on a PRAM.

We first remark about the intuition behind the array DI used in the DTEP algorithm. If entry (i, j) of this array is **true** then it is known that p_i implies p_j. This is certainly true at the beginning of the algorithm by way of the initialization of DI. Whenever, for an inference rule $p_i \ :- p_j p_k$ with two antecedents, one of them—say p_k—is known to be **true**, the rule can be simplified to $p_i \ :- p_j$. But this means that p_j implies p_i, as recorded in the assignment to $DI[j, i]$. Also, the net effect of squaring the matrix DI in the last step of the outer loop is that chains of implications are shortened. Note that entry (i, j) of the matrix becomes **true** through the squaring of the matrix only if there was a chain of implications from p_i to p_j before the squaring.

```
algorithm DTEP(N, P, I, Z);
int N; set of Boolean P, Z; set of inference rules I;
co N is the number of Boolean variables in P; Z is the subset of P distinguished
as axioms; I is a set of Horn clauses with at most two variables on the right-
hand side; P is implemented as an array of Boolean variables; the algorithm sets
to true exactly those elements of the array P corresponding to elements in the
minimal model
oc
begin
    array DI[1..N, 1..N];
    co initialization oc
    P[i] := true for all pᵢ ∈ Z, else false;
    DI[j, i] := true for i = j and all pᵢ :- pⱼ ∈ I, else false;

    do l times   co l will be specified below oc
        for i ∈ {1, ..., N} with P[i] = false do in parallel
            if ((P[j] = P[k] = true) ∧ (pᵢ :- pⱼpₖ ∈ I)) ∨ ((P[j] = true) ∧
            (DI[j, i] = true))
            then P[i] := true
            fi;

            if (P[k] = true) ∧ (pᵢ :- pⱼpₖ ∈ I) then DI[j, i] := true fi
        od;
        DI := DI · DI
    od
end DTEP.
```

Figure 2.9   PRAM algorithm for generic DTEP.

Lemma   1   *If p_i is in the minimal model M, and if there is a derivation tree for p_i in (P, I, Z) of size m then $P[i] =$ true after at most $l = 2.41 \log m$ iterations of the outer loop of the DTEP algorithm. $P[i]$ never becomes true for $p_i \notin M$.*

Proof   The proof is by induction on the number of iterations of the main loop. We use the following induction hypothesis:

> At the end of each iteration of the main loop, there is a derivation tree for p_i of size at most 3/4 of the size at the end of the previous iteration, allowing as inference rules the rules in I and the "direct implications" as given by the current values of the elements of DI, and as axioms the p_j with $P[j]$ currently **true**.

This induction hypothesis is trivially satisfied before the first iteration of the loop. For the rth iteration, let $m = |T|$ be the size of a derivation tree

T for p_i, using the current axioms and derivation rules.

We distinguish two cases:

(1) T has at least $m/4$ leaves: Since the parallel loop in the DTEP algorithm set $P[k] = \textbf{true}$ if the children of a vertex labeled p_k are all leaves, this case is trivial.

(2) Let r_k be the number of maximal chains in T with k internal vertices, and let b be the number of leaves of T, both at the start of the loop. A chain in T is part of a branch with all internal vertices having degree exactly 2. Then we have:

$$m \;=\; 2b - 1 + \sum_{k \geq 1} k \cdot r_k;$$

$$
\begin{aligned}
m' \;&\leq\; b - 1 + \sum_{k \geq 1} \lceil k/2 \rceil \, r_k \\
&\leq\; b - 1 + \frac{1}{2}\Big(\sum_{k \geq 1} k r_k + \sum_{k \geq 1} r_k\Big) \\
&\leq\; b - 1 + \frac{1}{2}(m + 1 - 2b) + \frac{1}{2} \cdot 2b \\
&<\; \frac{3}{4}m \, .
\end{aligned}
$$

Here, m' is the size of the derivation tree after execution of the loop, using the current set of axioms and inferences. The first inequality follows because squaring of the matrix DI halves the length of all chains. The third inequality stems from the observation that in a tree with b leaves, the number of (maximal) chains can be at most $2b - 1$, as can be seen by assigning each such chain to its lower endpoint.

There is a trivial derivation tree for p_i after at most $\log_{4/3} m \leq 2.41 \log m$ (note that all logarithms whose base is not mentioned explicitly are base 2). $\square$

Theorem   4   *Let (P, I, Z) be a derivation system with the property that each p in the minimal model M has a derivation tree of size*

$$\leq 2^{\log^c N}.$$

Then there is an $\mathcal{NC}$-algorithm to compute M.

Proof   It follows immediately from the above Lemma that under the stated conditions the DTEP algorithm runs in polylogarithmic time. More precisely, for the bound on the size of derivation trees given in the Theorem, the running time of the algorithm is $O(\log^{c+1} N)$, and it uses N^3 processors. $\square$

2.3.2   Applications of DTEP

Longest Common Substrings   As a very simple example of a DTEP application, we consider the *longest common substring* problem: Given two strings $a_1 \cdots a_n$ and $b_1 \cdots b_m$ over some alphabet Σ, we are supposed to find a longest common substring $a_i \cdots a_{i+r} = b_j \cdots b_{j+r}$.

To obtain a DTEP formulation of this problem, we introduce variables $p_{i,j,l}$, whose intended meaning is:

$$p_{i,j,l} \text{ is true if } a_i \cdots a_{i+l-1} = b_j \cdots b_{j+l-1}.$$

Thus, we have the following axioms:

$$p_{i,j,1} \text{ iff } a_i = b_j,$$

and these inference rules:

$$p_{i,j,l} \; : - \; p_{i,j,\lceil l/2 \rceil} p_{i+\lceil l/2 \rceil, j+\lceil l/2 \rceil, \lfloor l/2 \rfloor}.$$

An easy induction on the length l of a common substring shows that $p_{i,j,l}$ is true in the minimal model for the above derivation system if and only if the substring of length l starting at position i in the first string is equal to the substring of the same length at position j in the second string. A similar induction can be used to show that for every common substring of length l, there is a derivation tree of size linear in l in the above derivation system. These two observations together establish the longest common substring problem as an instance of DTEP.

Other Applications   It is possible to rephrase a number of other problems as instances of the Dynamic Tree Expression Problem. A few examples are contained in the following list:

- Transitive closure in graphs and digraphs

- The nonuniform word problem for context-free languages

- The circuit-value problem for planar-monotone circuits

- DATALOG programs with the *polynomial-fringe property*

- ALT($\log n$, expolylog n)

These are the problems recognizable by log-space bounded Alternating Turing Machines with a bound of $2^{\log^c n}$ for the size of their computation tree, for some constant $c > 0$. This class is the same as $\mathcal{NC}$.

For the DATALOG example, we refer the interested reader to [33]. We briefly discuss the third and the last example in the above list.

The Planar-Monotone-Circuit Value Problem

A *circuit* is a directed acyclic graph (dag). Its vertices are called *gates*, and the arcs correspond to *wires* going from outputs of gates to inputs. The gates are either input/output gates connecting the circuit to its environment, or they are combinational—mapping the values on the input wires of the gate to values on the output wires—according to the function represented by the gate. In a *monotone* circuit, all combinational gates are AND or OR gates. Therefore, the output of the circuit, as a function of the inputs, is a monotone function. A *planar* circuit is a circuit whose dag can be laid out in the plane without any wires crossing one another and with the inputs and output(s) of the circuit on the outer face.

We assume that circuits are given as follows. The description of a circuit with n gates is a sequence $\beta_0, \ldots, \beta_{n-1}$. Each gate β_i is

(1) 0-INPUT or 1-INPUT (also abbreviated as 0 and 1, respectively); or

(2) AND(β_j, β_k), OR(β_j, β_k), or NOT(β_j), where $j, k < i$. In this case, the gate computes the logical function indicated by its name, with the values on the input wires to the gate as arguments, and puts this function value on all of the gate's output wires.

The output wire of β_{n-1} carries the output of the circuit.

The *circuit-value problem* (CVP) consists in computing the output of a circuit, given its description. It is well known that CVP is complete for $\mathcal{P}$ under log-space reductions, and hence is a "hardest" problem for the class of polynomial-time algorithms [17]. Even if we restrict ourselves to monotone circuits (allowing only AND and OR gates) or to planar circuits, the corresponding restricted CVP remains $\mathcal{P}$-complete [13]. However, the circuit-value problem can be solved efficiently in parallel for circuits that are monotone *and* planar [11], [18]. While the original solutions were indirect and technically quite involved, the DTEP paradigm gives us a relatively simple approach.

For ease of presentation, we assume that the description $\beta_0, \ldots, \beta_{n-1}$ of a planar-monotone circuit satisfies the following additional properties:

- The circuit is arranged in *layers*. The first layer is constituted by the inputs to the circuit; the last layer consists of the output of the circuit. For every wire in the circuit, there is a (directed) path from an input to the output of the circuit, containing that wire.

- All wires run between adjacent layers without crossing one another.

- All gates on even layers are AND gates with at least one input.

- All gates on odd layers (except the first) are OR gates.

These restrictions are not essential. Given a general type description of a circuit, there are well-known $\mathcal{NC}$ algorithms to test whether it represents a monotone and planar circuit, and to transform it into a description of a functionally equivalent circuit satisfying the restrictions listed above.

To obtain a DTEP formulation, it is helpful to look at *intervals* of gates. Formally, let the triple (l, i, j) denote the interval from the ith through the jth gate (in the order in which they appear in the circuit description) on the lth layer of the circuit. We shall introduce Boolean variables $p_{l,i,j}$ with the intended meaning

$p_{l,i,j}$ is **true** iff the output of every gate in (l, i, j) is **true**.

We simply observe that an interval of AND gates has outputs all **true** if and only if all inputs to this interval of gates are **true**, and that these inputs come from an interval of gates on the next lower layer. For intervals of OR gates, the situation is a bit more complicated because, in order to obtain a **true** output from an OR gate, only one of its inputs needs to be **true**. We must therefore be able to break intervals of **true** OR gates down into smaller intervals that are **true** because a contiguous interval of gates on the next lower layer is **true**. Because wires don't cross in a planar layout, such a partition is always possible.

More formally, the inference rules for a given instance of the planar-monotone-circuit value problem are:

(1) $p_{l,i,j}$ $:- p_{l-1,i',j'}$ for every (nonempty) interval of AND gates. Here, i' denotes the first input of the first gate, and j' the last input of the last gate in the interval.

(2) $p_{l,i,j}$ $:- p_{l,i,k} p_{l,k+1,j}$, for all $i \leq k < j$.

(3) $p_{l,i,j}$ $:- p_{l-1,i',j'}$, for all layers l of OR gates and all pairs (i', j') where $i' \leq j'$ and $\beta_{i'}$ is an input of β_i, and $\beta_{j'}$ is an input of β_j.

The axioms are given by all intervals of **true** inputs to the circuit.

It is easy to see that the total length of the representation for the axioms and for all inference rules is polynomial in n. A straightforward induction, using the planarity of the circuit as outlined above, also shows that $p_{l,i,i}$ is **true** in the minimal model if and only if the ith gate on layer l has output **true** in the circuit.

Suppose $p_{l,i,j}$ is in the minimal model. To see that there is a linear size derivation tree for $p_{l,i,j}$ we consider the subcircuit consisting of all gates from which a gate in (l, i, j) is reachable. Whenever we use an inference rule of type 1 or of type 3 in a derivation tree for $p_{l,i,j}$ we charge the vertex in the derivation tree (corresponding to the left-hand side of the rule) to the interval (l, i, j). We observe that we need to use an inference rule of type 2 for an OR gate on some layer l' only if the interval of gates on layer $l' - 1$

feeding into this interval contains more than one maximal subinterval of **true** gates. In this case, we can split the interval (l', i, j) into subintervals (l', i, k) and $(l', k+1, j)$, in such a way that one of the inputs to gate k on layer l' is **false**. We charge the left-hand side of the type 2 inference rule to the interval of **false** gates on layer $l' - 1$ containing this input. Because of the planarity of the circuit, rules used to derive $p_{l',i,k}$ and $p_{l',k+1,j}$ are charged to disjoint sets of intervals. Therefore, the number of maximal intervals of gates with the same output value is an upper bound on the size of a smallest derivation tree for any element in the minimal model. We conclude

Theorem 5   *There is an $\mathcal{NC}$ algorithm for the planar-monotone-circuit-value problem.* $\Box$

Tree-size bounded, alternating Turing machines

Alternating Turing machines are a generalization of standard (nondeterministic) Turing machines [16], [6]. They are interesting because their time complexity classes closely correspond to (standard) space complexity classes, and because their computations directly correspond to first-order formulas in prenex form with alternating quantifiers.

For an informal description of alternating Turing machines (ATMs), we assume that the reader is familiar with the concept of standard (nondeterministic) Turing machines [16]. An ATM consists of a (read-only) input tape and some number of work or scratch tapes, as in the standard model, as well as a finite control. The states of the finite control, however, are partitioned into *normal, existential*, and *universal* states. Existential and universal states each have two successor states. Normal states can be *accepting, rejecting*, or *nonfinal*. An *instantaneous description*, or *id*, of an ATM consists of a string giving the state of the finite control of the ATM, the position of the read head on its input tape, and head positions on and contents of the work tapes. It should be clear that, given the input to the machine, an id completely determines the state of the machine.

A computation of an ATM can be viewed as follows. In any state, the machine performs whatever action is prescribed by the state it is in. If the state is a normal nonfinal state, the finite control then simply goes to its unique successor state. Accepting or rejecting states have no successors. If the state is existential or universal, the machine clones itself into as many copies as there are successor states, with each clone starting with an id identical to that of the cloning id—except that the state of the cloning id is replaced by that of the successor state. The cloning machine disappears. It is best to view the computation of an ATM as a tree whose nodes correspond to ids of the computation in such a way that children of a node in the tree represent successor ids.

Given this interpretation, each computation of an ATM can be mapped into a (binary) tree. This tree need not necessarily be finite, since there can be nonterminating branches of the computation. To define whether an ATM *accepts* or, dually, *rejects*, some input, we look at the corresponding computation tree as defined above. A *leaf* in this tree is called *accepting* iff the state in its id is accepting. An internal node whose id contains a normal state is accepting iff its (unique) descendent in the tree is accepting. An internal node with an existential state is accepting iff at least one of its immediate descendents is accepting, and a node with a universal state is accepting iff all of its immediate descendents are accepting. An id of the ATM is called *accepting* if it is attached to an accepting node in the computation tree. An ATM is said to accept its input if and only if the root of the corresponding computation tree is accepting by the above definition or, equivalently, if the initial configuration of the machine is accepting.

To cast ATM computations into the DTEP framework we note that if the space used by the ATM on its worktapes is bounded by $O(\log n)$, then the number of distinct ids for the machine is bounded by some polynomial in n. Assume that $id_0, \ldots, id_r$ is an (efficient) enumeration of these ids. Our instance of DTEP will have Boolean variables p_i, for $i = 0, \ldots, m$, with the intended meaning

$$p_i \text{ is \textbf{true} iff } id_i \text{ is accepting.}$$

The axioms are given by those id's containing accepting states, and we will have these inference rules :

- $p_i \; :- p_j p_k$ iff p_i contains a universal state with two successor states, and id_j and id_k are the two immediate successor ids of id_i; and

- $p_i \; :- p_j$ in all other cases where id_j is an (the) immediate successor id of id_i.

A straightforward induction shows that there is an exact correspondence between derivation trees for the p_i corresponding to the initial id of the ATM, and certain subtrees of the ATM's computation tree. These subtrees are obtained by removing, for each node whose id contains an existential state, all but one of the children of that node, together with their subtrees. Also, the leftover child has to be an accepting node.

Since the computation tree of the ATM is of expolylog size, so is any derivation tree obtained by the above construction. This concludes the (informal) proof for the following theorem:

Theorem   6   *Problems in the complexity class ALT(log n, expolylog n) are instances of the Dynamic Tree Expression problem. Since this class is the same as $\mathcal{NC}$, every problem in $\mathcal{NC}$ can be cast as an instance of DTEP.*

$\square$

2.3.3   Application to Algebraic Straightline Programs

Straightline programs consist of a sequence of assignment statements executed in order. Each variable referenced in a statement is either an input variable, or it has been assigned a value in an earlier assignment. In *algebraic straightline programs*, the expressions on the right-hand side of the assignment are basic unary or binary expressions involving the arithmetic operators $+$, $-$, $\times$, and $/$, as well as constants and input or computed variables. Each noninput variable is assigned a value in exactly one statement in the straightline program. In what follows, we are only concerned with straightline programs containing the operators $+$, $-$, and $\times$. Any such straightline program computes some (multivariate) polynomial of the input variables. Conversely, every multivariate polynomial can be expressed as an algebraic straightline program.

A straightline program corresponds, in a natural way, to a *directed acyclic graph*, or *dag*. In this dag, the vertices correspond to the assignment statements, the *sources* of the dag (i.e., vertices of indegree zero) correspond to the input variables of the program, and arcs signify the use of variables defined in one statement by some other statement. We define the *size* of a straightline program to be its number of assignment statements. Figure 2.10 shows a straightline program for the polynomial $((x_1 + x_2)^3 + x_1^2 x_3)(x_1 + x_2)$ with indeterminates x_1, x_2, and x_3, and Figure 2.11 gives the corresponding computation dag. In Figure 2.11, all edges represent arcs directed upward.

We assume without loss of generality that the operation in the last assignment in a straightline program is a multiplication.

To make straightline programs more uniform and easier to deal with, we first wish to transform them into an (algebraically equivalent) sequence of *bilinear forms* of the form

$$y_i = \left(\sum_{j<i} c_{ij}^{(l)} y_j \right) \cdot \left(\sum_{j<i} c_{ij}^{(r)} y_j \right).$$

Again, the y_j are either input variables or assigned a value in an earlier

---

$$
\begin{array}{llll}
x_4 &= x_1 + x_2 & x_5 &= x_4 \times x_4 \\
x_6 &= x_5 \times x_4 & x_7 &= x_1 \times x_1 \\
x_8 &= x_3 \times x_7 & x_9 &= x_6 \times x_8 \\
x_{10} &= x_4 \times x_9 &
\end{array}
$$

Figure 2.10   Straightline program for polynomial $((x_1 + x_2)^3 + x_1^2 x_3)(x_1 + x_2)$.

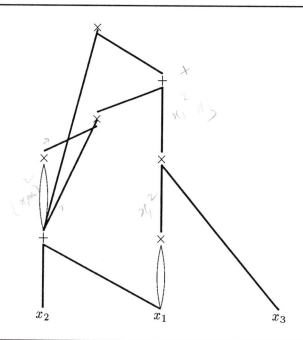

Figure 2.11   Computation dag for straightline program for polynomial $((x_i + x_2)^3 + x_1^2 x_3)(x_1 + x_2)$.

assignment. In fact, each noninput variable y_i is in exact correspondence with some x_{k_i} in the original straightline program, which is the result of a multiplication operation. Assuming that $i = k_i$ for the input variables, the coefficients $c_{ij}^{(l)}$ (resp., $c_{ij}^{(r)}$) give the number of distinct paths from x_{k_j} to the left (resp., right) multiplicand of x_{k_i} in the dag belonging to the original straightline program. These coefficients can be obtained in a straightforward way by repeatedly squaring a matrix A obtained as follows. We first construct, from the computation dag for the straightline program, an auxiliary dag by splitting every multiplication node v into three nodes, v^1, v^2, and v^3: v^1 receives the arc from the left multiplicand, v^2 the arc from the right multiplicand. All arcs originally leaving v are reattached to v^3. In the resulting digraph, the original input nodes and the type 3 nodes are sources, and the type 1 and type 2 nodes are the sinks. The matrix A mentioned above is the adjacency matrix of this digraph after we attach a loop (directed cycle of length one) to every sink. Further details are left to the reader. The coefficients in the bilinear forms can be computed on a PRAM in time $O(\log^2 n)$, using $M(n)$ processors. Here, n is the size of the

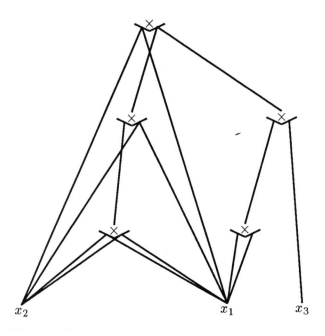

Figure 2.12   Graph of bilinear straightline program.

original straightline program, and $M(n) \leq n^3$ is the number of processors required to perform matrix multiplication in time $O(\log n)$.

Figure 2.12 shows a graphical representation of the transformed "bilinear straightline program" for the straightline program resp. computation dag given in Figures 2.10 and 2.11. Each bilinear form is represented by a flat "V," with the terms in the first (second) factor given by the arcs ending in the left (right) arm of the "V." The coefficient belonging to a term is attached to the arc unless it is one or zero. In the latter case, no arc is drawn at all.

For the digraph B of a bilinear straightline program, we can define operations analogous to the DTEP operations. The Boolean variables are replaced by pairs (c_i, v_i) for every node i in B. The first element, c_i, is a Boolean indicating whether the value of the polynomial corresponding to vertex i has been computed; if c_i is **true** then v_i holds this value. Initially, the values of all the input variables are known. We call a node of B a *1-node* iff it is not a sink (i.e., a node of outdegree zero) and the value of exactly one of its factors is known (i.e., the values of all nodes with arcs to the corresponding "leg" have been computed). All other nodes whose values are still unknown are called *2-nodes*.

We also have several matrices describing the arcs in the graph, and the coefficients attached to them. The set of arcs and coefficients, and consequently the entries in these matrices, change during the course of the algorithm.

- The (i,j)th entry of $C^{(l)}$ contains the coefficient of the arc from the ith node of B to the *left leg* of the jth node for all nodes i and j such that j is a 2-node; the other entries are zero.

- $C^{(r)}$ is defined correspondingly for the right-hand factors.

- The (i,j) entry of the matrix D, for j a 1-node, is the coefficient of the arc from node i to node j.

- The matrix $\bar{C}$ is a projection matrix; all its entries are zero except for those on the diagonal positions corresponding to 2-nodes.

Each phase of the adapted DTEP algorithm consists of the following steps:

co one phase of adapted DTEP **oc**

> **for** all i with $c_i = \textbf{false}$ and both factors known **do in parallel**
> compute v_i and set $c_i = \textbf{true}$;

> **for** all nodes i that just became new 1-nodes **do in parallel**
> multiply the coefficients of all arcs entering the unknown "leg" by the other, known factor;

> update all matrices so as to reflect the changes in the sets of 1-nodes and 2-nodes;

> $C^{(l)} := (\bar{C} + D) \cdot C^{(l)}; \quad C^{(r)} := (\bar{C} + D) \cdot C^{(r)};$
> $D := (\bar{C} + D) \cdot D;$

> update B to correspond to the newly computed matrices;

co end of phase **oc**

It is straightforward to verify that the bilinear program given by the modified graph after execution of one phase computes the same values as the original bilinear program. Also, one phase can be executed on a PRAM in time $O(\log n)$ employing $M(n)$ processors, where $M(n) \leq n^3$ is the number of processors required to multiply two $n \times n$ matrices in $O(\log n)$ parallel steps.

To obtain a bound on the required number of phases, it is not sufficient to unfold the dag B into a tree and then use our general results about DTEP because there are easy examples where the unfolded tree is of exponential size. However, the following *worst-term* argument works.

Define, for each node in the graph of a bilinear straightline program, its *formal degree* inductively as follows: The original input variables have formal degree 1, and each other variable has a formal degree equal to the sum of the formal degrees of its two factors. Here, the formal degree of a factor is defined to be the maximum of the formal degrees of all its terms.

For every factor of a noninput variable y_i in a bilinear straightline program, we define a *worst term*, derived from the execution of the above algorithm: Determine, for the corresponding "leg" of the node i in the sequence of graphs generated by the algorithm, which other node j with an arc to that leg is the latest to have its value computed. If j has an arc to i in the first graph of the sequence, y_j becomes the worst term for the factor of y_i under consideration, and we distinguish the arc from j to i. Otherwise, distinguish any path from j to the appropriate leg of i in the first graph of the sequence. Every node on this path corresponds to the worst term of the appropriate factor of the next node on the path.

Consider the subgraph B' induced by the distinguished arcs. Assume without loss of generality that B' has only one sink. Clearly, B' computes some monomial. If we unfold B' into a tree T, then T's size is exactly twice the formal degree of this monomial, minus one. The reason is that all nonleaf nodes in T are multiplication nodes, with exactly one term per factor.

A simple induction shows that the adapted DTEP algorithm computes the values of nodes in T in exactly the same phases as it computes them when run on B, the full graph. The analysis of the general DTEP algorithm applies to the adapted algorithm, executed on T. We obtain that the adapted DTEP algorithm computes the value of each y_i in a number of phases proportional to the logarithm of the formal degree of y_i. Since the initial transformation to a bilinear straightline program is basically a transitive closure computation, we obtain the following theorem:

Theorem 7 *Suppose S is a straightline program containing n variables and computing a polynomial of formal degree d. The adapted DTEP algorithm computes the value of the polynomial given by S, at a specified point, in time $O(\log n(\log n + \log d))$ using $M(n)$ processors on a unit cost PRAM.* $\square$

Thus, polynomials given by straightline programs can be evaluated by an $\mathcal{NC}$ algorithm as long as their formal degree is expolylog in n. This result was originally obtained in [22]. Our presentation is intended to give evidence for the wide applicability of the DTEP paradigm, and maybe to supply a somewhat simpler, less technical proof.

2.3.4   General Remarks About DTEP

The DTEP paradigm establishes an interesting connection between fast parallel computation and small proof or derivation trees. In more technical terms, the instances of DTEP and $\mathcal{NC}$ are characterized as $\text{ALT}(\log n, \text{expolylog} n)$ or the set of problems recognizable by log-space-bounded alternating Turing machines with "expolylogarithmically" bounded computation trees. DTEP is not necessarily intended to provide optimal parallel algorithms for a given problem. Instead, its primary function is to provide a convenient, high-level way for the specification of an efficient and highly portable parallel algorithm. The input to DTEP is in a format very similar to logic programs—declarative rather than procedural—and thus allows us to abstract away many peculiarities of any underlying parallel architecture.

The DTEP algorithm can be efficiently implemented on various "real-world" parallel architectures, such as binary hypercubes, butterfly networks, shuffle-exchange based networks, and multidimensional meshes of trees [21]. The algorithm can also be tailored and made more efficient (both in terms of time and number of processors) for more restricted instances of the dynamic tree expression problem. Some of these optimizations are also discussed in [21].

We have also shown how to extend the basic DTEP paradigm to problems in algebraic domains—in particular the parallel evaluation of algebraic computation dags, or straightline programs. Further interesting extensions include the possibility of handling non–Horn clauses in the basic formulation of DTEP, along the lines discussed in [33].

Another interesting line of research is the characterization of more classes of problems for DTEP in independent terms—in particular the derivation of conditions for the existence of small proof trees.

2.4 $\mathcal{P}$-Complete Algorithms

In this section, we are going to talk about hardness results for parallel computation, or at least some examples of them. We want to talk about the parallel complexity of Gaussian elimination. Formally, we shall prove that Gaussian elimination with pivoting—partial or full—is a $\mathcal{P}$-complete problem. This result was first obtained by Steve Vavasis, a student at Stanford, in a term paper for my course on Parallel Computation [34].

We have already met the parallel complexity class $\mathcal{NC}$, the class of problems solvable on a PRAM in polylogarithmic time using a polynomial number of processors. We also know that $\mathcal{NC}$ is contained in $\mathcal{P}$, the well-known class of problems solvable in polynomial time on a Turing machine. One of the big open questions in complexity theory is whether this containment is strict, and many—if not most—people think it is. If we accept this assumption (and the reader may notice the analogy with the $\mathcal{P}$ versus $\mathcal{NP}$ scenario), we can show that problems are not solvable efficiently in parallel by proving that they are $\mathcal{P}$-complete (or even harder, of course), a concept we have already discussed.

What do we know about Gaussian elimination? Sequentially, it's a well-known algorithm and has been analyzed backward and forward. Error analysis has been done. Gaussian elimination has been implemented in many places. It is numerically stable if you use pivoting, and it consists of a number of elimination steps where each step looks for a pivot element (of relatively large magnitude) and then uses the corresponding equation to eliminate the pivot variable from the remaining equations. By the way, each such elimination step is easy to parallelize. There are many systolic or mesh-oriented algorithms, and if we go to hypercubes it's even easier. But the elimination steps are performed one after another, basically one per row of the given matrix. Thus, if the matrix has n rows, the number of iterations would be something like n, which is a far cry from polylogarithmic in n.

We are going to show, in a formal sense specified below, that Gaussian elimination with pivoting (partial or full) is $\mathcal{P}$-complete, meaning that it is unlikely that it can be efficiently parallelized using only polynomially many processors. One could object that maybe the precision required for parallel computation is so big that no fast parallel algorithm exists. We show, however, that finite precision—say two decimals—is sufficient for our argument!

There are many other variants of Gaussian elimination that we are not going to talk about. However, we feel that for most if not all of them the proof given here can be adapted.

One important thing we want to introduce here is the concept of a $\mathcal{P}$-complete algorithm as opposed to a $\mathcal{P}$-complete problem. The difference is whether we look at a specific algorithm or method to solve a problem, or at the problem in general. Here, the problem is to solve linear systems of equations. There are many ways of doing this. One is Gaussian elimination, which is an algorithm—a rather specific method to do it. Another way would be to compute the inverse of the matrix by some method and then just multiply the right-hand side of the system by the inverse.

The concept of a $\mathcal{P}$-complete algorithm is the following: Suppose that we have some problem that we call P, and suppose we have some sequential polynomial-time deterministic algorithm $\mathcal{A}$ for this problem P. Thus, given any instance I of P, $\mathcal{A}$ will find a solution, and we call this (unique) solution $\mathcal{A}(I)$. In other words, $\mathcal{A}$ is a function—it maps inputs I to outputs $\mathcal{A}(I)$. To turn this whole thing into a decision problem, we may ask: Given an instance I of the problem, and some index i, is the ith bit of the output $\mathcal{A}(I)$

a 1 or a 0? That's a yes/no answer. So we call an algorithm $\mathcal{P}$-complete if this decision problem is $\mathcal{P}$-complete.

We want to stress that $\mathcal{P}$-completeness of algorithms as defined here is a relationship between the inputs and the outputs. It doesn't really say much about how the algorithm $\mathcal{A}$ proceeds. We just want to be able to reproduce what the sequential algorithm does in parallel in terms of its input/output behaviour. We are not saying that if you want to compute this you have to simulate the sequential algorithm step by step. There may in fact be—and often are—many other ways to come by the same answer. You may use all your intelligence and ingenuity to come up with a fast parallel solution. However, if the algorithm is $\mathcal{P}$-complete, then most likely we are out of luck.

Table 2.5   Examples of $\mathcal{P}$-complete algorithms

1. Greedy depth-first search
2. Greedy maximal path
3. Greedy maximal independent set
4. Greedy maximal set of augmenting paths
5. FFD bin packing

Table 2.5 lists a few examples for $\mathcal{P}$-complete algorithms. To explain a bit the meaning of "greedy" we look at the case of depth-first search. We are given an undirected graph—by means of adjacency lists for all of its vertices—and a starting vertex, and we are supposed to construct a depth-first search tree rooted at that given vertex. If we do this by the standard, greedy algorithm then—whenever the search reaches a vertex—it looks for the first vertex in its adjacency list that has not yet been visited, and it proceeds to that vertex. If there is no such vertex the search backs up to the vertex visited before. Of course, since the adjacency lists could be rearranged arbitrarily, there are in general many DFS trees. John Reif showed that if I is a representation of a graph in terms of adjacency lists, then computing the standard or greedy (as described above) depth-first search tree is $\mathcal{P}$-complete [29].

There are a few more examples of $\mathcal{P}$-complete algorithms in the table. The maximal-path problem, for instance, is to construct a path (like depth-first search) until you cannot continue anymore because you have seen every neighbour of the last node. For more details about this problem, the interested reader is referred to [2].

Here, we consider the case of Gaussian elimination with partial pivoting.

Formally, given a (nonsingular) $n \times n$ matrix M, we want to decompose M into

$$M = PLR$$

where L is a lower triangular matrix with 1's on the diagonal, R is an upper triangular matrix, and P is a permutation matrix determined by the pivot positions.

We'd just like to remark here, that, given the permutation matrix P, it is relatively easy to determine L and R. There is, in fact, an $\mathcal{NC}$ algorithm to determine them. However, we are not going to discuss the details here.

In order to show that Gaussian elimination is $\mathcal{P}$-complete, we are going to reduce a known $\mathcal{P}$-complete problem to Gaussian elimination with (partial) pivoting. This establishes that Gaussian elimination is $\mathcal{P}$-hard, i.e., it is at least as hard as everything in $\mathcal{P}$. But we know that it is in $\mathcal{P}$ itself because we all know an n^3 algorithm for Gaussian elimination.

The $\mathcal{P}$-complete problem we use is the circuit value problem. It is the generic $\mathcal{P}$-complete problem. We are given a directed acyclic graph consisting of gates (say AND, OR, and NOT gates) and of inputs, which are 0 or 1. Actually, in our case, we use NAND gates only, but it is easy to see that this does not restrict generality since each one of the other gates can be simulated by a small circuit built solely of NAND gates.

Thus, we are going to reduce the circuit value problem (for circuits consisting only of NAND gates) to the problem of determining the permutation matrix P resulting from the pivot positions in the Gaussian-elimination algorithm. Intuitively, what we want is to construct, from a given circuit, a linear system with matrix M such that performing Gaussian elimination on M somehow simulates the given circuit. In particular, we want to be able to read off from the choice of pivot elements whether the circuit evaluates to 0 or 1.

This reduction is a simple transformation. For every gate of the circuit it constructs a couple of rows in the matrix. If we have n gates in the circuit (including inputs), then we will have $2n$ rows in the matrix.

Suppose that the gates of the circuit, including the inputs, are numbered 1 through n, in such a way that whenever gate i is an input of gate j then $i < j$. Such a numbering is called *topological*. We denote the entries of the $2n \times 2n$ matrix M by $m_{i,j}$. They are all zero except for those determined explicitly by one of the following conditions:

(1) If gate i is a 1-input then $m_{2i-1,2i}$ and $m_{2i,2i}$ are

$$\begin{pmatrix} -3.9 \\ 0 \end{pmatrix}$$

(2) If gate i is a 0-input or a NAND gate then $m_{2i-1,2i}$ and $m_{2i,2i}$ are

$$\begin{pmatrix} -3.9 \\ 4 \end{pmatrix}$$

(3) If the output of gate i is an input of gate j then $m_{2i-1,j}$ and $m_{2i,j}$ are

$$\begin{pmatrix} 0 \\ 1 \end{pmatrix}$$

as are $m_{2j-1,i}$ and $m_{2j,i}$.

(4) Entry $m_{2i,n+i}$ is 1, for $i = 1, \ldots, n$.

As a small example, consider a circuit consisting only of a NAND gate (with two inputs). The inputs are numbered 1 (with value 1) and 2 (with value 0), the gate itself is numbered 3. Figure 2.13 shows the 6×6 matrix generated by the reduction. Entries that are zero by default are left blank in the figure.

We note that the numbers 3.9 and 4 are chosen somewhat arbitrary. They work, but one could certainly come up with a different set of numbers for the same job. We also note that the actual simulation of the circuit occurs during the first n elimination steps, with pivots in the left half of the matrix. The right half simply assures that the matrix is nonsingular.

Let's first verify that the matrix M constructed by the reduction is indeed nonsingular. This can easily be seen by permuting the rows as follows: first come all the odd-numbered rows, then the even-numbered rows. The resulting matrix is lower triangular, with nonzero elements on the diagonal: -3.9 in the left half and 1's on the right.

The following theorem states the fashion in which determination of the pivot elements (during the first n elimination steps, i.e., in columns 1 through n) simulates the underlying circuit.

Theorem 8   *For $i = 1, \ldots, n$, the element originally in position $(2i-1, i)$ (resp., $(2i, i)$) of matrix M will be the pivot element if the output of gate i is 1 (resp., 0).*

$$\begin{pmatrix} -3.9 & & 0 & & & \\ 0 & & & 1 & 1 & \\ & -3.9 & & 0 & & \\ & 4 & & 1 & 1 & \\ 0 & 0 & -3.9 & & & \\ 1 & 1 & 4 & & & 1 \end{pmatrix}$$

Figure 2.13   Matrix for NAND circuit.

Proof We only sketch the main ideas for the proof. The output of a NAND gate is 1 iff at least one of its inputs is 0. If a gate which could be an input has value 1, then the first of its two rows will contain the pivot element which is −3.9. If the output of a gate is 0, then the second of its rows will contain the pivot element, which, in this case, is 4. Only in the second case will the elimination procedure affect other rows in the matrix, namely, those belonging to gates for which the current gate is an input. A value of 0.25 will be subtracted from the second row of the pair belonging to each such gate, thus making sure that this row won't contain the pivot element. But since the corresponding NAND gate receives a 0 input, this is in line with our assumption stated at the beginning.

To give a formal proof, the cumulative effects of the off-diagonal elements have to be bounded. We omit these lengthy and technical arguments. We note, however, that more precise estimates can be used to show that using only two decimals for precision is sufficient to make the argument go through. □

To relate the above proof to our definition of a $\mathcal{P}$-complete algorithm, it should be clear that we have actually shown that it is $\mathcal{P}$-complete whether a specified entry of the permutation matrix P produced by the (partial) pivoting routine is 0 or 1.

As we noted, the $\mathcal{P}$-completeness argument also works for fixed precision. Thus, it is not differences in numbers that get smaller and smaller— or, for that matter, numbers that become larger and larger—that make Gaussian elimination hard for parallel computation; it is the underlying combinatorial structure, which insists on pivoting on the element with the largest absolute value in the current column. By the way, we have actually shown the somewhat stronger result that *pairwise pivoting* is $\mathcal{P}$-complete since (during the first n elimination steps) each pivot is chosen from just two candidate rows.

On the other hand, we have only shown a specific algorithm (or method) to solve linear systems to be bad for parallel speedup: There are parallel algorithms that solve linear systems of equations, and they are $\mathcal{NC}$ algorithms. One comes from work by Csanky who developed parallel methods to compute the determinant [10]. Once we have the determinant, we can use it to compute the inverse. Still other methods have been given in [7], [5], and [24].

We should also note that we only showed that Gaussian elimination most likely cannot be speeded up efficiently to obtain polylogarithmic running times. In practice, however, polynomial running times with a small polynomial (smaller than the obvious n^3) may already be quite interesting. And it shouldn't be too hard to observe that every single elimination step of the Gaussian elimination procedure can actually be done in polylogarithmic time, obtaining a total running time that is roughly linear in n.

2.5   Parallel Approximation Algorithms

In this section, we shall talk about approximation algorithms for bin packing. The first part, however, will be related to the topic of the last section on $\mathcal{P}$-complete algorithms: We shall show that the problem to construct an FFD packing is $\mathcal{P}$-complete. Then, we shall present an $\mathcal{NC}$ algorithm that produces a packing as good as FFD.

The *bin-packing problem* is as follows: We are given a set of n items, of varying sizes between 0 and 1, which we are supposed to pack into as few bins of unit size as possible. Bin packing is a well-known $\mathcal{NP}$-hard optimization problem. There are, however, heuristics—simple algorithms that provide solutions that are not too bad.

The first-fit decreasing method is the following: We first sort the items into nonincreasing order of size, and we arrange the bins in a row. Then, in order, we pick one item at a time and pack it into the first bin where it still fits. Intuitively, FFD is a good idea because we pack the large items first.

For the first-fit decreasing method, theoreticians were able to prove a performance bound of $\frac{11}{9}$ times the optimal number of bins plus some constant term. This proof is rather involved, and we won't mention any details. We'll just discuss the $\mathcal{P}$-completeness of the FFD heuristic, and we'll show, as mentioned before, another heuristic with the same performance bound, for which we know an $\mathcal{NC}$ algorithm. For more details on the topic, and for other approaches to the same problem, we refer the reader to [3] and the references given there.

First, we show the following:

Theorem   9   *Given a list of items, each of size between 0 and 1, in non-increasing order, and two distinguished indices i and b, it is $\mathcal{P}$-complete to decide whether the FFD heuristic will pack the ith item into the bth bin. This is true even if the item sizes are represented in unary.*

Proof   For the proof we use a reduction from the following variant of the circuit-value problem: A circuit consists only of AND and OR gates (i.e., is *monotone*) whose fan-out is at most two. This restricted version is clearly $\mathcal{P}$-complete as can be seen by an easy log-space reduction from the general circuit-value problem. The details of the construction are omitted here.

Table 2.6 Bins and item sizes for various types of gates

	Bins	*Items*
Fan-out one		
AND	$\delta_i, \delta_i+\delta_j-2\epsilon$	$\delta_i, \delta_i, \delta_i-2\epsilon, \delta_i-2\epsilon$
OR	$\delta_i+\delta_j-2\epsilon, \delta_i$	$\delta_i, \delta_i, \delta_i-2\epsilon, \delta_i-2\epsilon$
Fan-out two		
AND	$\delta_i, 2\delta_i-4\epsilon, \delta_i+\delta_j-3\epsilon, \delta_i+\delta_k-4\epsilon$	$\delta_i, \delta_i, \delta_i-2\epsilon, \delta_i-2\epsilon,$ $\delta_i-2\epsilon, \delta_i-3\epsilon, \delta_i-4\epsilon$
OR	$2\delta_i-4\epsilon, \delta_i, \delta_i+\delta_j-3\epsilon, \delta_i+\delta_k-4\epsilon$	$\delta_i, \delta_i, \delta_i-2\epsilon, \delta_i-2\epsilon,$ $\delta_i-2\epsilon, \delta_i-3\epsilon, \delta_i-4\epsilon$
Gate β_n	δ_n, δ_n	$\delta_n, \delta_n, \delta_n-2\epsilon, \delta_n-2\epsilon$

Our reduction is described in two stages. We first reduce the restricted monotone circuit-value problem to an FFD bin-packing problem with bins of variable sizes. The construction is then modified to give an FFD packing into unit capacity bins.

Let $\beta_1, \ldots, \beta_n$ be the gates of an n-gate monotone circuit, i.e., each β_i is either $\text{AND}(i_1, i_2)$ or $\text{OR}(i_1, i_2)$, with i_1 and i_2 the inputs of the gate. Each input can be a constant (**true** or **false**), or the value of some other gate β_j, $j < i$. In our first construction, we transform the sequence $\beta_1, \ldots, \beta_n$ into a list of items and a list of bins. The list of item sizes will be nonincreasing. For every gate β_i, we obtain a segment for each of the two lists. The segments for each list are concatenated in the same order in which the gates are given. For ease of notation, let

$$\delta_i = 1 - \frac{i}{n+1} \quad \text{and} \quad \epsilon = \frac{1}{5(n+1)}.$$

The list segments for each gate are determined by Table 2.6, where gate β_i is assumed to feed into gate β_j if it has just one output, and into gates β_j and β_k otherwise.

Let T_i denote any item of size δ_i, and F_i any item of size $\delta_i - 2\epsilon$. For every constant input of gate β_i, a T_i is removed from its list of items if the input is **false**, and an F_i if it is **true**.

We claim that packing the list of items (which is clearly nonincreasing) into the sequence of bins according to the FFD heuristic, emulates evaluation of the circuit in the following sense. Consider the bins in list order. When we start packing into the first bin of β_i's segment, for $i = 1, \ldots, n$, the remaining list of items starts with β_i's segment, and two of the first four items in this segment have already been removed. The other two of these four items encode the values of the two inputs to gate β_i: T_i stands for a **true** input, F_i stands for **false**. Suppose β_i is an AND gate with fan-out two. Then β_i's second bin receives a T_i if both of its inputs are **true** and an F_i, otherwise. In the first case, the second bin can further accommodate only the last item in β_i's list; whereas in the second case, it still has room for the third to last item in the list. As a result, packing β_i's items leaves space in the amount of $\delta_j - \epsilon$ and $\delta_k - \epsilon$ in β_i's last two bins if β_i evaluates to **true**. If the output of β_i is **false**, the corresponding amounts are δ_j and δ_k. Thus, in the first case, F_j and F_k will also be packed into the last two of β_i's bins since they are the largest items to fit. In the other case, T_j and T_k fit and will be packed. Therefore, after both inputs to β_j (similarly, β_k) have been evaluated, the two *remaining* of the first four items in β_j's (resp., β_k's) segment again properly reflect the values of the two inputs to the gate. Figure 2.14 shows the packings for two input combinations to a fan-out two OR gate. The OR gate functions quite similarly to the AND gate just described, with the role played by the first two bins more or less reversed. The details of the simulations performed by the other types of gates listed in Table 2.6 are left to the reader.

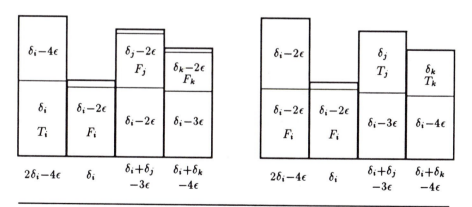

Figure 2.14   Packing for OR gate with (a) one **true** input and (b) two **false** inputs.

In the second part of the construction, we show how to use unit size bins. Let $u_1, \ldots, u_q$ be the nonincreasing list of item sizes, and let $b_1, \ldots, b_r$ be the list of variable bin sizes obtained in the first part. Define B to be the maximum of the b_i, and let $C = (2r+1)B$. We construct a list of decreasing items $v_1, \ldots, v_{2r}$ that when packed into r bins of size C, leave space b_i in the ith bin. Let

$$v_i = \begin{cases} C - iB - b_i, & \text{if } i \leq r; \\ C - iB, & \text{if } i > r. \end{cases}$$

When these items are packed according to the FFD heuristic, items v_i and v_{2r+1-i} end up in the ith bin, thus leaving b_i empty space. Also note that v_{2r} is at least as large as u_1. Let $w_1, \ldots, w_{2r+q}$ be the list of item sizes obtained by concatenating the v- and u-lists and normalizing the sizes by dividing each of them by C. Assume without loss of generality that the output gate β_n of the given circuit is an AND gate. An FFD packing of the items in the w list into unit-size bins will place the item corresponding to the second T_n in β_n's list into the last bin iff the output of the circuit is **true**.

The two parts of the construction described above can clearly be carried out on a multitape Turing machine using logarithmic work space. Since all numbers involved in the construction are bounded in value by a polynomial in the size of the circuit, we have shown that FFD bin packing is $\mathcal{P}$-complete in the strong sense (under log-space reductions), i.e., it remains $\mathcal{P}$-complete even if numbers are represented in unary (with fractions given by a pair of integers). $\square$

It turns out that it is actually the presence of arbitrarily small items that makes FFD inherently sequential ($\mathcal{P}$-complete). In fact, we can show the following theorem:

Theorem 10 *The packing obtained by the FFD heuristic can be computed by an $\mathcal{NC}$ algorithm for instances where all items have size at least $\epsilon > 0$. On a sorted list of n items, the algorithm uses $n/\log n$ processors and runs in time $O(\log n)$.* $\square$

Note that if we assume that the input list is already sorted, then the sequential time to do the FFD packing is really $O(n)$. This is the same number of operations as in the parallel algorithm claimed above in the theorem, and hence we call the latter an *optimal parallel algorithm*.

We use this theorem for $\epsilon = \frac{1}{6}$. Given an arbitrary instance of bin packing where we also have smaller items, we first pack the items that are at least $\frac{1}{6}$ using the algorithm of the theorem. Then we use the remaining smaller items to fill up whatever we have. However, we don't fill in a first-fit decreasing manner. Those bins that have been filled to $\frac{5}{6}$ or more are okay, and we disregard them. Where there is more space left than $\frac{1}{6}$, we take a contiguous bunch of the small items, group them all together, and put them in. Since all these items are less than $\frac{1}{6}$, the space that's left will also be less than $\frac{1}{6}$, unless we run out of items. This way we get a packing where one of two things happens: The first case is that the smaller items all fit into bins that were already partially filled by the large items. Since the large items were packed according to FFD, we get a guarantee of $\frac{11}{9}$ times optimal. In the second case, the smaller items really can be used to fill up all the partially filled bins to more than $\frac{5}{6}$, so there is less than $\frac{1}{6}$ left. There will be a few more bins that are used to put small items in, but only the last bin will be partially filled. All of the others will again be filled to at least $\frac{5}{6}$. Now, if every bin is filled to at least $\frac{5}{6}$, we have used at most 20 percent more bins than in the case where all bins are filled completely. Since $\frac{11}{9}$ is larger this finishes the argument.

Here is a (very) brief sketch of the algorithm whose existence was claimed in the theorem above. We know that all items are at least of size $\epsilon > 0$. We divide the algorithm into phases. In phase i, we pack items whose size is between $2^{-(i+1)}$ and 2^{-i}. So the interval of sizes goes down by $1/2$ each time. Since all the items have size at least ϵ, $\log(1/\epsilon)$ phases are sufficient. We do these phases sequentially, i.e., one phase after the other. We divide bins into *runs*, where a run is defined as a sequence of bins where available space is increasing (nondecreasing). Each run can be packed very fast in parallel. The details how this is done are much too involved to be presented here. They are, however, described in full in [3].

There are other approximation algorithms for bin packing, based on discretization. Given any $\epsilon > 0$, a discretization can be chosen such that the packing obtained is within $(1 + \epsilon)$ times optimal (plus some constant term). The packing algorithm is an approach based on dynamic programming. While this type of algorithm gets arbitrarily close to optimal (in the asymptotic sense), the constants involved depend on ϵ and are usually astronomically large, rendering the algorithm impractical.

We have seen that sometimes approximation algorithms can be used to deal with a $\mathcal{P}$-complete problem. Of course, in the sequential world, polynomial approximation algorithms for $\mathcal{NP}$-complete problems are also of interest and well studied. It turns out that some of these latter algorithms can even be made into $\mathcal{NC}$ algorithms resulting in fast parallel approximation. [20] gives a survey of $\mathcal{NC}$ approximation algorithms for some $\mathcal{P}$- and $\mathcal{NP}$-complete problems.

Bibliography

[1] Ajtai, M., J. Komlós, and E. Szemerédi. Sorting in $c \log n$ parallel steps. *Combinatorica* 3:1–19, 1983.

[2] Anderson, R. J., and E. W. Mayr. Parallelism and the maximal path problem. *Inf. Process. Lett.* 24:121–126, 1987.

[3] Anderson, R. J., E. W. Mayr, and M. Warmuth. Parallel approximation algorithms for bin packing. *Information and Computation*, September 1989. (To appear.)

[4] Anderson, R. J., and G. L. Miller. Deterministic parallel list ranking. *Proc. 3rd Aegean Workshop on Computing: VLSI Algorithms and Architectures, AWOC 88*, June/July 1988, Corfu, Greece.

[5] Borodin, A., J. von zur Gathen, and J. Hopcroft. Fast parallel matrix and GCD computations. In *Proc. 23rd Annual. IEEE Symposium on Foundations of Computer Science*, 65–71, Chicago, 1982.

[6] Chandra, A., D. Kozen, and L. Stockmeyer. Alternation. *J. ACM* 28(1):114–133, 1981.

[7] Chistov, A. L. Calculation of the characteristic polynomial of a matrix. *Proc. FCT*. Lecture Notes in Computer Science 199, 1985.

[8] Cole, R. Parallel merge sort. In *Proc. 27th Annual IEEE Symposium on Foundations of Computer Science*. 511–516, Toronto, Canada, 1986.

[9] Cole, R., and U. Vishkin. Optimal parallel algorithms for expression tree evaluation and list ranking. *Proc. Third Aegean Workshop on Computing: VLSI Algorithms and Architectures, AWOC 88*, 91–100, June/July 1988, Corfu, Greece.

[10] Csanky, L. Fast parallel matrix inversion algorithms. *SIAM J. Comput.* 5:618–623, 1976.

[11] Dymond, P. *Simultaneous resource bounds and parallel computation*. Ph. D. Thesis, Dept. of Computer Science, Univ. of Toronto, 1980.

[12] Fortune, S., and J. Wyllie. Parallelism in random access machines. *Proc. Tenth Annual ACM Symposium on Theory of Computing*, 114–118, San Diego, Calif., 1978.

[13] Goldschlager, L. The monotone and planar circuit value problems are log-space complete for $\mathcal{P}$. *SIGACT News* 9(2):25–29, 1977.

[14] Goldschlager, L. A space efficient algorithm for the monotone planar circuit value problem. *Inf. Process. Lett.* 10(1):25–27, 1980.

[15] Hochschild, P., E. Mayr, and A. Siegel. Techniques for solving graph problems in parallel environments. *Proc. 24th Annual IEEE Symposium on Foundations of Computer Science*, 351–359, Tucson, Ariz., 1983.

[16] Hopcroft, J., and J. Ullman. *Introduction to Automata Theory, Languages, and Computation*. Reading, Mass.: Addison-Wesley, 1979.

[17] Ladner, R. The circuit value problem is log-space complete for $\mathcal{P}$. *SIGACT News* 7(1):583–590, 1975.

[18] Ladner, R., and M. Fischer. Parallel prefix computation. *J. ACM* 27(4):831–838, 1980.

[19] Lipton, R., and J. Valdes. Census functions: An approach to VLSI upper bounds (preliminary version). *Proc. 22nd Annual IEEE Symposium on Foundations of Computer Science*, 13–22, Nashville, Tenn., 1981.

[20] Mayr, E. W. Parallel approximation algorithms. *Proc. International Conference of Fifth Generation Computer Systems*, 542–551, Tokyo: Institute for New Generation Technology, November/December 1988.

[21] Mayr, E., and G. Plaxton. Network Implementations of the DTEP Algorithm. Tech. Rep. STAN-CS-87-1157, Dept. of Computer Science, Stanford Univ., 1987.

[22] Miller, G., V. Ramachandran, and E. Kaltofen. Efficient parallel evaluation of straight-line code and arithmetic circuits. Tech. Rep. TR-86-211, Computer Science Dept., USC, 1986.

[23] Miller, G., and J. Reif. Parallel tree contraction and its application. *Proc. 26th Annual IEEE Symposium on Foundations of Computer Science*, 478–489, Portland, Ore., 1985.

[24] Pan, V., and J. Reif. Efficient parallel solution of linear systems. *Proc. 17th Annual ACM Symposium on Theory of Computing*, 143–152, Providence, R.I., 1985.

[25] Paterson, M. S. *Improved sorting networks with $O(\log n)$ depth.* Research Report RR89, Dept. of Computer Science, Univ. of Warwick, 1987.

[26] Pfister, G. The architecture of the IBM research parallel processor prototype (RP3). Tech. Rep. RC 11210 Computer Science, IBM, Yorktown Heights, 1985.

[27] Pippenger, N. On simultaneous resource bounds. *Proc. 20th Annual IEEE Symposium on Foundations of Computer Science*, 307–311, San Juan, P. R., 1979.

[28] Preparata, F., and J. Vuillemin. The cube-connected-cycles: A versatile network for parallel computation. *Proc. 20th Annual IEEE Symposium on Foundations of Computer Science*, 140–147, San Juan, P. R., 1979.

[29] Reif, J. Depth-first search is inherently sequential. *Inf. Process. Lett.* 20:229–234, 1985.

[30] Ruzzo, W. Tree-size bounded alternation. *Proc. 11th Annual ACM Symposium on Theory of Computing*, 352–359, Atlanta, Ga., 1979.

[31] Schwartz, J. Ultracomputers. *ACM Trans. on Programming Languages and Systems* 2(4):484–521, 1980.

[32] Seitz, C. The cosmic cube. *CACM* 28(1):22–33, 1985.

[33] Ullman, J., and A. van Gelder. Parallel complexity of logical query programs. *Proc. 27th Annual IEEE Symposium on Foundations of Computer Science*, 438–454, Toronto, Canada, 1986.

[34] Vavasis, S. Gaussian elimination with pivoting is $\mathcal{P}$-complete. Dept. of Computer Science, Stanford Univ. Preprint 1987.

CHAPTER 3

WILLIAM DALLY

Massachusetts Institute of Technology

Network and Processor Architecture for Message-Driven Computers

Introduction

Parallel computing is essentially a problem of scheduling and resource management. Any computation can be represented as a graph in which the vertices represent operations to be performed and the edges represent the movement of information between these operations. To perform the computation, we must decide where each operation happens in space and time and how information moves between the operations—either in space over a communication network or in time through a memory. These scheduling decisions are subject to the constraint of limited processor, memory, and communication resources.

The four sections in this chapter are based on four lectures on parallel computing. Each touches on one aspect of the scheduling and resource management limitations described above. The first three sections concentrate on moving information in space; they discuss how to design efficient communication networks. Chuck Seitz has given an example of a communication network in his discussion of the Ametek 2010 (see Chapter One).

This chapter explores the design of such networks in more generality. Section 3.1 describes how to build wire-efficient networks. Given a maximum wire density, how do we design a network that gives us the lowest latency at a given operating load? Section 3.2 deals with the analysis of networks. A queuing model of these networks is constructed and applied to determine network latency under load. Network design is covered in Section 3.3. Many engineering issues of router architecture and driver design are discussed.

This section also deals with the problem of scheduling operations. Once we have networks that can deliver messages between processing elements, how does the processor perform an operation in response to the arriving messages? To execute concurrent programs, the processor must be able to create processes and switch between processes with a minimum of overhead. Mechanisms to solve this problem that are built into the message-driven processor, currently under development at MIT, are described. These mechanisms are used to construct operating system services that implement a global virtual address space, very fast management of processes, and migration of objects.

3.1   Wire-Efficient Communication Networks for Multicomputers

What makes a network efficient? To answer this question we must look at the technology used to construct networks and determine what limits network performance. An efficient network is one that makes the best use of the limiting resource. Figure 3.1 shows the bottom three levels of packaging in an electronic system: chip, module, and board. The chip is a torus routing chip (TRC) [7], the first chip that demonstrated the type of network discussed in this chapter. The TRC chip measures about 4mm on a side. Large microprocessor and memory chips are typically 8–10mm on a side. One or more chips are packaged into a ceramic carrier similar to the one shown in the figure. This second-level packaging technology is used to match the pitch of the pads on the routing chip (250μ centers) with the pitch of the PC board ($\approx 1000\mu$ centers). Even with a large chip, only ten PC board wires can cross under the area of the chip. At the board level as well as at the chip level, the system interconnection is limited by wire density. Wires are a limiting factor because of power and delay as well as density. Most of the power dissipated in these networks is CV^2f power used to drive the wires, and most of the delay is propagation delay over wires or RC delay in driving wires. Thus, to make an efficient network, we must make efficient use of wires.

Figure 3.1   Three levels of packaging: chip, module, and board. One or more chips (8-10mm on a side with a wire pitch of 4μ and a pad pitch of 250μ) are attached to a ceramic module (wire pitch of 250μ). The modules are then mounted on PC boards (wire pitch of 1000μ).

Analyzing a network in terms of wires is a departure from traditional measures of cost. In the past wires were neglected, and switches and pins were considered the limiting factors. Networks designed to make the best use of switches and pins often made inefficient use of wire. Designing networks to give the best performance for a given wire density results in considerably better performance.

The problem of designing a communication network is illustrated in Figure 3.2. A number of nodes communicate by passing messages over the network. Each node injects messages with average length L into the network at an average rate of λ bits per cycle. The message rate, λ, represents the duty factor of a node, the fraction of time the node is injecting a message into the network. If traffic is nonuniform, we may specify a separate λ_{ij} for each pair of nodes i,j. The area available to wire the network limits the number of wires crossing a cut of the network to B, the bisection width of the network.

Bisection width, B, gives us a lower bound on wire density. B is the minimum number of wires that, when cut, divide the network into equal halves. The wire density may be higher across some other physical cut of the network [8]. For example, a tree has a bisection width of one, but its wire density is considerably higher. For the cases we consider here, however, bisection width is a good approximation of wire density.

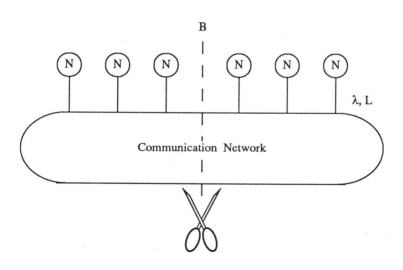

Figure 3.2   The Problem.

The objective of the network designer is to minimize latency and maximize throughput. Latency, $T(\lambda, L)$, is the average time required to deliver a message in the network. Latency is measured from the time the first bit of a message leaves the source node until the last bit of the message arrives at the destination node. Throughput, σ, is the maximum message rate, λ, that can be sustained by the network: $\sigma = \max \lambda \ni \exists T(\lambda, L)$. At offered message rates greater than σ, the network saturates and latency goes to infinity.

To meet the objective of minimum latency and maximum throughput, the designer manipulates three independent variables: *topology*, *routing*, and *flow control*.

- **Topology** refers to the interconnection graph of the network $I = G(N, C)$. The vertices of this graph are the nodes of the network, N, and the edges are the physical channels that connect the nodes, $C \subset N \times N$. If we think of parallel computing as a resource allocation problem, topology is the earliest form of resource allocation: fabrication time binding. Figure 3.3 shows three interconnection graphs: a mesh, a butterfly, and a tree.

We will see shortly that many of the details of the topology (e.g., whether we put a twist in here or add an end-around channel there) do not matter very much. Performance in direct networks is largely determined by the choice of radix, k, and dimension, n. Adjusting these two parameters allows us to trade off the two components of latency.

- **Routing** is the method used to choose paths in the network. It is a relation, $C \times N \times C$, that maps the channel occupied by the head of a message and the destination node for the message into a set of

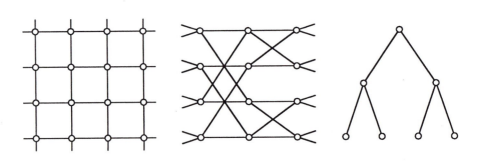

Figure 3.3   Three examples of network topology, (a) a 4 × 4 mesh, (b) a 2-ary 3-stage butterfly, and (c) a 2-ary 3-deep tree.

channels that can be used next by the message. Routing is a form of run-time resource allocation. Given a topology, a present position, and a destination, the routing relation determines how to get a message from here to there.

- **Flow Control** is the method used to regulate traffic in the network. It prevents messages from running over each other and controls how fast each advances to ensure an orderly progression of messages through the network. If two messages want to use the same channel at the same time, flow control determines (1) which message gets the channel and (2) what happens to the other message.

To put these three variables in perspective, imagine an automobile trip. The topology is defined by the road map. Routing requires picking a route from your source city to your destination city. Finally, flow control is analogous to the traffic signals and rules that prevent your car from occupying the same piece of road as another car.

3.1.1   Topology

A topology is evaluated in terms of the following five parameters:

- **Bisection Width**   As described above, the channel bisection, B, is the minimum number of channels that, when cut, separate the network into two equal parts. The wire bisection, B_W, is the number of wires crossing this cut of the network. $B_W = BW$ where W is the width of a channel in bits. Bisection width measures how much wiring density is required by a network. For design purposes, wire bisection is fixed, and the bisection width constrains how wide each channel can be made: $W = B_W/B$.

- **Degree**   The degree of a node, d, is the number of channels incident on a node. The number of channels into the node is the *in degree* of the node, d_{in}. The number of channels out of a node is the *out degree*, d_{out}. The total degree is the sum, $d = d_{\text{in}} + d_{\text{out}}$. The degree affects the cost of a node. The number of pins on each node is Wd, and the complexity of the router's control logic is usually related to d.

- **Diameter**   The diameter of a network is the longest shortest path between two nodes.

$$D = \max_{i,j \in N} \left(\min_{p \in P_{ij}} \text{length}(p) \right), \tag{3.1}$$

where P_{ij} is the set of paths from i to j. Diameter has traditionally been used as the primary figure of merit for networks. Hence the popularity of networks such as binary n-cubes with low diameter.

To find the diameter, consider all pairs of nodes $i, j \in N$ and find the shortest path from i to j. Then take the maximum of this path length over all pairs of points. In a mesh, the i and j that give you the maximum are any two opposite corners. All the paths between these corners have the same length $2(k-1)$. For example, in the 4 × 4 mesh shown in Figure 3.4, the diameter is 6.

- **Wire Length** The length of the wires in the network determines the speed at which the network can operate and the amount of power dissipated driving the wires.

- **Symmetry** A network is symmetric if it is isomorphic to itself with any node labeled as the origin. In a symmetric network the network *looks* the same from every node. Rings and tori are symmetric, while linear arrays and meshes are not. Symmetric networks simplify many resource management problems. For example, given a uniform traffic pattern, they result in uniform loading of network channels, while asymmetric networks do not.

We will examine both indirect networks (k-ary n-flys), and direct networks (k-ary n-cubes) as examples of interactions between these parameters and see how choices of k and n affect bisection, B, degree, d, and diameter, D.

Indirect Networks In indirect networks, the switching nodes are distinct from the processing nodes [32]. They are called *indirect* because messages between processing nodes are routed indirectly through the switching nodes. Figure 3.5 shows two examples of this class of network. The nodes on the left and right sides of the network are the processing nodes. In some machines the network is folded back on itself so that the nodes on the left

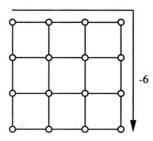

Figure 3.4   The diameter of a 4 × 4 mesh.

and right are the same. In other machines, processor nodes are placed on one side of the network, and memory nodes are placed on the other side. The nodes in the center of the network are the switching nodes.

Indirect networks have been used on machines such as the BBN butterfly [4] and IBM RP3 [29]. Many variations of indirect networks have been proposed in the literature [25] [28] [16]. It has been shown that most of these variations are equivalent [40]. The details of the network are relatively unimportant. The key decision in designing an indirect network is choosing the network radix k and dimension n.

The networks shown in Figure 3.5 are radix k, dimension n butterflys. We will refer to them as k-ary n-flys. A k-ary n-fly has $N = k^n$ processing nodes connected by n stages of k^{n-1} $k \times k$ switch nodes. Processing node i is connected to an input of node $\lfloor i/k \rfloor$ in the first stage of the switch and an output of the corresponding node in the last stage of the switch. Output port j of node i in the sth stage of the switch is connected to an input port of node $(i + jk^{(n-s-1)})$ mod N/k. A 2-ary 3-fly (8 nodes) is shown in Figure 3.5a. A 16-node 4-ary 2-fly is shown in Figure 3.5b. Both the BBN butterfly and the IBM RP3 use 4-ary n-fly networks.

The parameters of these networks are given by the following equations:

$$B_I = N/2, \tag{3.2}$$

$$B_{WI} = NW/2, \tag{3.3}$$

$$d_{\text{in}I} = d_{\text{out}I} = k, \tag{3.4}$$

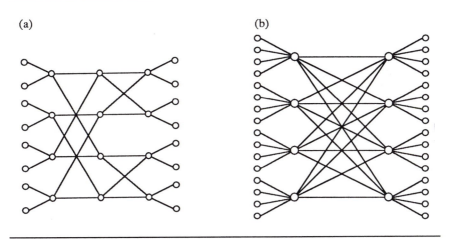

(a)

(b)

Figure 3.5   Indirect Networks: (a) a 2-ary 3-fly, and (b) a 4-ary 2-fly.

$$d = 2k, \tag{3.5}$$

$$D = n + 1. \tag{3.6}$$

In Figure 3.5a, the minimum bisection cuts the network horizontally through its center, $B = N/2 = 4$. The degree of each node is $d = 2k = 4$, and the diameter of the network is $D = n + 1 = 4$. The corresponding numbers for Figure 3.5b are 8, 8, and 3.

Indirect networks are characterized by a high bisection width ($N/2$), low degree ($2k$), low diameter ($n + 1$), and long wires. These networks are symmetric.

The bisection width $B = N/2$ does not reflect the actual maximum wire density for this class of networks. Slicing the networks vertically at any point cuts N wires. This vertical partition more accurately reflects the wiring problems encountered in implementing this class of networks. The area required to wire an indirect network in a plane is proportional to N^2, making these networks very costly in terms of wires. Indirect networks are as difficult to build as crossbar switches. They have fewer switches but require the same amount of wire.

As one varies k and n with the number of processing nodes, N, and the wire bisection, B_W, fixed, the degree and diameter of the indirect network are directly controlled. The channel width, however, remains fixed at $W = B_W/B = 2B_W/N$.

Indirect networks have a bisection width that is independent of the choice of k and n. This is a major disadvantage in that it prevents the designer from trading off the bandwidth of a channel against the diameter of the network. Direct interconnection networks overcome this limitation.

Direct Networks   In a direct network, all switching is performed in the processing nodes [32]. The prototypical direct network is the k-ary n-cube. Figure 3.6 shows two examples of this class of network.

A k-ary n-cube is a radix k cube with n dimensions. The radix implies that there are k-nodes in each dimension. A k-ary 1-cube (Figure 3.6a) is a k node ring. A k-ary 2-cube (Figure 3.6b) is constructed by taking k k-ary 1-cubes and connecting like elements into k-rings. In general, a k-ary n-cube is constructed from k k-ary $(n-1)$-cubes by connecting like elements into k-rings. Every node has an address that is an n digit, radix k number. Nodes are connected to all nodes with an address that differs in only one digit. As with the flys, a k-ary n-cube has $N = k^n$ nodes.

Many networks are included in the family of k-ary n-cubes. At the extreme of $k = 2$, we have a binary n-cube. At the extreme of $n = 1$ we have a ring. For $n = 2$ we have a torus. These networks have been used in several message-passing computers [33] [35].

(a)

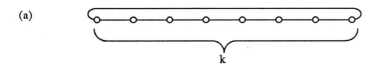

(b)

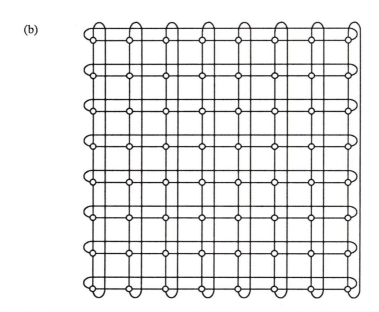

Figure 3.6   Direct Networks: (a) an 8-ary 1-cube, and (b) an 8-ary 2-cube.

The parameters of a k-ary n-cube are given in the following equations:

$$B_D = 2N/k, \tag{3.7}$$

$$B_{WD} = 2NW/k, \tag{3.8}$$

$$d_{\text{in}D} = d_{\text{out}D} = n, \tag{3.9}$$

$$d = 2n, \tag{3.10}$$

$$D = nk/2. \tag{3.11}$$

In Figure 3.6a, the bisection width is 2, the degree is 2, and the diameter is 4. The corresponding numbers for Figure 3.6b are 16, 4, and 8.

The bisection is given by cutting the network across the highest dimension. This cuts two channels in each ring of this dimension. Since there are k channels in each ring, this is $2/k$ of the N channels in this dimension for a total of $B_D = 2N/k$.

For small n, direct k-ary n-cubes are characterized by a low and controllable bisection width ($2N/k$), low degree ($2n$), high diameter ($nk/2$), and short wires. The toroidal versions of these networks (with ends connected as in Figure 3.6) are symmetric. The mesh versions of these networks (without the end-around connections) are asymmetric.

Low dimensional k-ary n-cube networks are very easy to wire. Their wiring complexity is proportional to N, compared to N^2 for multistage indirect networks. If we choose n less than or equal to 3, we have the added advantage that most wires are very short. Short wires allow the network to operate at higher speed and dissipate less power.

The bisection width of a k-ary n-cube depends on the choice of k and n. This property allows the designer to trade off bandwidth for diameter. When long messages are being sent, bandwidth dominates latency. In this case, making k large and n small will increase bandwidth and reduce latency. If shorter messages are being sent, a different choice of k and n can reduce diameter at the expense of bandwidth. As we will see shortly, latency is minimized when k and n are chosen to make the two components of latency nearly equal. Before examining the optimal choice of k and n, however, we must first discuss the remaining variables of network design: routing and flow control.

3.1.2   Routing

Routing is the method used for a message to choose a path over the network channels. Routing can be thought of as involving a relation, $\mathbf{R}$, and a function, ρ.

$$\mathbf{R} \subset C \times N \times C, \tag{3.12}$$

$$\rho : P(C) \times \alpha \mapsto C. \tag{3.13}$$

The routing relation $\mathbf{R}$ identifies the permissible paths that may be used by a message to reach its destination. Given the present position of a message, C, and its destination node, N, $\mathbf{R}$ identifies a set of permissible channels, C, that can be used as the next step on the route. $\mathbf{R}$ is a relation rather than a function because there may be more than one permissible path for a message to follow.

The function ρ selects one path from the set of permissable paths. At each step of the route, ρ takes the set of possible next channels, $P(C)$, some additional information about the state of the network, α, and chooses a particular channel, C. The additional information, α, may be constant, random, or state information based on traffic in the network.

This routing discipline considers the current position of a message to be a channel rather than a node [9]. $\mathbf{R}$ is $C \times N \times C$ rather than $N \times N \times C$. We will see later how considering the present position of a message to be a channel allows us to construct deadlock-free routing relations. If position is considered to be a node, deadlock must be avoided by storage allocation.

Routing methods can be classified as *deterministic*, *oblivious*, or *adaptive*. With deterministic routing, the path a message follows depends only on its source and destination nodes. In this case $\mathbf{R}$ is a function and α is constant (no additional information is provided). The *e-cube* or dimension order routing [24] used by the torus routing chip [7] and the mesh routing chip [15] is an example of deterministic routing.

An oblivious routing method may choose different paths through the network, but may use no information about the network state in choosing a path. The message is oblivious to any other traffic in the network. With oblivious routing, $\mathbf{R}$ is a relation (there may be many permissible paths). To be oblivious, α must contain no information about the state of the network. It may be random, a function of time, or a function of the contents of the message.

The most general case is adaptive routing, where the router may use information about the state of the network. In this case, α may be any function.

Figure 3.7 shows two examples of deterministic routing. In a k-ary n-fly (Figure 3.7a), messages can be routed using the digits of the destination address to select the output port at each switch. At stage i, the $n - i - 1^{st}$ digit of the destination address is examined and the corresponding output port is selected. The figure shows a route from node 2 to node 5 in a 2-ary

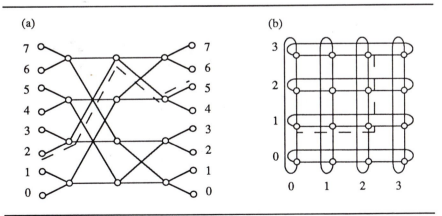

Figure 3.7   Routing Examples: (a) destination tag routing in a k-ary n-fly, and (b) e-cube routing in a k-ary n-cube.

3-fly. The destination address is 101 (binary), giving a route out the 1-port (upper port) of the switch in the first stage, the 0-port of the switch in the second stage, and the 1-port of the switch in the third stage.

In a k-ary n-cube (Figure 3.7b), messages can also be routed using the digits of the destination address in order. Each digit of the address corresponds to a dimension. In the 4-ary 2-cube of the figure there are two dimensions, X and Y, corresponding to the two digits in the address. The e-cube routing algorithm examines the destination address one digit at a time [24]. For the ith digit, a route in the ith dimension to the proper coordinate is chosen. In the figure a message is delivered from node (0,1) to node (2,3) by routing first in X (to (2,1)) and then in Y. Despite the differences in the underlying networks, the algorithm is almost identical to destination tag routing in the k-ary n-fly.

3.1.3   Flow Control

Flow control refers to the method used to regulate traffic in a network. It determines when a message or a part of a message can advance. Flow control is the resource management policy that is used to allocate communication resources, wires, and buffers; to information units, messages, packets, and flits.

Figure 3.8 shows the three information units important to understanding flow control [9].

- **Message** The logical unit of communication. Two objects communicate by sending a message. This is the only unit seen by clients of the network service.

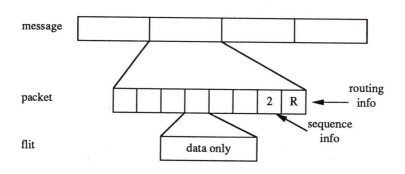

Figure 3.8   Information units: A message is the logical unit of communication. It is subdivided into **packets**, the smallest unit containing routing information. Packets may in turn be subdivided into **flits**, the smallest physical unit on which flow control is performed.

- **Packet** A message is divided into one or more packets. A packet is the smallest unit that contains routing information—e.g., the destination address. If there is more than one packet in a message, each packet also contains a sequence number to permit reassembly. When messages are short, they are typically sent as a single packet. Long messages must be broken into many packets to avoid degrading network performance.

- **Flit** A packet can be further divided into flow control digits or *flits*, the smallest unit on which flow control is performed. That is, communication resources, wires and buffers, are allocated on a flit basis. In general, a flit contains no routing information. Only the *leading* flit of a packet knows where it is going. The remaining flits must follow the flit ahead to determine their route.

A simple analogy illustrates the difference between packets and flits. Packets are like automobiles. Since they know where they are going, they can be interleaved freely. Flits, on the other hand, are like railroad cars. They must follow the flit ahead of them to find their destination. They cannot be interleaved with flits of another packet, or they would lose their only contact with their destination. This analogy will be helpful when deadlock is discussed.

In store-and-forward networks, packets are handled as a single flit— a single buffer is allocated to one packet. In multicomputer networks, however, the latency of store-and-forward routing is unacceptable. Newer multicomputers use *wormhole routing*, where wires and buffers are allocated to flits significantly smaller than an entire packet.

Performing flow control on units smaller than packets reduces latency, as shown in Figure 3.9. With store-and-forward routing[1] a packet is received in its entirety before being transmitted to the next channel. If a packet of L bits is transmitted across a channel of W bits/cycle with a cycle time of T_C and must cross D channels, the zero-load ($\lambda = 0$) latency of store-and-forward routing is the product of the time to transmit the packet across a single channel, L/W, and the number of transmissions required to reach the destination, D.

$$T_{SF} = T_C \left(\frac{L}{W} \times D \right). \tag{3.14}$$

With wormhole routing, packets are divided into flits. Channels and buffers are allocated flit-by-flit. A flit can advance as soon as it is allocated the resources it requires. It need not wait for the entire packet to be received. With this pipelined routing, the latency becomes the sum of

---

[1]The word routing here is a misnomer. Both store-and-forward routing and wormhole routing refer to *flow control* methods rather than routing methods.

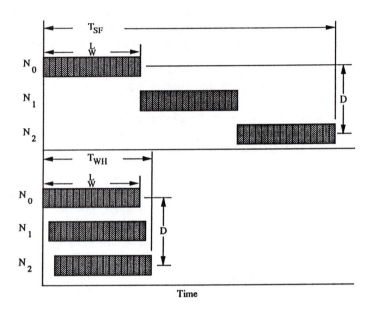

Figure 3.9   Store-and-forward routing (top) and wormhole routing (bottom). With store-and-forward routing, latency is the product of aspect ratio, L/W, and distance, D. Wormhole routing reduces latency to the sum of these two terms.

amount of time required to transmit a packet across a single channel, L/W, and the amount of time required for each flit to reach the destination, D.

$$T_{WH} = T_C \left(\frac{L}{W} + D \right). \tag{3.15}$$

A hybrid strategy, *virtual cut-through* [21], allocates storage buffers to packets as in store-and-forward, but pipelines the transmission of flits as in wormhole. It has the latency properties of wormhole routing, T_{WH}, but requires that deadlock avoidance be handled as in store-and-forward routing.

As will later become apparent, reducing latency to the sum of D and L/W allows us to trade off the two components of latency by appropriate choice of topology parameters k and n for a direct network. Wormhole

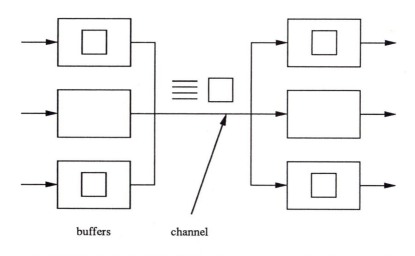

buffers   channel

Figure 3.10   A channel consists of the **physical channel:** a communication medium (e.g., wires or fibers); **source buffers,** buffers holding flits awaiting use of the channel; **destination buffers,** buffers for flits that have just used the channel, and some channel state.

routing also has the advantage of requiring very little storage, resulting in a small, fast communication controller that does not consume node memory bandwidth.

Figure 3.10 illustrates the resources that flow control has to allocate. Actual communication is performed by the physical channel. Buffers are used to store flits before and after they use the channel. Some channel state is required to remember the route for nonleading flits traversing the channel. A source buffer, a destination buffer, and some channel state make up a *virtual channel.* It is a channel from the point of view of resource allocation, but may share a physical communication channel with other virtual channels. Virtual channels, because they deal with routing, are allocated packet-by-packet. The buffers associated with a virtual channel remain allocated to a packet for its duration. They are allocated to the individual flits of the packet on a flit-by-flit basis. The physical channel is also allocated flit-by-flit.

When the leading or head flit of a packet arrives at a channel, it requests a virtual channel. Once the channel is allocated, it will remember the next step of the route (i.e., which outgoing channel to use) until the tail flit of the packet is encountered. Starting with the head flit, each flit in turn must now compete for buffer and channel resources. For a flit to advance, three conditions must hold: (1) It must be resident in a source buffer; (2) it must be allocated an empty destination buffer; and (3) it must be allocated use of the physical channel.

Flow control methods are distinguished by how they resolve collisions between packets. When two packets request the same channel at the same time, two questions arise: (1) Which packet receives the channel? and (2) what happens to the other packet? Figure 3.11 shows four common flow control disciplines for dealing with the "other" packet.

- **Buffering** (Figure 3.11a) One packet is granted the channel. The other packet is allowed to continue advancing and is stored in a buffer. Also known as *virtual cut-through* [21], buffering has the advantage of wasting none of the communication resources shown in Figure 3.10. It has two disadvantages, however: (1) It requires a large amount of buffer storage—enough for the largest packet; and (2) the buffers must be allocated in an acyclic manner to avoid deadlock. The required buffer storage is often more than can be built into a router and must be allocated in the processing node's memory, which slows the router down to the memory cycle time. Also, valuable node memory bandwidth is spent on through messages.

- **Blocking** (Figure 3.11b) The other packet is stopped in place. Also known as *wormhole routing* [9], blocking idles the resources that are allocated to the blocked packet. Blocking requires very little storage, resulting in a small, fast communication controller. Blocking flow control with a small number of flit buffers usually gives the best performance for a given set of storage and communication resources. Blocking with limited buffering is used on the Ametek 2010 [36].

- **Dropping** (Figure 3.11c) The other packet is dropped (eliminated). It is allowed to continue advancing, but its flits are not stored as they arrive at the node. The information is lost and must be retransmitted. Dropping results in a severe waste of resources, because all of the channel and buffer time used to get the dropped packet to the point of conflict has been wasted. As offered network traffic, λ, increases, the delivered traffic increases until the network becomes congested. Then, rather than saturating at the peak network throughput (as buffering and blocking disciplines do), the waste in a dropping network causes delivered traffic to be reduced. The traffic reaches an asymptote much lower than peak traffic. Only a few messages are being delivered, and most of the network resources are being wasted on messages that are dropped and retransmitted. Beyond peak throughput the network is "unstable."

  Dropping also has the disadvantage of requiring a positive acknowledgement on packet arrival that further wastes communication resources. In addition, a retransmission protocol is required that wastes

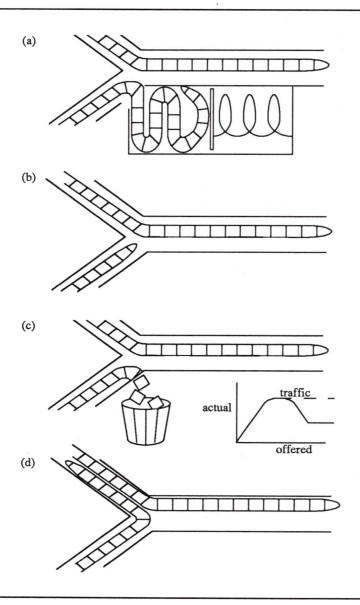

Figure 3.11   Packet collisions may be resolved by (a) buffering, (b) blocking, (c) dropping, or (d) misrouting.

compute and storage resources and increases latency with timeouts on dropped packets. Dropping is used on the BBN butterfly, for example.

- **Misrouting** (Figure 3.11d) The other packet is routed to an idle but incorrect channel. Sometimes called *desperation routing*, this method of resolving packet collisions wastes resources by using channels and buffers to take a packet further from its destination. Also, misrouting may result in *livelock* unless both a minimum number of buffers and an appropriate protocol are provided. After being misrouted for n cycles, a set of messages may be in exactly the same positions—the cycle then continues until the machine is powered down. Misrouting is used on the Denelcor HEP and on the Connection Machine.

Most networks combine these flow control disciplines. For example, several multicomputers use a hybrid of buffering and blocking. They have limited buffer storage, less than a full packet. They buffer the first few flits of a message and then block the remaining flits. A study of how latency depends on the number of buffers is presented in [26].

3.1.4 Deadlock Avoidance

The flow control discipline must allocate resources to packets in a manner that avoids deadlock. Deadlock can occur when there is a cyclic dependency for resources. If two packets each hold resources required by the other to move, both packets will be blocked indefinitely.

Two examples of deadlock are shown in Figure 3.12. In (a) neither packet, destined for the node three hops away, is able to move because there is no free packet *buffer* on the next node of the route. In (b) neither packet is able to advance because the *channel* it requires next is already in use. In both cases deadlock is caused by resource dependencies. In (a) the resource is packet buffers, while in (b) it is channels (flit buffers and routing state). To avoid deadlock, we must first identify the resources packets are competing for and then introduce a mechanism for breaking cyclic dependencies on the resources.

Structured Buffer Pools A method of breaking dependencies among packet buffers (Figure 3.12a) is to structure the buffer pools so they form a partial order [17]. With store-and-forward or cut-through routing, storage resources are allocated entirely in terms of packets. Once the first flit of a packet is accepted, sufficient storage must be available to hold the entire packet. Packet buffers are associated with nodes, so once a packet is buffered there is no history of the channel it arrived on. The routing relation is $N \times N \times C$.

The buffer pool is structured so that a packet in a given buffer, A, can be stored only in a restricted set of buffers, S, on the next node of the

(a)

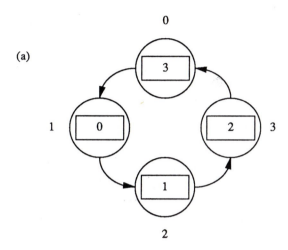

(b)

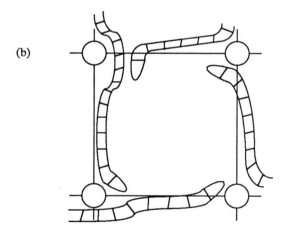

Figure 3.12   Two examples of deadlock (a) 4-ary 1-cube with store-and-forward routing is deadlocked by a cycle of buffer allocations, and (b) a k-ary 2-cube with wormhole routing is deadlocked by channel dependency.

route. If buffer $B \in S$, we say that there is a dependency from A to B, $dep(B, A)$. Buffer allocation is restricted so the buffer dependency graph, the directed graph defined by the dependency function, dep, is acyclic.

For example, allocating buffers to packets depending on the distance from their destination node results in an acyclic buffer dependency graph. If the diameter of a network is $D_{\max}$, we create $D_{\max}$ packet buffers on each node. At each step of a route, if a packet is h hops from its destination, it can route only into the hth buffer on the next node of the route.

Although structured buffer pool deadlock avoidance works well for local area and wide area networks, the large amounts of storage required and the high latency make it inappropriate for multicomputers. The use of virtual channels overcomes these problems.

Virtual Channels   With virtual channels we avoid deadlock by making the routing relation acyclic [9]. The resources being allocated are virtual channels—flit buffers at either end of a physical communication channel and some associated routing state. If several virtual channels are associated with a single physical channel, several packets can use the same physical communication resource without interacting from the point of view of deadlock avoidance.

To prevent deadlock we will restrict the routing relation to make the channel dependency acyclic. Recall that $\mathbf{R}$ is a relation $C \times N \times C$. Remembering the channel from which each packet arrived at a node allows us to define a dependency relation between channels. We define the channel dependency relation, $dep \subset C \times C$ to be the projection of $\mathbf{R}$ onto $C \times C$. A pair of channels, (a, b) is in dep iff $\exists n \in N \ni (a, n, b) \in \mathbf{R}$.

In general a routing relation will have cycles in its channel dependency graph (the graph induced by dep). We restrict $\mathbf{R}$ to eliminate these cycles. In doing so we may disconnect the network so there no longer exist routes between pairs of nodes. To reconnect the network we add new virtual channels. These channels are added in such a way that dep remains acyclic.

Figure 3.13 illustrates the application of virtual channels to a 4-cycle. Because there is only one way to route around the network, a cycle exists in the channel dependency graph, dep. The existence of this cycle makes it possible for the network to deadlock [9]. To eliminate this cycle we must remove some triples from $\mathbf{R}$ to eliminate one of the edges from the channel dependency graph. In this case we remove the triples $(c_3, n_i, c_0) : i = 1, 2$. This removes the edge (c_3, c_0) from dep.

In breaking this cycle, however, we have disconnected the network. It is no longer possible to route between node pairs $(n_3, n_1), (n_3, n_2)$, and (n_2, n_1). To reconnect the network we add two new virtual channels, c_0' and c_1'. These channels are in parallel with and share physical channels with c_0 and c_1. Using these virtual channels, we can add the triples $(c_3, n_i, c_0') : i = 1, 2$ and (c_0', n_2, c_1') to $\mathbf{R}$ to restore connectivity.

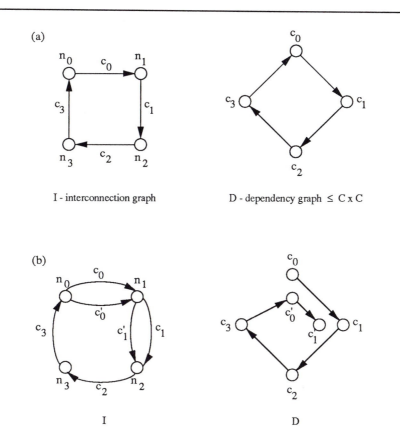

(a)

I - interconnection graph   D - dependency graph $\leq$ C x C

(b)

I   D

Figure 3.13   (a) Routing in a 4-cycle (4-ary 1-cube) is prone to deadlock because its dependency graph contains a cycle. (b) Removing two triples from the routing relation and adding two virtual channels to reconnect the network converts the cycle into a "spiral," thus breaking the deadlock.

Another view of this application of virtual channels to deadlock avoidance in a 4-cycle is shown in Figure 3.14. The 4-cycle is shown as a traffic circle with four entrances. The flit buffers correspond to the areas of roadway between the entrances. To indicate that wormhole routing is used, packets are shown as worms. When two packets enter the circle, each is allocated two flit buffers. "Wormlock" then occurs because the next packet buffer requested by each worm is occupied by the other. Recall our previous analogy and imagine two railroad trains entering the circle. Since they cannot interleave their cars (flits), they are deadlocked. Adding two virtual channels (Figure 3.14 (b)) converts the traffic circle into a spiral. Because

(a)

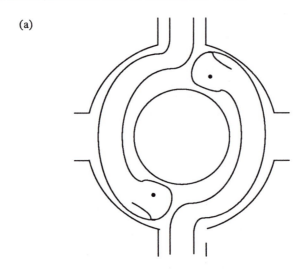

(b)

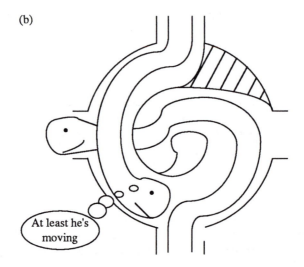

At least he's moving

Figure 3.14   Virtual channels break deadlock by converting cycles in a network into spirals. (a) Two packets (worms) enter a cycle and become deadlocked because of channel dependence. (b) Adding virtual channels allows one packet to make progress.

there is no dependency cycle, one packet is able to progress, but the other packet remains blocked.

When used in conjunction with e-cube routing to order channels by dimension, the technique illustrated in Figures 3.13 and 3.14 can be used to avoid deadlock in any k-ary n-cube network.

Adaptive Routing   Figure 3.15 illustrates the use of virtual channels to implement deadlock-free adaptive routing. Consider a k-ary 2-cube network (torus or mesh) such as the one used on the Ametek 2010 [36]. If three nodes in the same row (same x-coordinate) attempt to send messages at the same time (Figure 3.15a), only one will be able to progress. While there exist many noninterfering routes, e-cube routing restricts the messages to take a route that results in interference. A similar problem occurs when many messages are sent to nodes in the same column (same y-coordinate).

If we relax the dimension order routing and allow messages to switch from the y-dimension back to the x-dimension, we create a channel dependency that may result in deadlock. In Figure 3.15b, each of the four packets (worms) requires the channel occupied by the worm ahead in the cycle, and the network is therefore deadlocked.

The cyclic dependency can be broken by restricting routing so that eastbound packets (positive x-direction) are not allowed to travel on northbound (positive y-direction) channels. With this restriction, deadlock is no longer possible, but the network has become disconnected. We restore connectivity by introducing a virtual northbound channel for eastbound messages (Figure 3.15c). The resulting network performs deadlock-free adaptive routing [12].

The technique illustrated in Figure 3.15 can be applied to any k-ary n-cube network. As each dimension, i, is added to the network, traffic in the previous dimensions is divided into 2^{i-1} groups according to its directions in the $i-1$ previous dimensions. One direction of travel in the ith dimension is then partitioned into 2^{i-1} virtual channels, one for each group. The technique is very costly for networks with many dimensions (large n) because the number of virtual channels grows exponentially with n. A comparison of network performance with adaptive and nonadaptive routing is given in [26].

Virtual Channel Summary   The development of virtual channels [9] made possible the use of wormhole routing in cyclic networks with $k > 2$. Before virtual channels it was not understood how to avoid deadlock in such networks. The application of virtual channels extends beyond deadlock avoidance, however. They are a versatile mechanism that can be applied to solve a number of network problems. The properties of virtual channels and a few of their applications are summarized here. By no means is this list exhaustive.

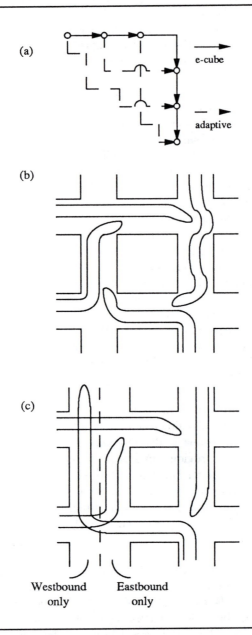

Figure 3.15   Using virtual channels to implement adaptive routing. (a) *e*-cube routing may result in poor use of channel resources. (b) Relaxing dimension order can lead to deadlock from cyclic channel dependencies. (c) This deadlock can be broken by adding a virtual northbound channel.

- Virtual channels can be used to construct a deadlock-free routing in any strongly connected network.

- Only a few flits per node of buffering are required. Because of this small memory requirement, no node memory space or bandwidth is needed for through messages, and the network can operate at a speed independent of the node memory.

- Virtual channels can be applied to

  — Deadlock avoidance: As in the two examples shown in Figure 3.14, channel pairs are removed from the routing relation to break deadlock, and then virtual channels are added to restore connectivity.

  — Adding network priorities or parallel networks: Virtual channels can be used to duplicate a network logically by doubling the number of flit buffers. The resulting two networks are logically separate—traffic in one will not block traffic in the other. They can be used to implement two priorities of message delivery.

  — Adding network services: In addition to duplicating the network, one may add services to one of the logical networks sharing physical channels. For example, the original proposal for the IBM RP3 [29] called for two networks—a normal network and a *combining network*—to handle accesses to synchronization variables. Rather than duplicating the costly physical channels, these two networks could be implemented with virtual channels sharing the same set of physical channels.

3.1.5   Wire-Efficient Topology

Now that we understand the variables of network design, let's return to the question of selecting a network that gives us the lowest latency, $T(0, L)$, for a given wiring density, B_W, number of nodes, N, and message length, L. We will make use of two major ideas: (1) Using direct networks allows us to trade off channel bandwidth W against diameter D, and (2) performing flow control on a flit-by-flit basis (wormhole routing) makes latency the sum of two components D and L/W.

We will consider only k-ary n-cube networks. We saw earlier that k-ary n-flys require too much wiring density to be competitive and do not permit us to trade off diameter against bandwidth. The question becomes one of choosing k or n (there is only one independent variable) to minimize latency. Small variations in the network topology are relatively unimportant. For example, choosing a mesh rather than a torus gives at most a factor of 1.5 improvement in throughput at the expense of asymmetry and uneven channel loading. The proper choice of k and n can improve performance by an order of magnitude [14].

The zero-load latency of the network for random traffic is given by

$$T = D_{\text{avg}} + L/W. \qquad (3.16)$$

Equation (3.16) gives latency in terms of cycles and neglects the fact that propagation delay T_p (the coefficient of D) may be different from channel period T_c (the coefficient of L/W).

Assuming a constant B_W and a torus-connected k-ary n-cube, the average distance is $(k-1)n/2$ and the channel width is $k/2$ giving

$$T = \frac{(k-1)n}{2} + \frac{2L}{k}. \qquad (3.17)$$

Equation (3.17) illustrates how the choice of radix trades off latency between the component due to distance and the component due to aspect ratio. Increasing k increases distance and decreases aspect ratio.

The lowest latency occurs when $D \approx L/W$ or $\frac{(k-1)n}{2} \approx \frac{2L}{k}$. Because $n = \frac{\log N}{\log k}$, it is not possible to solve for a minimum in closed form. However, it has been shown [14] that the minimum latency occurs at a smaller n (larger (k)) than the point where $D = L/W$.

Figure 3.16 shows how latency varies with dimension for B_W and N fixed. The figure shows three curves for networks of size $N = 256$, 16K, and 1M nodes. The points marked with an X indicate fixed-radix networks of each size. Other locations on the curve can be realized using mixed-radix networks.

For networks with few dimensions (small n), the latency due to distance dominates. For example, a 2-D 1M node network has a diameter of 1,024. As n is increased, latency is reduced until distance and aspect ratio are equal ($D = L/W$). As n is increased beyond this point, the component of latency due to message length or aspect ratio L/W dominates.

For the 256 node network, minimum latency occurs for $n = 2$ (16-ary 2-cube). A $32 \times 32 \times 16$ ($n = 3$) mixed-radix network (marked with an O) gives the best performance for a 16K network. A 32-ary 4-cube and 16-ary 5-cube both give good performance for the 1M node network. Figure 3.16 shows that low-dimensional networks ($2 \leq n \leq 4$, $8 \leq k \leq 32$) outperform high-dimensional networks when wire density is the limiting factor in constructing the network.

Low-dimensional networks are even more attractive than this comparison indicates for the following reasons:

- It is not feasible to scale B_W linearly with N. Because the wires must be packaged in two or three physical dimensions, a bisection width of $B_W = N^r : 0.5 \leq r \leq 0.66$ is more reasonable. With a smaller bisection width, the minimum latency point shifts to the left.

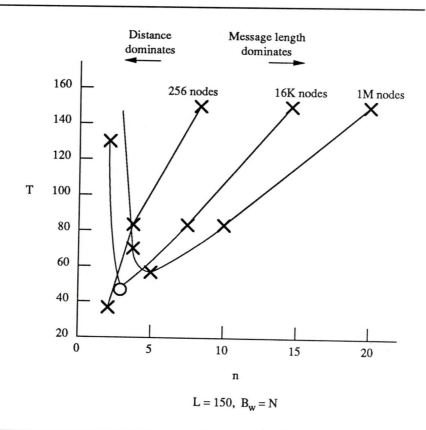

Figure 3.16   Latency versus dimension for networks of 256, 16K, and 1M nodes, $L = 150$ bits, $B_W = N$.

- This comparison assumes that all channels are the same speed. Low-dimensional networks have shorter wires (for $n \leq 3$ the wire length is the node-to-node spacing) and hence operate faster and dissipate less power.

- Low-dimensional networks better exploit locality. Because their latency is more strongly affected by distance, low-dimensional networks work better for traffic patterns where distance is reduced (e.g., nearest neighbor communication). Because the latency of high-dimensional networks is dominated by aspect ratio, they benefit very little from locality.

Table 3.1 Latency calculation for two 1,024-node networks: a binary 10-cube and a 32-ary 2-mesh. $(L = 200, B_W = 1024, T_c = T_p = 50ns)$

	2-ary 10-cube	32-ary 2-mesh	
W	1	16	bits
D_{avg}	5	20	hops
L/W	200	12.5	cycles
T_{aspect}	10	0.625	μs
T_{dist}	0.25	1.0	μs
T_{SF}	50	12.5	μs
T_{WH}	10.25	1.625	μs

- Router complexity is proportional to the number of dimensions because the control logic is duplicated for each dimension.

Table 3.1 shows the latency calculation for two 1,024-node networks with equal wire bisection. The low-dimensional 32 × 32 mesh gives a latency that is 16 percent of the binary 10-cube latency. For nearest neighbor communications, the latency (T_{aspect}) is 6 percent of the 10-cube latency.

The table also shows the advantages of wormhole routing as compared with store-and-forward routing. A binary 10-cube with store-and-forward routing[2] has a latency of 50μs. Switching to wormhole routing while retaining the binary 10-cube topology gives a factor of 5 performance improvement to 10.25μs. Changing the topology to a 32 × 32 mesh gives a factor of 6 (combined factor of 30) performance improvement to 1.625μs. Performance is increased a factor of 30 while keeping the same wire density and operating at the same frequency.

---

[2] A machine manufactured by N-Cube corporation has almost exactly these specifications. Because it routes messages using its processor, however, its latency is much greater than shown here.

3.1.6   Summary

In this section we introduced the variables involved in designing a multi-computer interconnection network and developed a method for choosing a network that gives the minimum latency for a given wire density. The variables we discussed are **topology**, the interconnection graph of channels and nodes, **routing**, the method used to find a path through this graph, and **flow control**, the policy used to allocate storage and communication resources to messages in transit and to resolve collisions between messages.

We considered both direct and indirect network topologies. **Direct networks**, such as k-ary n-cubes, allow us to trade off the diameter of the network against the bandwidth of individual channels. The important decision in choosing a topology is the choice of k and n that determines this tradeoff. Minor differences in topology—the difference between a torus and a mesh, for example—are relatively unimportant.

Wormhole routing is a flow control discipline that allocates communication buffers on a flit-by-flit basis and resolves collisions by blocking. It reduces latency to a sum of two components, one due to message length and the other distance. With store-and-forward routing, latency is a product of these two components.

Networks using wormhole routing can be made deadlock-free by using **virtual channels**. This deadlock-avoidance technique eliminates dependencies by restricting the routing relation rather than by restricting buffer allocation. A virtual channel consists of a few flit buffers and some routing state. Several virtual channels may share the same physical communication channel. Virtual channels allow us to build deadlock-free communication networks with very modest storage requirements at each node (a few flits). They can also be used to implement several logically separate networks sharing the same set of wires.

Finally, we saw that for networks limited by wire density, low-dimensional k-ary n-cube networks offer the minimum latency. For a fixed number of nodes and a fixed bisection width, as dimension is increased latency shifts from being dominated by distance to being dominated by message length. The minimum latency occurs when the two components are nearly equal.

3.2   Analysis of Multicomputer Communication Networks

Section 3.1 introduced the three variables of networks: topology, routing, and flow control. We went on to do a zero load ($\lambda = 0$) analysis that showed that low-dimensional k-ary n-cubes give minimum latency for a given wiring density. In this section we carry our analysis one step further and investigate the behavior of these networks under non-zero loading.

Network performance under load can be characterized by two curves, as shown in Figure 3.17. If we run the network for a very long time with a constant rate of offered traffic, λ, its queue lengths or, if $\lambda > \lambda_{\text{sat}}$, its accepted traffic, λ_a, reach stable values. These values reflect the steady-state performance of the network.

As the message rate λ is increased, *collisions* occur between *messages*. These collisions increase the network latency from its zero-load value, $T(0, L)$. The latency curve generally takes the shape shown in Figure 3.17a. Latency increases slowly at first and then more rapidly, approaching a vertical asymptote at **network saturation**. Beyond a saturation message rate, λ_{sat}, no steady-state solution for network latency exists. At higher message rates, latency grows without bound.

Beyond the saturation point, $\lambda > \lambda_{\text{sat}}$, network performance is described by the rate of *accepted* traffic, that is, the rate of traffic that reaches its destination. Figure 3.17b shows a typical plot of accepted versus offered traffic. Below the saturation point a steady-state solution for latency exists, and all traffic is accepted. In this region, the curve is the straight line $\lambda_a = \lambda$. Beyond the saturation point the shape of the curve depends on the flow control discipline. Networks that use blocking or buffering flow control (no resources wasted) typically remain saturated with $\lambda_a = \lambda_{\text{sat}}$ as traffic is increased (solid curve). In networks that use dropping flow control (resources wasted on collision), accepted traffic drops off beyond the saturation point and eventually reaches an asymptote at $\lambda_x << \lambda_{\text{sat}}$ (dotted curve).

Clearly we would never operate a network above λ_{sat} for long periods of time. The importance of the accepted versus offered traffic curve is that it gives us an idea of how the network responds to transient loads that exceed the saturation point. Networks that saturate with $\lambda = \lambda_{\text{sat}}$ recover easily from a transient load $\lambda_T > \lambda_{\text{sat}}$ once traffic is again reduced to a stable value $\lambda_S < \lambda_{\text{sat}}$. Since traffic is being removed from the network at λ_{sat}, the queue lengths will return to their λ_S values with some time constant. Networks that saturate to a lower asymptote, λ_x, however, can become stuck in a congested state. If $\lambda_S > \lambda_x$, queue lengths will continue to grow, and the network will not return to its stable λ_S state. The time constant at which the network responds to changes in load is related to $\tau = Q/\lambda_{\text{sat}}$, where Q is the total queuing along a typical path through the network.

For a given network, performance curves (similar to Figure 3.17) are usually produced using simulation or from tests of the actual network. However, to evaluate design alternatives, an analytic method of arriving at the performance curves is invaluable. In this section we will develop some methods for analyzing the performance of multicomputer networks using queuing theory.

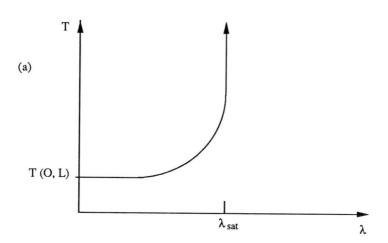

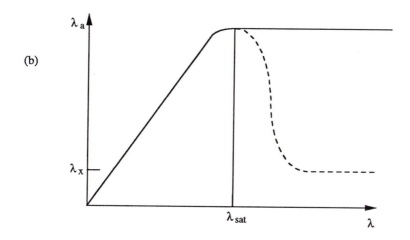

Figure 3.17   Steady-state network performance under load can be characterized by two curves (a) latency, $T(\lambda, L)$, versus load, λ, and (b) accepted traffic λ_a versus offered traffic, λ.

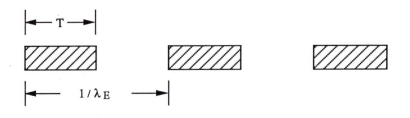

Figure 3.18   Traffic on a channel is described by its message rate λ_E and its service time T.

3.2.1   Parameters

Figure 3.18 illustrates the variables used in analyzing networks: message rate and service time. We deal with message rate both as λ (bits/cycle) and as $\lambda_E = \lambda/L$ (messages/cycle). As shown in the figure, the channel is busy for a period T every $1/\lambda_E$ cycles, giving a duty factor of $T\lambda_E$. The message rate determines the duty factor of a channel while the service time determines the granularity of resource allocation.

3.2.2   Latency Calculation

Latency is calculated as shown in Figure 3.19. We begin with the latency at the destination and work our way back to the source, adding the additional latency and increase in service time due to queuing delays. Assume that a destination node removes a message from the network as soon as it arrives (the consumption assumption) [9]. The service time at the destination is

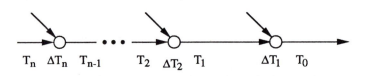

Figure 3.19   Latency is calculated starting with the service time at the destination, T_0, and working back toward the source, adding the additional latency due to collisions, ΔT_i, at each switch.

the number of cycles required to remove the message from the network or the aspect ratio of the message,

$$T_0 = \frac{L}{W}. \tag{3.18}$$

At the ith switch from the destination, we will calculate an additional delay, ΔT_i, due to queuing and collisions, and a new service time, T_i. To make these calculations we need to know the service time and message rate (T and λ) on every arc in the graph. To make the calculation tractable, we assume that all message rates describe independent random processes. This procedure is repeated until we reach the source and calculate T_n. The procedure becomes more complicated in cyclic networks where we have to solve for a number of service times simultaneously. The latency is given by

$$T(\lambda, L) = \sum_{i=1}^{n} \Delta T_i + T_0. \tag{3.19}$$

The two terms correspond exactly to the D and L/W terms in the zero-load latency equation (3.15).

The relationship between the increase in service time, $T_i - T_{i-1}$, and the increase in latency, ΔT_i, is determined by the flow control discipline—or more specifically, by the amount of queuing in the network. In a network with complete buffering (e.g., virtual cut through [21]), messages are never blocked on a channel, and thus there is no increase in service time; $T_i = T_0 : i = 1, \ldots, n$. In a message that uses strict blocking (no queuing), all blocking time adds to the busy time of the channel as well, and $T_i = T_{i-1} + \Delta T_i$. With partial queuing, service time increases by the amount that the incremental latency, ΔT_i, exceeds the queue length. The addition of queuing also changes the expression for ΔT_i.

The latency at each stage is calculated as shown in Figure 3.20. We model each point of contention as a two-to-one merge. One input of the merge is the channel for which we are calculating ΔT (in this case A). We lump all other channels competing for the resource into the other channel (B). A rate is given for each input channel, and a service time is given for the output channel. We calculate the additional latency on the channel of interest, ΔT_A, by multiplying the probability of a collision by the expected waiting time of a collision and adding the pipeline delay of the merge, T_p. The T_p terms sum up to the D term in our zero-load latency equation (3.15); the other term is the additional latency due to a collision.

$$\Delta T_A = P(\text{collision})E(\text{delay}) + T_p. \tag{3.20}$$

The probability of a collision is the probability that the output channel is busy handling the other input, which is the duty factor of the other channel, $T_0 \lambda_B$. Given that we have a collision, the waiting time has a

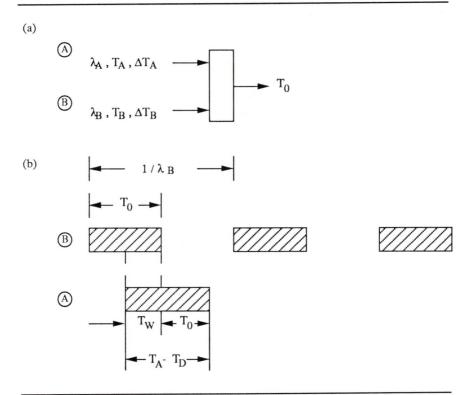

Figure 3.20   The increase in latency at a node is given by multiplying the probability of a collision with the expected wait time. (a) Model of contention at a single node. (b) Gantt chart showing a collision.

uniform distribution from 0 to T_0. With no queuing, the average waiting time is half the output service time, $T_0/2$. Thus we can rewrite (3.20) as

$$\Delta T_A = (T_0 \lambda_B)(\frac{T_0}{2}) + T_p, \qquad (3.21)$$

or

$$\Delta T_A = \frac{\lambda_B T_0^2}{2} + T_p. \qquad (3.22)$$

Two observations can be made about (3.22):

- Given a fixed total message rate, $\lambda_A + \lambda_B$, latency is reduced when the channel is unbalanced. Because the latency of the A channel depends on the rate of the B channel, having $\lambda_A \gg \lambda_B$ causes more messages to see less interference. In the extreme case $\lambda_B = 0$, there is no contention at all. As we will see shortly, an appropriate choice of a routing relation can unbalance network rates to increase performance.

- Latency depends quadratically on *service time*, T_0. If service time is increased, we both increase the probability of a collision and increase the expected delay due to a collision. For this reason, it is important that (1) service time increase as little as possible in each stage, and (2) that the maximum packet length be kept small.

Adding queuing to a network reduces the increase in service time at each stage and changes the expression for additional latency, ΔT_A. Strictly speaking, with queuing, the simple expression for latency in equations (3.20–3.22) is no longer valid since an arriving message on channel A is now competing with other messages on channel A as well as with messages on channel B. If we assume an infinite M/M/1 queue [22], the equation for additional latency with queuing ΔT_{AQ} becomes

$$\Delta T_{AQ} = \frac{T_0}{1 - (\lambda_A + \lambda_B)T_0} + T_p. \tag{3.23}$$

We will make the approximation of retaining equation (3.20) for two reasons. First, our queues are not infinite. Queues in many networks are shorter than a single message. Second, at message rates of interest, the probability of more than a single *message* being in the queue is very small.

With no queuing, the service time at each stage increases by the average waiting time, T_W. The waiting time is the increase in latency less the fixed propagation delay:

$$T_W = \Delta T_A - T_p. \tag{3.24}$$

If l flits of queuing are added to the input of the stage, waits of up to l cycles can be tolerated with no increase in service time. Waits of time $T > l$ cause an increase of $T - l$ cycles. Since waiting time is uniformly distributed from 0 to T_0, given the occurrence of a collision, the average increase in service time, T_y, is

$$T_y = \frac{(T_0 - l)^2}{2T_0}. \tag{3.25}$$

Multiplying by the probability of a collision gives the average increase in service time per stage.

$$T_i - T_{i-1} = \frac{\lambda_B(T_{i-1} - l)^2}{2}. \tag{3.26}$$

If sufficient queuing is provided to hold an entire packet ($l \geq T_0$), there will be no increase in service time. For smaller queue sizes, equation (3.26) overestimates service time because it considers each stage independently. In fact, the queues of different stages interact, as shown in Figure 3.21.

The effect of partial queuing is illustrated in Figure 3.21. The figure shows activity on channels from stage i (top) down to the source (bottom). Since we are determining the effect of later channels on earlier channels, time decreases from left to right. A collision at stage i causes an increase in latency time of $\Delta T_i = T_x + T_p$ and a corresponding increase in service time to $T_i = T_0 + T_x - l$. The queuing in subsequent stages decreases the service time. At stage $j > i$, the service time is $T_j = T_0 + \max(T_x - (j - i + 1)l, 0)$. Because of this effect, the network approaches the performance of complete queuing whenever there is sufficient queuing to eliminate the increase in service time between collisions. This occurs when

$$l > l_x = \frac{\lambda_B T_0^2}{2}. \qquad (3.27)$$

Typically a queue of several times l_x is used. For example, in the J-machine [13], an average message is 24 flits (6 words of 4 flits each), and thus $T_0 = 24$ cycles. And an average load on a competing channel is $\lambda_B = .001$ messages/cycle. For these parameters, $l_x = .6$. The network uses 4 flits of queuing per node. A simulation study of the effect of queuing on network performance is given in [26]. Because of the difficulty of

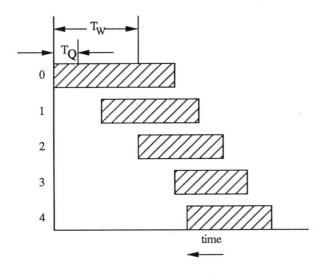

Figure 3.21   Partial queuing causes the increase in service time to be reduced by l in each preceding stage. Time in this figure is shown advancing from right to left.

analyzing partial queuing, the analyses below assume either no queuing or complete queuing.

In calculating latency, it is important to include the delay due to source queuing. Some networks push much of their delay due to contention back to the source. If the time spent waiting to enter the network is not considered a component of latency, these networks would appear to have very little increase in latency with applied load.

3.2.3   Latency of an Indirect Network

As an example of latency calculation we will consider a k-ary n-fly (an indirect network) and calculate its latency as a function of traffic under the following usual assumptions:

(1) A uniform distribution of destinations

(2) A deterministic destination service time

(3) The message sources are considered to be independent Poisson processes with rate λ_E messages/cycle

Results computed using these assumptions agree well with simulation and test data in which the assumptions 2 and 3 are replaced with actual message lengths and sources. Relaxing assumption 1, however, can change the results drastically. For example, some networks have *hot spots*—peaks in the destination distribution function that can cause considerable congestion in the network [30].

Since destinations are uniformly distributed, traffic is evenly spread over the network. Each arc carries traffic of rate λ_E. At a given switch, $k - 1$ channels are competing with a given channel for an output. On the average, $1/k$ of these channels wants the same output, so the competing traffic is $\lambda_B = (k - 1/k)\lambda_E$. Thus at each stage the increase in latency is given by equation (3.22) to be

$$\Delta T_i = \frac{(k-1)\lambda_E T_{i-1}^2}{2k} + T_p. \tag{3.28}$$

If the network has complete queuing, there is no increase in service time: $T_i = T_0$. If we assume no queuing, equation (3.26) gives us the following recurrence for calculating the service times:

$$T_i = T_{i-1} + \frac{(k-1)\lambda_E T_{i-1}^2}{2k}. \tag{3.29}$$

The average latency of the network can be solved for using equation (3.19).

Finally, we must include the effect of source queuing:

$$T = \frac{T_n}{1 - \lambda_E T_n}. \tag{3.30}$$

Figure 3.22 shows latency as a function of traffic for 1,024-node indirect networks with no queuing. The curves are calculated from equations (3.29) and (3.30). The contribution from nT_p is ignored. The figure shows that low-radix indirect networks suffer considerable latency due to contention. The 2-ary 10-fly saturates at 30 percent capacity and has an increase in latency of 200 percent at 20 percent capacity. The 32-ary 2-fly, on the other hand, does not saturate until 55 percent capacity and has an increase in latency of 60 percent at 20 percent capacity.

Calculating the performance of an indirect network is straightforward because there are no cycles in the traffic graph. We can construct a partial order of switches and solve them one at a time. Mesh-connected direct networks can be solved in the same manner. Also, if a network has complete queuing, there is no dependence on service times, and the equations can be

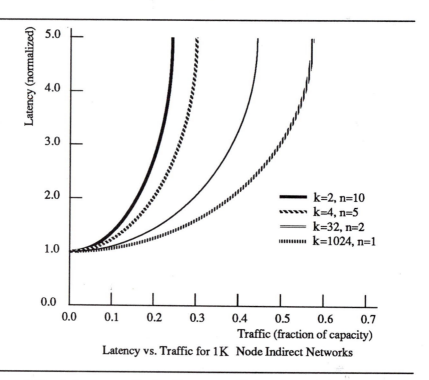

Latency vs. Traffic for 1K Node Indirect Networks

Figure 3.22   Latency normalized to T_0 versus traffic for 1,024-node indirect networks with no queuing.

solved one at a time. In the next section we will develop some techniques for dealing with torus-connected direct networks without queuing that have cyclic dependencies on service times.

3.2.4   Latency of a Direct Network

In this section we analyze a torus-connected k-ary n-cube using e-cube routing. We analyze this network hierarchically. We start with the entire network and partition it into dimensions. Each dimension is then replaced with a model that accounts for messages that route in or skip the dimension. For the messages that route in a dimension, a multistage model is used to calculate incremental latency and service time.

Using e-cube routing, the network at the level of dimensions is acyclic. Once a message travels to a lower numbered dimension, it never again visits a higher numbered dimension. Ignoring for now the messages that may *skip* dimensions, we can consider the k-ary n-cube to be a **multistage network**, as shown in Figure 3.23. Each dimension is a stage. Once again we start at the destination and work backward, calculating the latency and service times in each dimension.

Figure 3.24 shows the model used to calculate the latency and service time of a dimension. When messages enter a dimension, a fraction $\gamma = 1/k$ of them are already at the appropriate coordinate in that dimension. These messages skip over the dimension and later merge with messages that pass through the dimension. Since this merging introduces contention latency that depends on the relative fraction of traffic skipping versus routing, messages that skip a dimension must be handled separately [14].

Traffic arriving from a previous dimension has a total rate of λ_E. This traffic consists of two components: a component that skipped the previous dimension, $\lambda_S = \gamma\lambda_E$, and a component that routed through the previous dimension, $\lambda_R = (1 - \gamma)\lambda_E$. The components of these two streams that route in the current dimension, $\lambda_{SR} = \gamma(1 - \gamma)\lambda_E$ and $\lambda_{RR} = (1 - \gamma)^2\lambda_E$, compete for access to the channel. Similarly, the components that skip the

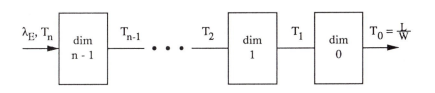

Figure 3.23   Dimension level model for calculating the latency of a k-ary n-cube.

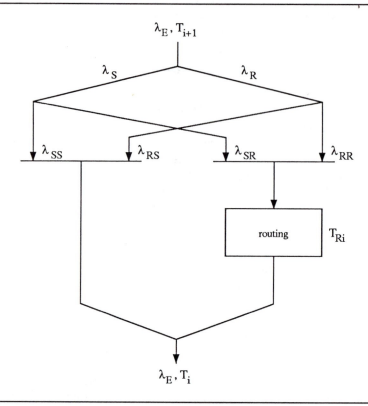

Figure 3.24   Model of a single dimension. Messages that skip either the previous dimension or the current dimension are considered separately because they must compete for a channel when they reenter a dimension.

dimension, λ_{SS} and λ_{RS}, must compete for access to the path around the channel.[3] The latency of these two contention points is calculated using (3.22). The latency due to routing through the channel, T_{Ri}, is added to the contention latency of the components that route. Details of these calculations are given in [14].

Two important special cases are easy to analyze. For a binary n-cube, $k = 2$, $\gamma = 0.5$, and since there is only a single hop in a dimension, there is no routing contention, $T_{Ri} = 0$. In this case, assuming no queuing, the service time is given by the recurrence

$$T_{i+1} = T_i \left(1 + \frac{\lambda_E T_i}{8} \right). \tag{3.31}$$

---

[3] If multiple bypass paths are available, there is no competition.

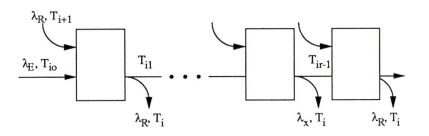

Figure 3.25   Model of routing latency in a single dimension.

When k becomes very large (for practical purposes $k > 10$ is large), $\gamma \approx 0$, and we can ignore the merging contention of messages skipping the dimension. In this case the service time becomes

$$T_{i+1} = T_i + T_{Ri}.\tag{3.32}$$

The routing latency in a single dimension, T_{Ri}, is calculated using the model of Figure 3.25. Each stage of the network handles two sources of traffic, entering traffic with rate λ_R and continuing traffic with rate λ_C. Because the network is symmetric, there is uniform traffic on the continuing channels $\lambda_C = \sigma\lambda_R$, where $\sigma = \frac{k-2}{2}$ is the average number of continuing hops in one dimension.[4] Let T_{ij} denote the service time seen by a message on the $j+1^{\text{st}}$ continuing channel it encounters after arriving in a dimension. Service time on the jth continuing channel is given from (3.22) to be

$$T_{i(j-1)} = T_{ij} + \frac{\lambda_R T_{i0}^2}{2}.\tag{3.33}$$

Repeating this equation σ times gives the service time on the first continuing channel in terms of itself:

$$T_{i0} = T_i + \frac{\lambda_C T_{i0}^2}{2}.\tag{3.34}$$

Solving for T_{i0} gives

$$T_{i0} = \frac{1 - (1 - 2\lambda_C T_i)^{\frac{1}{2}}}{\lambda_C}.\tag{3.35}$$

---

[4]This is an approximation. If virtual channels are used to avoid deadlock, the logical network (where the contention delays occur) is a spiral and does not have uniform traffic. The physical network does have uniform traffic. See [26] for more details.

Adding the latency incurred on the entering channel gives

$$T_{i+1} = T_{i0} + \frac{\lambda_C T_{i0}^2}{2}. \qquad (3.36)$$

The delay due to contention on the entering channel is the same as the delay due to contention on all the continuing channels combined. The e-cube routing causes messages to go an average of σ hops on the continuing channels before they take a hop on an entering channel. As a result, the contention is asymmetric. Performance is improved because most of the traffic sees very little contention. With a routing discipline that allowed more *turns*, the traffic on the entering and continuing channels would be closer together, and the total delay due to contention would increase.

Figure 3.26 shows the latency and accepted traffic performance curves for a 32-ary 2-cube [14]. Traffic is shown as a fraction of network capacity.[5] The latency curve (Figure 3.26a) shows very good agreement between theory and experiment. It also indicates that for this network, saturation occurs when λ is between 40 percent and 50 percent capacity. The increase in latency up to the 30 percent point is only about 10 percent of the zero-load latency. This shows that zero-load latency is in fact a good indicator of performance.

The accepted traffic versus offered traffic curve (Figure 3.26b) shows experimental data. It illustrates the saturation property of blocking flow control. The accepted traffic reaches an asymptote of λ_{sat} and remains there as offered traffic is increased.

3.2.5   Summary

In this section we have used a simple queuing model to analyze the performance of multicomputer networks. This model predicts performance to within 10 percent of experimental results. It is useful in that it allows a designer to evaluate many alternatives without the need for costly simulations.

We analyzed both an indirect and a direct network. The indirect network suffered from a considerable increase in latency with applied traffic. A 32-ary 2-fly showed a 60 percent increase in latency at 20 percent capacity.

For the direct network, we saw that zero-load latency was in fact a good indicator of network performance. For a 32-ary 2-cube, the additional latency due to contention was less than 10 percent of total latency at 30 percent capacity. Networks of this class saturate at between 30 percent and 50 percent capacity. Once the expected traffic is determined, a network

---

[5] At capacity, every channel in the network is advancing a message toward its destination every cycle.

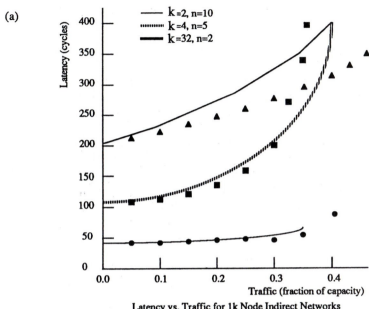

Latency vs. Traffic for 1k Node Indirect Networks

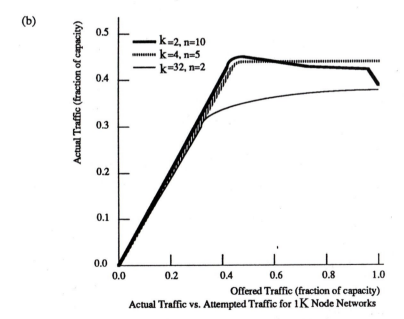

Actual Traffic vs. Attempted Traffic for 1K Node Networks

Figure 3.26   Network performance curves for a k-ary n-cube (a) latency versus applied traffic (fraction of capacity) solid line is analysis, points are experiment, (b) accepted traffic versus applied traffic.

should be designed with a capacity that is at least three times the expected average traffic.

3.3   Design of Communication Controllers

Multicomputer networks are composed of a *communication fabric* (e.g., wires) and a set of communication controllers or *routers*. The routers buffer, synchronize, and route messages over the physical channels provided by the communication fabric. They provide the routing and flow control functions for the network. In the preceding two sections we considered the cost and performance of the network in terms of the communication fabric. In this section we will examine the design of the routers. In particular, we will discuss router organization (how primitive switches and storage elements are connected to make up a router), synchronization (the timing of data transfer), and the electrical design of transmitters and receivers.

3.3.1   Router Organization

Primitive Components   Routers are constructed from three primitive components (Figure 3.27).

- **Switch** A primitive switch selectively connects its input to its output as directed by a control line.

- **Latch** A latch stores information. It samples its input when directed by a control line.

- **Control** A control block is a state machine that examines some input lines and generates control lines to direct other functions.

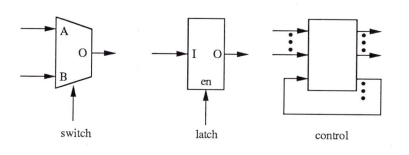

Figure 3.27   Routers are constructed from switches, latches, and control state machines.

Routing Modules   We combine these components to build simple routing modules such as a queue stage (Figure 3.28a), a 1:2 switch (Figure 3.28b), or a 2:1 merge (Figure 3.28c). The *switch* and *merge blocks* can be generalized to be 1:*n* and *n*:1, or higher degree switches and merges can be constructed from two-way modules. The inputs and outputs of these modules carry both the data being routed and flow control information (requests and acknowledgements). All inputs and outputs have identical formats and protocols, and thus the data and control lines of any input port can be connected to the corresponding data and control lines of any output port.

The queue module (Figure 3.28a) contains a single latch and a control box that performs flow control to make the latch act as a stage of a queue. The switch module (Figure 3.28b) contains two switches, a latch, and a control box. The control box examines some routing information and selects one of the two outputs. The control box also performs flow control to decide when to latch the next flit into the latch. The merge module contains a control box that arbitrates between requests from its two inputs. Chuck Seitz has termed these simple routing modules *routing automata* [34] [15].

In the networks we examined in Section 3.1, our switching nodes were either $n \times n$ switches (in the case of *k*-ary *n*-cubes) or $k \times k$ switches (in the case of *k*-ary *n*-flys). These complex switches are constructed from the routing modules, as shown in Figure 3.29. The modules are connected into a network that realizes the complex switch. In effect we are building a network within each node of a larger network.

If a switching node is to perform all possible $n \times n$ permutations, it must be constructed from a *non-blocking* network [6]. For example, the crossbar shown in Figure 3.29a constructed from *n* 1:*n* switches and *n* *n*:1 merges is a non-blocking network. The torus routing chip [7] is an example of a router implemented using a crossbar network. Constructing a crossbar, however, results in a large, slow router.[6] Since network performance depends on router performance (T_p and T_c), a crossbar router results in a slow network.

To overcome the performance limitation of the crossbar, we can construct a simpler *blocking* network. A blocking network may restrict our choice of routing relation (**R** and ρ) and may suffer from internal message collisions, but if the network is chosen to optimize the most common paths, the overall performance is significantly improved.

For example, the linear array of Figure 3.29b both restricts routing and is subject to internal collisions. It cannot route from a lower dimension to a higher dimension, and an attempt to route from *p* to *y* and *x* to *p*

---

[6]There are less expensive non-blocking networks such as the Batcher sorting network [3] or the Clos network [6], but they are still quite expensive compared to the simple linear array.

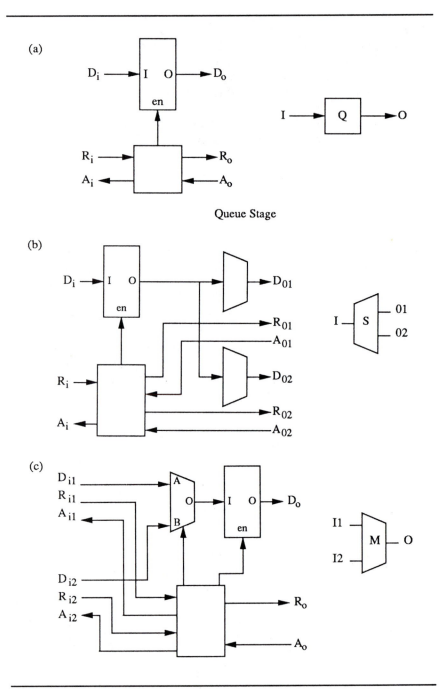

Figure 3.28   Primitive components are combined to construct simple routing modules such as (a) a queue stage, (b) a 1:2 switch, or (c) a 2:1 merge.

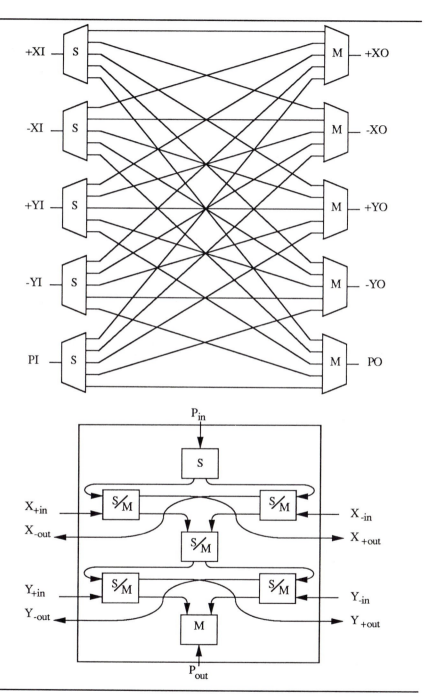

Figure 3.29 Anatomy of a router: (a) an $n \times n$ crossbar switch and (b) an $n \times n$ switch is constructed from a linear array of routing modules.

simultaneously may result in a collision in the y dimension. This router, however, is perfectly suited to e-cube routing. With e-cube routing, messages never travel to higher dimensions, and most messages ($\frac{\sigma}{\sigma+1}$ of them) travel straight through in one dimension. Because the single dimension transit requires only a simple 2:2 switch/merge, it can be made to operate very quickly. The network design frame [11] and the mesh routing chip [15] are examples of routers that use the linear array organization.

Design of a Routing Module   As shown in Figure 3.30, a routing module is constructed from three stages: input, switching, and output. Simple routers can be designed entirely at this level without the need to

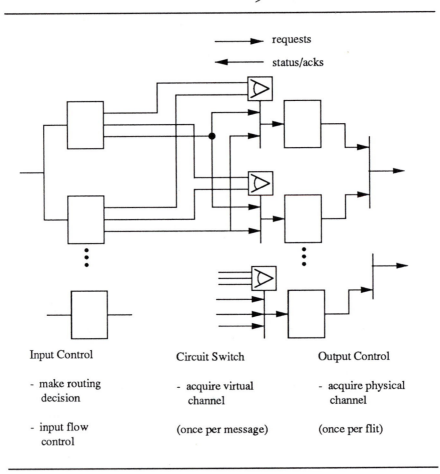

Input Control

- make routing decision

- input flow control

Circuit Switch

- acquire virtual channel

(once per message)

Output Control

- acquire physical channel

(once per flit)

Figure 3.30   A routing module consists of one or more of the stages shown here: *input control*, *switching*, and *output control*.

define intermediate modules. More complex routers are constructed from a network of routing modules, as described above. The function of each stage is as follows:

- The **input** control performs input flow control and routing. It determines the next channel of the route and makes a request for that channel. This arbitration is performed once per message. Once the channel is acquired, the input control sequences the flits of the message through the router, signaling the previous stage when the next flit is to advance. When the last flit of the message has passed, the channel is released.

- The **switch** stage performs virtual channel arbitration and switching on a per message basis. At the beginning of each message a request is made by input control. An arbiter for each output virtual channel grants one request and refuses the others. The switch is configured to route data and flow control information for the granted request to the appropriate output buffer.

- The **output** control performs flit-by-flit arbitration for use of the physical communication channel. As it receives flits from the switch stage, it competes with other virtual channels for access to the physical channel. When the physical channel is granted and the data transmitted, it signals the switch to send the next flit.

A routing module is a circuit switch. A path from input to output is established for the duration of the message. The input and output control perform flit-by-flit arbitration and flow control to pipeline messages through the circuit switch.

The Torus Routing Chip   The photograph of a torus routing chip in Figure 3.31 shows how routing components fit together to construct a router [7]. The large rectangular region in the center of the chip is a 5 × 5 10-bit wide (8-data + 2-flow control bits) crossbar switch. Five 5-way arbiters are built into the switch. Along the top and bottom of the switch are five input controllers. These units perform the routing calculation, compete for access to the switch, and perform input flow control. To the left and right of the switch are five output controllers that perform output flow control and arbitrate for use of the physical channel. The output controllers also include four stages of queuing each.

Network Design Frame   The network design frame (NDF), shown in Figure 3.32, uses the linear array router organization [11]. The chip consists of separate x and y dimension routers connected as shown in Figure 3.29b. Both routers are implemented in a single data path at the center of the

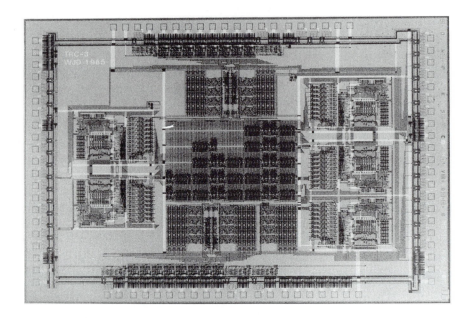

Figure 3.31   Photomicrograph of a torus routing chip.

chip. A router organized in this way is both smaller and faster than a crossbar router.

The NDF implements a bidirectional torus or mesh-connected network with both directions sharing the same set of bidirectional data wires. A token-passing arbiter is used by the output stages of the two chips sharing the wires to control access. The chip reduces power consumption, noise, and delay by using low voltage (1V) signaling on the communication channel.

3.3.2   Synchronization

Synchronization refers to ordering events in time. For a router, the events are requests and acknowledgements for resources (a virtual channel, a flit buffer, or a physical channel). Synchronization orders the acquisition of the channel, the sequencing of flits through the channel, and the release of the channel. The inputs and outputs of the routing modules carry flow control

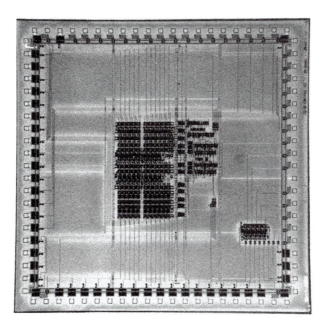

Figure 3.32   Photomicrograph of a network design frame.

information along with their data. The format of this information and
the flow control protocol depend on the synchronization technique used.
There are two approaches to synchronizing routers (or any digital system):
synchronous or self-timed (also called asynchronous).

Self-Timed Design   With self-timed design [31], a minimum amount
of synchronization is performed. Most events are allowed to be unordered;
only interacting events are ordered. Each module input or output carries
its own timing information in the form of request or acknowledge lines. An
edge on one of these lines corresponds to an event. Only when two module
outputs interact is it necessary to synchronize these lines. When two asynchronous data streams or sources of events are combined, a synchronizer
or arbiter is required to avoid metastability problems that can occur when
two competing events occur at nearly the same time.

Figure 3.33 shows the control part of a self-timed merge element. Only

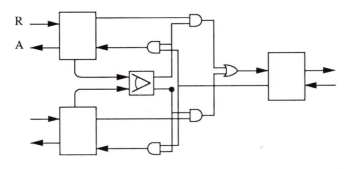

Figure 3.33   The control path of a self-timed merge element.

one of the two input controllers (left) is granted access to the output by the arbiter (center). The arbiter output gates the request and acknowledge signals between the enabled input and the output. The arbiter acts as a synchronizer as well as an arbiter. It includes an *excluded middle* circuit [31] that prevents either output from being asserted until the arbiter has made a definite decision. The arbiter guarantees that even if the two requests occur at nearly the same time, only one input will be granted the output.

The protocol used to associate events with edges of the request and acknowledge lines is called a *signaling convention*. Three self-timed signaling conventions are illustrated in Figure 3.34. With the 4-cycle signaling convention (Figure 3.34a), transmission of a flit proceeds in the following steps:

(1) Data is applied to the data lines and request goes high to signal valid data.

(2) Acknowledge goes high to signal that the data has been accepted and the input lines may change state.

(3) The request line goes low, indicating that the data lines are no longer driven.

(4) Acknowledge goes low to signal that the receiver is ready for the next data item.

Steps 3 and 4 in the list above are not absolutely necessary. The events they signal are not required to sequence the data transmission. A 2-cycle signaling convention (Figure 3.34b) eliminates these two transitions and uses only steps 1 and 2. Every toggle of the request line (that is, every

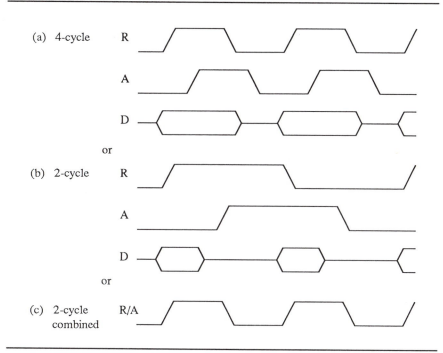

(a) 4-cycle   R

A

D

or

(b) 2-cycle   R

A

D

or

(c) 2-cycle   R/A
combined

Figure 3.34   Self-timed signaling conventions: (a) 4-cycle, (b) 2-cycle, and (c) 2-cycle combined.

change of state) is considered to be an instance of transition 1. Every transition on the acknowledge line is considered to be an instance of transition 2. By eliminating two transitions, 2-cycle signaling significantly reduces the delay associated with signaling. For this reason, it was used externally on the torus routing chip (Figure 3.31), while 4-cycle signaling was used internally.

When virtual channels are being used, it is unacceptable to retain access to the physical channels while awaiting an acknowledge. For this reason, the 2-cycle protocol is modified. The transition on the request line signals that data is valid. The receiver must be ready for this data because it is asserted for a fixed *hold* time only, and then the channel is released. A transition on the acknowledge line signals that the receiver is ready for the next flit. In effect this is a 2-cycle protocol where the signal transitions correspond to events 1 and 4 above. Events 2 and 3 occur by convention within a small, fixed time (the hold time) of event 1.

Because request and acknowledge transitions never occur simultaneously, it is possible to combine these two signals into a single line (Figure 3.34c).

The transmitter pulls the single request/acknowledge line high to request (transition 1). The receiver pulls the same line low to acknowledge (transition 2). This 2-cycle, combined signaling was used on the NDF (Figure 3.32). It reduces pin count but requires careful circuit design to work properly. Input thresholds and timing must be adjusted so the transmitter (receiver) is guaranteed to see that it has pulled the line high (low) before the receiver (transmitter) is allowed to pull the line back low (high).

Synchronous Routers In a synchronous system, all events occur on transitions of a single global signal called the *clock*. Synchronous systems have a number of advantages that have led to their almost universal use in conventional computing systems. For example, efficiency is gained because only a single source of events need be distributed. Noise immunity is gained; because signals are sampled by a clock, glitches occurring outside the sampling aperture are rejected. Also, since all signals are synchronized to the same source of events, they can be combined logically without fear of metastability, and the design of arbiters is thus simplified—no excluded middle circuit is required for synchronization. Synchronous state machines are also simpler than their asynchronous counterparts because races and hazards are not a concern.

A disadvantage is that synchronous systems are often slower than self-timed systems because every circuit operates at a clock period set by the slowest circuit in the machine, with margins added for clock skew and other variations. If a router is pipelined in three stages, for example, all three stages operate at a rate set by the slowest stage on the slowest chip in the machine. In a self-timed circuit, each circuit operates at its own fastest speed.

Synchronous systems also introduce the problem of distributing a high-frequency clock with a minimum of *clock skew*. Because of variations in delay, the clock event is not seen simultaneously at all points in the system. The largest difference in observed times for one clock event is the clock skew. If the earliest time at which the event is observed is t_{min} and the latest time is t_{max}, the skew is $t_{skew} = t_{max} - t_{min}$. Small clock skews require the clock period to be increased by t_{skew} to allow signals that start propagating at t_{max} a full cycle before they are sampled at $t_{cy} + t_{min}$. Clock skews larger than the shortest propagation path through the logic will cause the circuit to malfunction at any speed as variables from two different clock cycles interact.

Clock distribution systems can deal with skew in two steps, as shown in Figure 3.35. To distribute a low clock skew to the pins of a chip, series terminated drivers feed matched transmission lines (Figure 3.35a). Each line is made to have exactly the same length and impedance, and thus all clock edges reach the chip pins with the same fixed delay. For very large fanouts, a clock driver tree is used with matched drivers as well as

(a) Matched transmission lines to chip inputs.

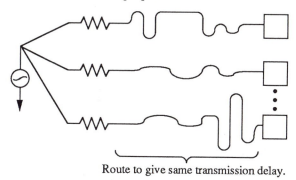

Route to give same transmission delay.

(b) PLL to compensate for orr chip skew.

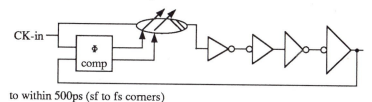

to within 500ps (sf to fs corners)

Figure 3.35   Methods for distributing a clock without skew. (a) Matched transmission lines and drivers deliver an aligned clock to the chip inputs. (b) A phase-locked loop is used to equalize internal clock driver delays.

transmission lines. Very often there is a manual delay adjustment on each driver that is used to *align* the system.

A harder problem is to eliminate the skew caused by the on-chip clock buffer. A large CMOS chip has a clock capacitance of several hundred pico-Farads and for reliable operation requires that the clock be decoupled from external power supply noise. To meet these requirements, an on-chip clock driver is required. Driving the large load from off chip results in slow edges, excessive ringing, and a noisy clock.

These on-chip clock drivers are a significant source of clock skew. A typical CMOS process may have parametric variations that cover a 2:1 speed range. To eliminate the skew caused by fast drivers on some chips and slow drivers on others, a phase-locked loop circuit may be used to equalize the delay of the on-chip driver (Figure 3.35b). The phase comparator compares the on-chip clock to the off-chip clock. Phase differences other than 180° generate error signals that adjust the delay to be exactly one half clock period. The phase-locked loop used with a divider also allows the external clock frequency to be reduced. We have simulated such a phase-locked loop skew compensator using SPICE. These simulations showed on-chip clock skews between the *fast-n fast-p* and *slow-n slow-p* process corners were reduced to < 500ps.

Synchronous systems have advantages for adaptive routers where information from many channels must be combined to make the routing decision. In an asynchronous system, each of these information sources must be synchronized separately. In a synchronous router they are all synchronized to the same event source and may be combined with simple combinational logic.

3.3.3   Driver Circuit Design

A large part of the time and power used by a router goes into driving the wires that make up the physical communication channels. Careful design of the channel drivers and receivers can result in more efficient use of this time and power. Most routers built today operate with 5V signal levels because the TTL standard power supply voltage is 5V. Routers, however, drive signals only to other routers and thus have no need to use standard signal levels. Low dimensional routers ($n \leq 3$) have the advantage of driving very short lines and can operate with higher output impedance than a driver designed to drive a long transmission line.

Lowering router signal levels reduces power dissipation quadratically because $P = V^2/R$. This reduction comes at the expense of reduced noise margins. Most of the noise in digital systems is due to inductive and capacitive coupling between signal lines and self-inductance of power lines. If delay is held constant while signal voltage levels are reduced, these noise sources scale linearly with voltage. As V is reduced from V_1 to αV_1, dV/dt is also reduced by α. Thus noise due to capacitive coupling scales linearly with voltage. Assuming a triangular current waveform, the di/dt required to change a signal from 0 to αV_1 into a load C_L in time t_r is given by

$$\frac{di}{dt} = \pm \frac{4\alpha V_1 C_L}{t_r^2}. \tag{3.37}$$

Since di/dt also scales linearly with α, noise due to mutual and self-inductance is reduced at the same rate as the noise margins are reduced. The only significant noise source that does not scale with the signal voltage is the power supply noise due to the parts of the chip that still operate at a high voltage (including the receivers). Careful design of the power distribution system (e.g., separate grounds for pads and chip internals) is required to prevent this noise from coupling into the communication channels.

Alternatively, if noise margins are larger than required, low-voltage pads can be used to reduce delay at the expense of noise margins. If we reduce the output impedance of our pad, the output current, i, and its derivative increase. From equation (3.37), di/dt and thus noise scale as $1/t_r^2$.

Figure 3.36 shows the schematic of a low-voltage driver and receiver used on the NDF chip [11]. It is based on a design proposed by Knight [23]. The driver consists of two n-type transistors sized to give the desired output impedance (100Ω in the NDF). Both devices operate in the resistive region. The upper device is made 4/3 the size of the lower device to equalize their resistances because when it is on it operates with a $(V_{GS} - V_T)$ of between 3 and 4V, while the lower device (when on) always has $(V_{GS} - V_T) = 4V$.

The receiver is a cascode amplifier. When the signal falls below $V_R - V_T$,

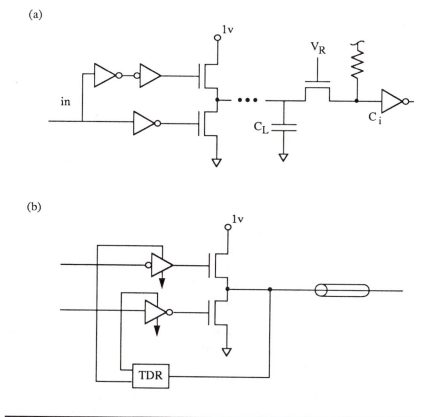

Figure 3.36   Low-voltage driver and receiver design. (a) Design used in the NDF. (b) Self-series-terminating design.

the input transistor conducts, pulling the input to the inverter below its switching point. When the signal rises, the input transistor shuts off, and the resistor (actually a current source) pulls the input of the inverter up to the supply voltage.

One shortcoming of this design is that a significant amount of static power is dissipated in the receiver. If the receiver input node (the input of the inverter) has a capacitance C_i, and a rise time of t_r is required, the bias current for the receiver must be at least $I = C_i V_i / t_r$, and thus the power is

$$P_i = \frac{C_i V_i^2}{t_r}.$$

(3.38)

If the cycle time is t_c, the power dissipated driving the line is

$$P_L = \frac{C_L V_L^2}{t_c}.$$

(3.39)

Thus the low-voltage drivers save power only when the capacitance ratio C_L/C_i is great enough that the savings in driver power more than make up for the receiver dissipation. That is when

$$P_{L1} - P_{L2} > P_i,$$

(3.40)

$$\frac{C_L(V_{L1}^2 - V_{L2}^2)}{t_c} > \frac{C_i V_i^2}{t_r},$$

(3.41)

$$\frac{C_L}{C_i} > \frac{t_c}{t_r} \frac{V_i^2}{(V_{L1}^2 - V_{L2}^2)}.$$

(3.42)

The term $\frac{V_i^2}{(V_{L1}^2 - V_{L2}^2)}$ is close to unity. Thus the savings in power dissipation occurs whenever the capacitance ratio is significantly greater than the ratio of cycle time to rise time. For the NDF, the capacitance ratio is 50 (10pF/200fF), while the cycle to rise ratio is 10 (20ns/2ns).

If the system is synchronous, the power dissipation in the receiver can be made to depend on $1/t_c$ by using a clocked amplifier. In this case, power dissipation is reduced for any $C_L/C_i > 1$.

In networks with some long wires, the output driver can be made self-series-terminating, as shown in Figure 3.36b. A reference pad adjusts the output impedance of the drivers to match the impedance of the line by varying the supply voltage of the inverters driving the final stage. The reference circuit drives a step into the line and samples the line before the wave has reflected back. If the voltage is less (greater) than half the step voltage, the pad output impedance is increased (decreased) until the pad is matched to the line.

3.3.4   Summary

In this section we touched on three engineering issues that are significant in the design of a router: organization, synchronization, and driver circuit design. This is just a small sample of the engineering problems that must be solved to build a high-performance router.

Routers are constructed from primitive switches, latches, and control blocks. These components are connected together into simple routing modules that switch, merge, or buffer channels. Routing modules have inputs and outputs that include flow control information and conform to a standard signaling convention so that any input can be connected to any output. A router is constructed from a network of routing modules. For restricted routing relations (as with e-cube routing) a very simple router organization (such as a linear array) can outperform a non-blocking network if it optimizes the frequently used paths.

Routers control the movement of flits with a sequence of events. In a self-timed or asynchronous router, events on channels that do not interact are independent and need not be ordered. Each channel carries its own timing signal. In a synchronous router, all channels share a single global source of events.

Much power and time is spent transmitting signals over the physical communication channel. We examined the design of a low-voltage driver that improves the efficiency of signal transmission.

3.4   The Message-Driven Processor (MDP)

An efficient multicomputer will have the processors on each node tightly coupled into the network. In the preceding three sections we have discussed methods for building networks that deliver messages between the nodes of a multicomputer in a few microseconds. In addition to this network latency, T_{net}, we must also consider the latency of the processing node, T_{node}. The total latency, T, is the sum of these two components:

$$T = T_{net} + T_{node}. \tag{3.43}$$

After the network has delivered a message to the node, the processor must buffer the message, interpret it, and begin the task of processing the message. With a conventional processor running a conventional operating system, these operations require $> 100\mu s$. In this section we will discuss mechanisms that reduce this message-handling overhead to $< 1\mu s$ in the message-driven processor (MDP) [10].

The MDP is being designed as part of the J-Machine project at MIT [13]. The project has three main goals:

(1) To reduce the overhead associated with dispatching a task in response

to a message, T_{net}, to a level that permits the execution of fine-grain (< 10 instruction) concurrent programs.

(2) To improve the efficiency of computing systems by increasing memory bandwidth and the fraction of area devoted to processing.

(3) To build a general purpose parallel computer capable of supporting many models of computation. The strategy is to provide efficient primitive mechanisms for communication, synchronization, and naming.

The remainder of this section describes how the MDP addresses each of these goals.

3.4.1 Processors are Inexpensive

We build computer systems, parallel or serial, from VLSI chips. VLSI provides a uniform medium for comparing the cost and performance of alternative architectures and designs. Computer memories, processors, and interface units are all built from the same technology. Their cost is a function of area.

A VLSI chip is very inexpensive to fabricate. Figure 3.37 shows a 5-inch CMOS wafer. Over 250 chips are fabricated on the wafer. Of these, about 30 percent (approximately 80 chips) are expected to be defect-free. Wafers of this type cost about \$250 in quantity for an unpackaged cost of about \$3 per chip. The chips on this wafer are 5mm × 8mm or $160M\lambda^2$ (λ is half the minimum line width [27]). We can use these cheap parts to build fast, inexpensive processing elements.

In VLSI a 1M-bit DRAM chip has an area of $256M\lambda^2$. In the same area we can build a single-chip processing node, as shown in Figure 3.38. The chip includes the following elements:

A 32-bit processor	$16M\lambda^2$
A floating-point unit	$32M\lambda^2$
A communication controller	$8M\lambda^2$
512Kbits RAM	$128M\lambda^2$

Such a single-chip processing node would have the same processing power as a board-sized node, but significantly less memory. We refer to a machine built from these nodes as a *jellybean machine* because it is built with commodity part (jellybean part) technology.

A fine-grain processing node has two major advantages: density and memory bandwidth. Several hundred single-chip nodes can be packaged on a single printed circuit board, permitting us to exploit hundreds of times the concurrency of machines with board-sized nodes. With on-chip memory

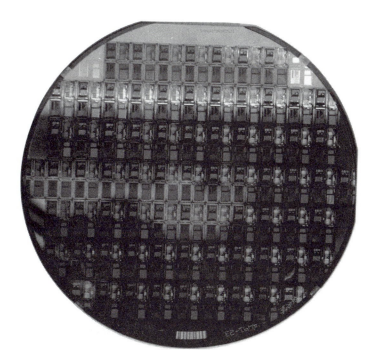

Figure 3.37   A 5-inch silicon wafer fabricated on a 1μ CMOS line. This wafer contains over 250 5mm × 8mm die sites and can be expected to yield over 80 working dice.

we can read an entire row of memory (128 or 256 bits) in a single cycle without incurring the delay of several chip crossings. This high-memory bandwidth allows the memory to buffer messages from a high bandwidth network while simultaneously providing the processor with instructions and data.

Fine-grain machines are quite efficient. We measure efficiency as

$$e_A = A_1 T_1 / A_N T_N \tag{3.44}$$

(where A_i is the area of i processors and T_i is the time to solve a problem on i processors) rather than as

$$e_N = T_1 / N T_N \tag{3.45}$$

(where N is the number of processors). Proponents of coarse-grain machines argue that a machine constructed from several thousand single-chip

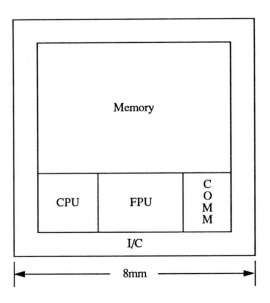

Figure 3.38   In the area of a 1Mbit DRAM chip, one can construct a processing node with a 32-bit processor, a floating point unit, a communication controller, and 512Kbits of memory.

nodes will be inefficient because many of the processing nodes will be idle. N is large, hence e_N is small. A user, however, is not concerned with N but rather with what the machine costs and how long it takes to solve a problem. Fine-grain machines have a very high e_A because they are able to exploit more concurrency in a smaller area.

Because processors are inexpensive, it is not important that all the processors be kept busy. Processor utilization (3.45) is a poor measure of efficiency. The critical resources in a multicomputer are the communication bandwidth and memory capacity. To see how well these resources are applied, the efficiency of the machine as a whole must be measured as the ratio of its relative speedup, $S = T_1/T_N$, and its relative cost, $C = A_N/A_1$.

How can we apply silicon area to building more powerful computers? It is easy to apply silicon area to making more processors, but it is quite difficult to apply area to make a single processor more powerful. Figure 3.39 illustrates the diminishing returns in the performance of a uniprocessor. The figure shows performance of processors normalized to the speed of the technology (milli-Instructions/τ) as a function of area ($M\lambda^2$). Normalizing to τ eliminates the ≈ 15 percent/year performance improvement that

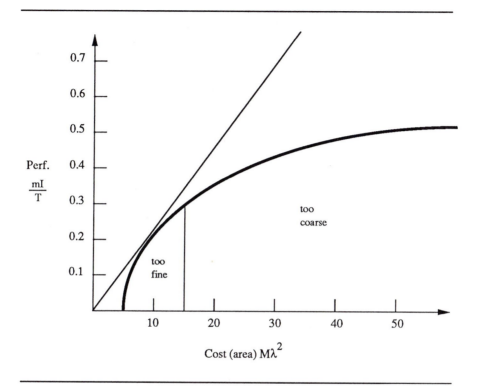

Figure 3.39   Performance (milli-Instructions/τ) as a function of chip area ($M\lambda^2$). As the area of a chip is increased, there is a diminishing return in performance.

results from better technology and allows us to concentrate on the change in performance due to architecture. The figure is intended only to illustrate a concept and does not reflect data from any actual processors.

There is a fairly large overhead for building even a 1-bit processor ($\approx 5M\lambda^2$), and thus there is a large incremental improvement $\partial P/\partial A$ in adding a second bit to the processor. Similar returns result from adding additional bits, some registers, and support for common arithmetic operations. However, once the processor has 32-bit data paths and support for the most common operations, there is very little return on additional expenditures of area. $\partial P/\partial A$ is reduced. At the point where $\partial P/\partial A = P/A$, $\approx 15M\lambda^2$ in the figure, we get a larger incremental return on our area investment by building additional processors than by building a larger single processor.

The performance of a multicomputer built from processors of this optimum size is illustrated by the straight line in Figure 3.39. This linear increase in performance as we add processors is achieved only in the most ideal cases. In most real applications, uneven distribution of processing load and overhead associated with communication and synchronization will cause the actual speedup curve to be somewhat lower. Operating at a fine grain tends to even out the processing load at the expense of increased overhead. The MDP includes mechanisms to achieve fine-grain concurrency without excessive overhead.

The J-Machine, under design at MIT [13], will connect up to 65,536 of the jellybean parts (Figure 3.38) with the networks described in sections 3.1 through 3.3. Our initial prototype will contain 4,096 processors. Figure 3.40 is a sketch of such a 4K-node workstation. The machine is a memory system in that each board contains a large number of identical chips containing mostly memory. The difference is that each memory chip includes a processor and communication unit that improve the bandwidth and efficiency of the computing system.

3.4.2 Fine-Grain Programs Expose More Concurrency

Many computationally demanding problems have an abundance of concurrency. This concurrency exists at many levels: for example, at the coarsest grain we iterate over the gridpoints of a problem. For each gridpoint we may perform some vector operations that can be carried out in parallel. Each operation may require the evaluation of some expressions or method that can be performed simultaneously. Within one expression, several arithmetic operations can be performed in parallel.

Figure 3.40   The J-Machine achieves high efficiency by combining many processing nodes, each containing a powerful processor and a small amount of memory.

The *grain size* of a program refers to the level at which concurrency is exploited—that is, the size of the tasks and messages that make up the program. Coarse-grain programs have a few long ($\approx$ 10ms) tasks, while fine-grain programs have many short ($\approx$ 5μs) tasks. Because more tasks can execute at a given time—that is, there is more concurrency—fine-grain programs (in the absence of overhead) result in faster solutions than coarse-grain programs. Also, load balancing is more uniform in fine-grain programs because work is allocated in smaller units.

Figure 3.41 shows the computation graph of a program. Each vertex corresponds to a task at the level of method (subroutine) evaluation. In our Concurrent Smalltalk system [19], we have observed that invoking methods on objects averages twenty instructions. The average data object is eight words in length, and the average message is six words in length. In the extreme case of dataflow [39], a task may be a single instruction. Before a task can execute, it must receive its input data from preceding tasks. The edges in the computation graph reflect the communication and synchronization actions required to resolve these dependencies.

Existing message-passing computers have communication and synchronization overheads in excess of 500 instruction times. As a result, programs for these machines are required to aggregate many fine-grain tasks into a single large-grain task of several thousand instructions. Conceptually, 100 vertices of the computation graph are grouped together to amortize the communication and synchronization overhead. By reducing communication and synchronization overhead to permit efficient execution at a grain size of ten instructions, we can exploit 100 times as much concurrency.

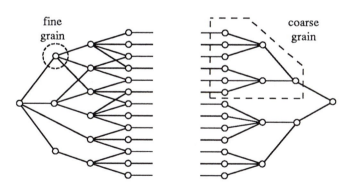

Figure 3.41   Computation graph of a parallel program. Each vertex is a task. Edges indicate dependencies between tasks. Communication and synchronization actions are usually associated with each edge.

Table 3.2

Machine	T_{net}	T_{node}	Task Size	Concurrency
iPSC-1	5ms	500μs	10ms	10^2
J-Machine	2μs	1μs	5μs	10^5

Table 3.2 compares the communication and synchronization overhead (T_{net} and T_{node}) of a first-generation, message-passing concurrent computer, the Intel iPSC-1 [20] with the expected performance of the J-Machine. Using the techniques described in the first three sections of this chapter, the network latency has been reduced to 2μs from 5ms. Using the mechanisms described in the following section, the MDP reduces the node overhead from 500μs to 1μs. In combination, these reductions in overhead permit the execution of much shorter tasks—5μs versus 10ms—resulting in substantially more concurrency—10^5 versus 10^2.

3.4.3   The Message-Driven Processor (MDP)

The Message-Driven Processor (MDP) [10] provides low overhead mechanisms for communication, synchronization, and translation. These mechanisms can be used to support many models of computation, making it unnecessary to hardwire a specific model of computation.

The MDP reduces overhead by coupling its execution mechanism into the network. Most processors are instruction-driven processors that fetch an instruction from memory. An instruction control unit looks at the opcode of that instruction and dispatches a sequence of operations to interpret the instruction. At an abstract level, the MDP treats messages like instructions. The MDP fetches a message from the network, examines the opcode of the message, and dispatches a sequence of operations to interpret the message.

Rather than hardwire this message interpretation mechanism, however, the MDP performs most of its message interpretation by executing instructions. Nevertheless, three critical areas require hardware support:

- **Communication** The MDP uses a SEND instruction to transmit messages over the network. Hardware queue management is provided to allocate buffers to incoming messages. Row buffers are used to increase the memory bandwidth to handle incoming message streams.

- **Synchronization** The MDP schedules and creates processes in hardware by dispatching on message arrival. Messages in the queue are also considered to be processes on the ready list.

- **Translation** The MDP uses a translate instruction to translate virtual addresses into physical segment descriptors.

Send Instruction   The MDP injects messages into the network using a send instruction that transmits one or two words (at most one from memory) and optionally terminates the message. The first word of the message is interpreted by the network as an absolute node address (in x,y format) and is stripped off before delivery. The remainder of the message is transmitted without modification. A typical message sent is shown in Figure 3.42. The first instruction sends the absolute address of the destination node (contained in R0). The second instruction sends two words of data (from R1 and R2). The final instruction sends two additional words of data, one from R3 and one from memory. The use of the SEND2E instruction marks the end of the message and causes it to be transmitted into the network.

A first-in-first-out (FIFO) buffer is used to match the speed of message transmission to the network. In some cases, the MDP cannot send message words as fast as the network can transmit them. Without a buffer, *bubbles* (absence of words) would be injected into the network pipeline, thereby degrading performance. The SEND instruction loads one or two words into the buffer. When the message is complete or the eight-word buffer is full, the contents of the buffer are launched into the network.

Early in the design of the MDP we considered making a message send a single instruction that took a message template, filled in the template using the current addressing environment, and then transmitted the message. Each template entry specified one word of the message as being either a constant, the contents of a data register, or a memory reference offset from an address register (like an operand descriptor). The template approach was abandoned in favor of the simpler one or two operand SEND instruction

```
SEND    R0          ; send net address
SEND2   R1,R2       ; header and receiver
SEND2E  R3,[3,A3]   ; selector and continuation - end msg.
```

Figure 3.42   MDP assembly code to send a four-word message uses three variants of the SEND instruction.

because the template did not significantly reduce code space or execution time. A two operand **SEND** instruction results in code that is nearly as dense as a template and can be implemented using the same control logic used for arithmetic and logical instructions.

Previous concurrent computers have used direct-memory access (DMA) or I/O channels to inject messages into the network. First an instruction sequence composed a message in memory. DMA registers or channel command words were then set up to initiate sending. Finally, the DMA controller transferred the words from the memory into the network. However, this approach to message sending is too slow for two reasons. First, the entire message must be transferred across the memory interface twice, once to compose it in memory and a second time to transfer it into the network. Second, for very short messages, the time required to set up the DMA control registers or I/O channel command words often exceeds the time required to simply send the message into the network.

Message Reception The MDP maintains two message/scheduling queues (one for each priority level) in its on-chip memory. The queues are implemented as circular buffers. As messages arrive over the network, they are buffered in the appropriate queue. To improve memory bandwidth, messages are enqueued by rows. Incoming message words are accumulated in a row buffer until the row buffer is filled or the message is complete. The row buffer is then written to memory.

It is important that the queue have sufficient performance to accept words from the network at the same rate at which they arrive. Otherwise, messages would back up into the network, causing congestion. The queue row buffers, in combination with hardware update of queue pointers, allow enqueuing to proceed using one memory cycle for each four words received. Thus a program can execute in parallel with message reception with little loss of memory bandwidth.

Providing hardware support for allocation of memory in a circular buffer on a multicomputer is analogous to the support provided for allocation of memory in push-down stacks on a uniprocessor. Each message stored in the MDP message queue represents a method activation, much as each stack frame allocated on a push-down stack represents a procedure activation.

Scheduling and Dispatch Each message in the queues of an MDP represents a task that is ready to run. When the message reaches the head of the queue, a task is created to handle the message. At any time, the MDP is executing the task associated with the first message in the highest priority non-empty queue. If both queues are empty, the MDP is idle—that is, executing a background task. Sending a message implicitly schedules a task on the destination node. This simple, two-priority scheduling mechanism removes the overhead associated with a software scheduler.

More sophisticated scheduling policies may be implemented on top of this substratum.

Messages become *active* either by arriving while the node is idle or executing at a lower priority, or by being at the head of a queue when the preceding message *suspends* execution. When a message becomes active, a task is created to handle it. Task creation, changing the thread of control and creating a new addressing environment, are performed in one clock cycle, as shown in Figure 3.43. Every message header contains a message *opcode* and the message *length*. The message opcode is loaded into the instruction pointer to start a new thread of control. The length field is used along with the queue head to create a message segment descriptor that represents the initial addressing environment for the task. The message handler code may open additional segments by translating object IDs in the message into segment descriptors.

No state is saved when a task is created. If a task is preempting lower

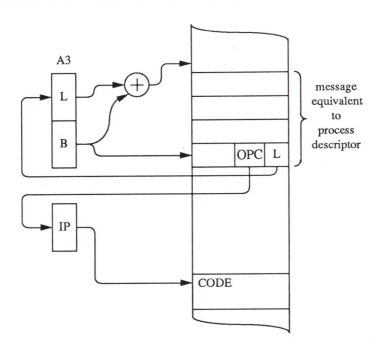

Figure 3.43   Message dispatch. In one clock cycle, a new task is created by (1) setting the instruction pointer to change the thread of control and (2) creating a message segment to provide the initial addressing environment.

priority execution, it executes in a separate set of registers. If a task, A, becomes active when an earlier task, B, at the same priority suspends, B is responsible for saving its live state before suspending.

The dispatch mechanism is used directly to process messages requiring low latency (e.g., combining and forwarding). Other messages (e.g., a remote procedure call) specify a handler that locates the required method (using the translation mechanism described below) and then transfers control to it.

For example, a remote procedure call message is handled by the call handler code, as shown in Figure 3.44. The execution of this handler is depicted in Figure 3.45. The first instruction gets the method ID. The next instruction translates this method ID into a segment descriptor for the method. The final instruction transfers control to the method. The method code may then read in arguments from the message queue. The argument object identifiers are translated to segment descriptors using the translate instruction. If the method needs space to store local state, it may create a context object. When the method has finished execution, or when it needs to wait for a reply, it executes a SUSPEND instruction, passing control to the next message.

An early version of the MDP had a fixed set of message handlers in microcode. An analysis of these handlers showed that their performance was limited by memory accesses, and thus there was little advantage in using microcode. The microcode was eliminated, the handlers were recoded in assembly language, and the *message opcode* was defined to be the physical address of the handler routine. Frequently used handlers are contained in an on-chip ROM. This approach simplifies the control structure of the machine and gives us flexibility to redefine message handlers to fix bugs, for instrumentation (e.g., to count the number of sends), and to implement new message types.

Synchronization with Tags   Every register and memory location in the MDP includes a 4-bit tag that indicates the type of data occupying the location. The MDP uses tags for synchronization on data availability in addition to their conventional uses for dynamic typing and run-time type checking. Two tags are provided for synchronization: future, and

```
MOVE    [1,A3],R0    ; get method id
XLATE   R0,A0        ; translate to segment descriptor
LDIP    INITIAL_IP   ; transfer control to method
```

Figure 3.44   MDP assembly code for the CALL message.

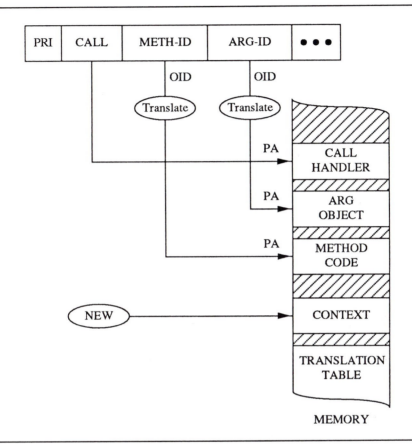

Figure 3.45   The CALL message invokes a method by translating the method identifier to find the code, creating a context (if necessary) to hold local state, and translating argument identifiers to locate arguments.

c-future. A future tag is used to identify a named placeholder for data that is not yet available [2] [18]. Applying a strict operator to a future causes a fault. A future can, however, be copied without faulting. A c-future tag identifies a cell awaiting data. Applying any operator to a c-future causes a fault. As they are unnamed placeholders, they cannot be copied.

The c-future tag is used to suspend a task if it attempts to access data that has not yet arrived from a remote node. When a task sends a message requesting a reply, it marks the cell that will hold the reply as a c-future. Any attempt to reference the reply before it is available will fault and suspend the task. When the reply arrives, it overwrites the c-future and resumes the task if it was suspended.

Figure 3.46 illustrates the use of c-futures. When a program executes the statement C := A + B, it discovers that object B is remote. It sends a message to B to compute the sum and tags C with a c-future to indicate that it is not present. When the value is returned from B, it overwrites the c-future. An attempt to reference C before the value arrives results in a trap.

Translation   The MDP is an experiment in unifying shared-memory and message-passing parallel computers. Shared-memory machines provide a uniform global name space (address space) that allows processing elements to access data regardless of its location. Message-passing machines perform communication and synchronization via node-to-node messages. These two concepts are not mutually exclusive. The MDP provides a virtual addressing mechanism intended to support a global name space while using an execution mechanism based on message passing.

The MDP implements a global virtual address space using a general translation mechanism. The MDP memory allows both indexed and set-associative access. By building comparators into the column multiplexer of the on-chip RAM, we are able to provide set-associative access with only a small increase in the size of the RAM's peripheral circuitry.

The translation mechanism is exposed to the programmer with the ENTER and XLATE instructions. ENTER Ra,Rb associates the contents of Ra (the key) with the contents of Rb (the data). The association is made on the full thirty-six bits of the key so that tags may be used to distinguish different keys. XLATE Ra,Ab looks up the data associated with the contents of Ra and stores this data in Ab. The instruction faults if the lookup *misses* or if the data is not a segment descriptor. XLATE Ra,Rb can be used to look up other types of data. This mechanism is used by our system code to cache ID to segment descriptor (virtual to physical) translations, to cache ID to node number (virtual to physical) translations, and to cache class/selector to segment descriptor (method lookup) translations.

Tags are an integral part of our addressing mechanism. An ID may translate into either a segment descriptor for a local object or a node address for a global object. The tag allows us to distinguish these two cases, and a fault provides an efficient mechanism for the test. Tags also allow us to distinguish an ID key from a class/selector key with the same bit pattern.

Most computers provide a set-associative cache to accelerate translations. We have taken this mechanism and exposed it in a pair of instructions that a systems programmer can use for any translation. Providing this general mechanism gives us the freedom to experiment with different address translation mechanisms and different uses of translation. We pay very little for this flexibility because performance is limited by the number of memory accesses that must be performed.

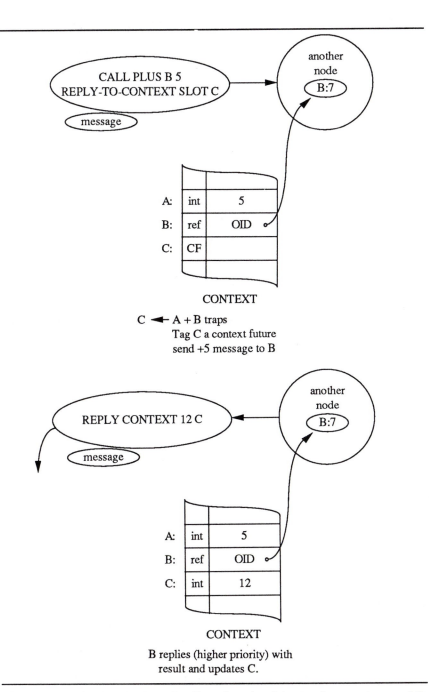

Figure 3.46   Synchronization using Tags: Location C is tagged as a **c-future** while a message is sent to object B to compute its value. An attempt to reference C before the value is returned results in a trap.

3.4.4   JOSS: The Jellybean Operating System

Fine-grain concurrent computers present some interesting challenges to the operating system. The small amount of memory in each node and the small task size executed require very different strategies from those used in conventional operating systems [1].

To support the J-Machine we have built an operating system, JOSS [38], that is designed to deal efficiently with the small tasks, small objects, and small memories of a fine-grain concurrent machine. It is also tailored for an environment where local computation is inexpensive. Communication bandwidth and memory capacity are the limiting resources.

JOSS provides

- **Memory Management** services to allocate and name objects and to locate an object, once given its name (ID).

- **Task Management** services to create, suspend, resume, and destroy tasks.

- **Code Distribution** services that maintain a method cache on each node and perform combining to distribute copies of method code on demand.

- **Message Services** include remote procedure call, late-binding send, mulitcast forwarding, and combining.

A few of the most novel aspects of JOSS are described in detail below.

Memory Management Services   Most message-passing multicomputers have a separate memory address space on each node. Nodes interact only by sending messages between processes [37]. A partitioned address space makes distributed data structures difficult to construct [8], limits the size of a process's address space to the memory size of a node, and requires entire processes to be relocated to balance memory use. Also, because storage on remote nodes cannot be directly accessed, these machines replicate the operating system and application code on each node.

JOSS overcomes these limitations by providing a global object namespace. All data and code are stored in objects. Each object is assigned a unique global ID. Given an object ID, a task on any node can reference the corresponding object. Objects are free to migrate between nodes. Accesses to objects are bounds checked and protected. The system supports distributed objects [19]. Large distributed objects are implemented as a collection of small constituent objects accessed via a single ID. A one-to-many translation service prevents the single ID from becoming a bottleneck.

A global object namespace provides many of the advantages of a shared-memory multicomputer while retaining the scalability of a message-passing machine. Distributed data structures are easily constructed by linking objects on different nodes using IDs. Processes have an address space limited only by the size of an ID. Also, code need not be replicated on each node since it can be referenced through its ID.

To support this global object space, JOSS provides (1) services to allocate and deallocate objects, and (2) services to translate object names (IDs) into object locations. Both functions are layered, with one component providing the service locally within a node and a second component extending the service across the network.

Objects are created locally by allocating a contiguous region of memory off the top of the heap and assigning a unique name (ID) to the object. The global ID space is partitioned so that nodes may assign unique global IDs autonomously. Object creation is extended across the network by providing a NEW message that creates an object of a specified class on a remote node and returns its ID.

If nodes are creating new objects frequently, the 32-bit ID space can become exhausted in just a few minutes. To avoid system failure when IDs are exhausted, JOSS recycles IDs. Before reusing an ID, the system attempts to dereference it. The ID is reused only if the object to which it was previously assigned no longer exists.

Objects are deleted by marking. Their storage is reclaimed by a compactor that copies down objects in memory to fill unused holes. As segments are relocated during compaction, the local translation table is updated and all segment registers are invalidated. Compaction is very fast because local memory is small and fast and because the operation is completely local—no communication is required.

Given an object ID, an object is located in two steps. First, a distributed global name table is accessed to find the node on which the object is resident. A message is then sent to the node, where the ID is translated into a segment descriptor for the object. The segment descriptor for an object is strictly local information, and thus each node may relocate objects locally without interacting with any other nodes.

As Figure 3.47 shows, both steps are performed by translating the ID using the MDP's XLATE instruction backed up by a hash table. If the object is local, the segment descriptor is immediately returned. If the object is remote, a *hint* as to its present location may be present in the translation cache. If not, a message is sent to the node on which the object was created, its *hometown node*, to access the object's distributed global name table entry. The use of hints saves a message send for most object references. The global name table must be consulted, however, if the hint becomes stale or is discarded to free up space.

In a fine-grain multicomputer, segment-based memory management is preferred to paging because fine-grain relocation and protection is required. The ability to compact all of memory in a few milliseconds eliminates

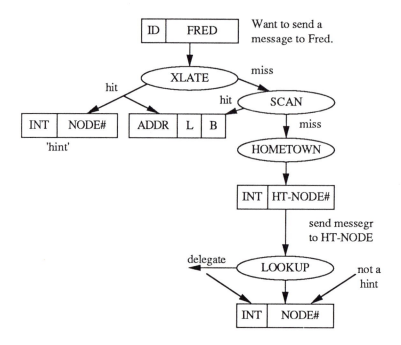

Figure 3.47   An object is located by translating its object ID.

concerns over external fragmentation, but internal fragmentation remains an issue. Objects are small and must be protected and relocated individually. To support fine-grain computation, a paging system would either have to have a very small page size (eight words), or have to sacrifice protection by packing unrelated objects into the same page.

Code Distribution   With only a small amount of memory on each node of the J-Machine, it is not possible to replicate the program and system code on each node, as is done in most message-passing computers. Instead, the J-Machine maintains a single, distributed global copy of all code and makes local copies of methods on demand. Only a small portion of the operating system, a few hundred words of code, is replicated on each node.

A method cache is used to hold local copies of code for execution. If a task requires a method that is not in the cache, it is suspended, and a message is sent requesting the method. On the average, twenty instructions are fetched on each *miss*, and therefore the method cache would achieve a 95 percent hit ratio even if there were no reuse. For a modest sized cache and typical reuse, warm-start hit ratios exceed 99 percent [5].

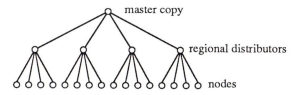

Figure 3.48   A combining/distribution tree is used to combine requests for a method when there is a miss in the method cache.

Many nodes of the machine often need the same method at the same time. If they all sent a message to the node containing the master copy of the method, their requests would be served serially, slowing execution. To avoid this bottleneck, JOSS performs combining of requests for methods. Each node sends its code request up a tree of *distributors*, as shown in Figure 3.48. At each level of the tree the following action is performed:

- If the method is cached in the distributor, a copy is sent back down the tree immediately.

- If another request is already pending for the same method, the two requests are *combined*. No immediate action is taken. When the method arrives in response to the pending request, it will be sent to all requesters.

- Otherwise, the distributor records the pending request and forwards it up the tree.

To avoid a root bottleneck when many different methods are requested, different combining/distribution trees are used for different methods. The tree nodes used for a particular request are determined by combining bits from the requesting node number with bits from the hometown node number of the method. The method's hometown node is always the root of the tree.

The use of combining/distribution trees to handle method requests increases the latency of such requests by a factor of $\approx \log_r N$, where r is the radix of the tree. For example, a radix $r = 8$ tree in an $N = 4,096$-node machine will have four levels and take about four times as long to handle a method request than if the request were sent directly to the root. This increased latency is more than offset by eliminating the serial access bottleneck.

3.4.5   Summary

This section has described the message-driven processor (MDP). The MDP provides low overhead mechanisms for communication, synchronization, and naming. Communication is supported by means of a send instruction, message buffering, and automatic message buffer allocation. Task scheduling and dispatch on message arrival are provided for synchronization along with tags to indicate data presence. A translate instruction in concert with segmented memory access is used to support naming. The MDP's mechanisms are primitive and can be used to support many different models of computation. The brief discussion of the JOSS operating system illustrates how a programming system can be built on top of these mechanisms.

These mechanisms have very low overhead. The MDP can create a task in response to an arriving message in less than a microsecond. By reducing the node's component of message overhead, T_{node}, the MDP is able to take advantage of the low latency networks discussed in this chapter. An ensemble of MDPs connected by a low latency network can efficiently execute concurrent programs with a grain size as small as ten instructions. Operating at such a fine grain size exposes sufficient concurrency in many problems to keep thousands of MDPs busy.

Bibliography

[1] Bach, M. J., *The Design of the UNIX Operating System,* Prentice Hall, 1986.

[2] Baker, H. and Hewitt, C., "The Incremental Garbage Collection of Processes," *ACM Conference on AI and Programming Languages,* Rochester, New York, August, 1977, pp. 55–59.

[3] Batcher, K. E., "Sorting Networks and Their Applications," *Proceedings AFIPS FJCC,* Vol. 32, 1968, pp. 307–314.

[4] BBN Advanced Computers, Inc., *Butterfly Parallel Processor Overview,* BBN Report No. 6148, March 1986.

[5] Chien, Andrew, "The Particle-in-Cell Code (PIC): An Application Study for CST," *MIT Concurrent VLSI Architecture Memo,* December 1988.

[6] Clos, Charles, "A Study of Non-Blocking Switching Networks," *Bell Systems Technical Journal,* Vol. 32, March 1953, pp. 406–424.

[7] Dally, William J. and Seitz, Charles L., "The Torus Routing Chip," *J. Distributed Systems,* Vol. 1, No. 3, 1986, pp. 187–196.

[8] Dally, William J., *A VLSI Architecture for Concurrent Data Structures*, Kluwer, Hingham, MA, 1987.

[9] Dally, William J. and Seitz, Charles L., "Deadlock-Free Message Routing in Multiprocessor Interconnection Networks," *IEEE Transactions on Computers*, Vol. C-36, No. 5, May 1987, pp. 547–553.

[10] Dally, William J. et al., "Architecture of a Message-Driven Processor," *Proceedings of the 14th ACM/IEEE Symposium on Computer Architecture*, June 1987, pp. 189–196.

[11] Dally, William J., and Song, Paul, "Design of a Self-Timed VLSI Multicomputer Communication Controller," *Proc. International Conference on Computer Design, ICCD-87*, 1987, pp. 230–234.

[12] Dally, William J., "Fine-Grain Concurrent Computers," *Proc. 3rd Symposium on Hypercube Concurrent Computers and Applications*, ACM 1988.

[13] Dally, William J. et al., "The J-Machine: A Fine-Grain Concurrent Computer," *Information Processing 89*, Elsevier, 1989, pp. 1147–1153.

[14] Dally, William J., "Performance Analysis of k-ary n-cube Interconnection Networks," *IEEE Transactions on Computers*, to appear.

[15] Flaig, Charles, M., *VLSI Mesh Routing Systems*, Technical Report 5241:TR:87, Dept. of Computer Science, California Institute of Technology, 1987.

[16] Goke, L. R. and Lipovski, G. J., "Banyan Networks for Partitioning Multiprocessor Systems" *First Symposium on Computer Architecture*, 1973, pp. 21–28.

[17] Gunther, Klaus D., "Prevention of Deadlocks in Packet-Switched Data Transport Systems," *IEEE Transactions on Communications*, Vol. COM-29, No. 4, April 1981, pp. 512–524.

[18] Halstead, Robert H., "Parallel Symbolic Computation," *IEEE Computer*, Vol. 19, No. 8, Aug. 1986, pp. 35–43.

[19] Horwat, Waldemar, Chien, Andrew A., and Dally, William J., "Experience with CST: Programming and Implementation," *Proceedings of the ACM SIGPLAN '89 Conference on Programming Language Design and Implementation*, 1989, pp. 101–109.

[20] Intel Scientific Computers, *iPSC User's Guide*, Order No. 175455-001, Santa Clara, CA, Aug. 1985.

[21] Kermani, Parviz and Kleinrock, Leonard, "Virtual Cut-Through: A New Computer Communication Switching Technique," *Computer Networks*, Vol. 3., 1979, pp. 267–286.

[22] Kleinrock, Leonard, *Queueing Systems, Volume 1: Theory*, Wiley, New York, 1975.

[23] Knight, T. and Krymm, A., "A Self-Terminating Low Voltage Swing CMOS Output Driver," *IEEE Journal of Solid State Circuits*, Vol. 23 No. 2, April 1988, pp. 457–464.

[24] Lang, C. R. Jr., *The Extension of Object-Oriented Languages to a Homogeneous, Concurrent Architecture*, Dept. of Computer Science, California Institute of Technology, Technical Report 5014, May 1982.

[25] Lawrie, Duncan H., "Alignment and Access of Data in an Array Processor," *IEEE Transactions on Computers*, Vol. C-24, No. 12, December 1975, pp. 1145–1155.

[26] Mailhot, John, *A Comparative Study of Routing and Flow Control Strategies in k-ary n-cube Networks*, S.B. Thesis, Massachusetts Institute of Technology, May 1988.

[27] Mead, Carver A. and Conway, Lynn A., *Introduction to VLSI Systems*, Addison-Wesley, Reading, Mass., 1980.

[28] Pease, M. C., III, "The Indirect Binary n-Cube Microprocessor Array," *IEEE Transactions on Computers*, Vol. C-26, No. 5, May 1977, pp. 458–473.

[29] Pfister, G. F. et al., "The IBM Research Parallel Processor Prototype (RP3): Introduction and Architecture," *Proc. International Conference on Parallel Processing, ICPP*, 1985, pp. 764–771.

[30] Pfister, G. F. and Norton, V. A., "Hot Spot Contention and Combining in Multistage Interconnection Networks," *IEEE Transactions on Computers*, Vol. C-34, No. 10, October 1985, pp. 943–948.

[31] Seitz, Charles L., "System Timing" in *Introduction to VLSI Systems*, C. A. Mead and L. A. Conway, Addison-Wesley, 1980, Ch. 7.

[32] Seitz, Charles L., "Concurrent VLSI Architectures," *IEEE Transactions on Computers*, Vol. C-33, No. 12, December 1984, pp. 1247–1265.

√[33] Seitz, Charles L., "The Cosmic Cube," *Comm. ACM*, Vol. 28, No. 1, Jan. 1985, pp. 22–33.

[34] Seitz, Charles L., Private communication, April 1986.

[35] Seitz, C. L., et al., "The Architecture and Programming of the Ametek Series 2010 Multicomputer," *Proc. Third Conference on Hypercube Concurrent Computers and Applications*, ACM, Jan. 1988, pp. 33–37.

[36] Seitz, C. L., Chapter 1 of this book.

[37] Su, W., Faucette, R., and Seitz, C., *C Programmer's Guide to the Cosmic Cube*. Technical report 5203:TR:85, California Institute of Technology, September 1985.

[38] Totty, Brian, *An Operating Environment for the Jellybean Machine*, MIT Artificial Intelligence Laboratory Memo No. 1070, 1988.

√[39] Veen, Arthur H., "Dataflow Machine Architecture," *ACM Computing Surveys*, Vol. 18, No. 4, December 1986, pp. 365–396.

[40] Wu and Feng, "On a Class of Multistage Interconnection Networks," *IEEE Transactions on Computers*, C-29(8), 1980, pp. 694–702.

CHAPTER 4

S. LENNART JOHNSSON
Yale University

Communication in Network Architectures

Acknowledgement

Most of the material presented here is drawn from joint publications with Ching-Tien Ho, IBM Research Center, and Abhiram Ranade, Department of Computer Science, U.C. Berkeley. The material was developed during their graduate studies. Their contributions to the understanding of optimal communication in highly concurrent systems are indeed significant.

Introduction

Supercomputers are expected to have a performance of at least 1 trillion instructions per second by 1995, and a primary storage of tens to hundreds of gigabytes (GB) [25]. At this rate of computation and memory size, the operation code, the operand addresses, and the operands require from 300 to 400 bits for a single instruction. The functional units must be supplied from 300 to 400 trillion bits per second, or about 16 million bits per cycle at a 25 MHz clock rate. This clock rate is somewhat conservative for MOS technologies, but system clock rates are not expected to become higher by more than a factor of 2 or 3. The width of the storage needs to be several million bits. Assuming each processor can deliver 25 Mflops/sec, a total of 40,000 processors is required for a system with a total capacity of a trillion floating-point instructions per second. A system of this complexity is entirely feasible to build. In half micron technology 40,000 chips with on-chip floating-point units and memory are projected to have a total of approximately 64 GB of primary storage. With the required storage bandwidth, and with tens of thousands of processing units, a network is the only feasible alternative for passing data between processors and storage units. Even if a technology that was an order of magnitude faster than MOS technologies were used, such as bipolar GaAs technology (used for the CRAY-3), thousands of processing units would still be required for an architecture with a performance of a trillion floating-point operations per second.

The determination of address maps and routing algorithms for the effective use of the communication system in (highly) parallel computers is critical in high performance computing systems based on current (packaging) technologies. The data transfer rate on a chip is one to two orders of magnitude higher than the transfer rate at the chip boundary. The transfer rate at the board boundary is typically within a factor of 5 of the rate at the chip boundary. Again, the rate at which data can be moved on a board is one to two orders of magnitude higher than the rate at the board boundary. With 500–1,000 pins on the board boundary, 100–200 boards, and a 25 MHz clock rate, the peak data motion rate across board boundaries is about 1 Tbit/sec, which is two to three orders of magnitude below the required rate for a system with a performance in the Tflop/sec range. A sustained performance of this magnitude with current packaging technologies is not possible without locality of reference.

In the highly concurrent architectures considered here, tens of thousands of operations can be performed concurrently. Architectures of this kind are often referred to as *data parallel* to emphasize massive parallelism and to distinguish the architectures from *control parallel* architectures, which usually offer a considerably lower degree of concurrency. In a data parallel programming model, algorithms are designed based on the structure and representation of the problem domain. It is considered to consist of sets of *elementary objects*, where objects in the same set are subject to the same transformations, at least most of the time. The objects in the same set can be operated on concurrently. Different sets of elementary objects are subject to different transformations, but they, too, may be operated on concurrently. The elementary object contains the state as well as the object description. An algorithm is expressed both as a sequence of transformations of the state of an elementary object, and interactions between elementary objects. For instance, in the finite element method, the physical domain is discretized by a set of finite elements. The solution on each element is approximated by a set of interpolation functions. The values of the interpolation functions at a number of nodal points is either predetermined to be zero or computed in the solution process. The number of nodal points per element is determined by the desired interpolation order. Two possible choices of elementary objects are apparent: finite elements and nodal points. For this particular application, the advantages and disadvantages of these two choices of elementary objects are evaluated in [69].

In the data parallel programming model, sets of elementary objects are represented by a higher level data type, such as the vector data type in some Fortran dialects or the array extensions of Fortran 90 [85], or arrays in APL, poly in C* [3], and parallel variables in *Lisp. The number of elementary objects that can be present in the primary storage is simply determined by the storage required by an object and the total storage. The elementary objects are often referred to as *virtual processors* [39, 117]. In general, several virtual processors are mapped to the storage of each physical processor. The number of virtual processors per physical processor is called the *virtual processor ratio* [117]. The storage of a physical processor is divided among as many virtual processors as is given by the virtual processor ratio. That many virtual processors timeshare a physical processor.

With higher level data types present, in data parallel languages operations on sets are also natural parts of the language, such as reduction, copy, parallel prefix, and certain permutations. With the data structure distributed across tens of thousands of storage modules interconnected by a network, the choice of *address map* and the management of the communication inherent in these operations are important for performance. Given the communications limitations of the technology, the communication may be the dominating factor in executing operations on sets distributed over a large number of storage units.

In order to provide some insight into the problems of, and techniques for, preserving locality by a judicious choice of address map, and effective communications management by carefully devised routing algorithms, we give some examples of typical supercomputer applications and the algorithms used therein. We classify these algorithms with respect to the data interaction, modeling the computations with a *computation graph*, where nodes represent data and where edges represent data interaction. Many of the basic algorithms have either regular computation graphs or graphs that can be regularized without loss of efficiency. There are also many computations that require irregular and dynamic computation graphs, however, such as event-driven simulation, the solution of partial differential equations in complicated geometries, or equations having widely varying solutions. For highly regular graphs, address maps and routing schemes for optimal communication under a variety of conditions are known for many types of networks. Much less is known regarding optimal address maps and communication for irregular computation graphs.

Key issues in determining the address map are the communication and the load balance it implies. In optimizing the address map with respect to communication, it is often desirable to preserve *locality* of reference. Each of the many possible measures of distance has its own merit. In solving partial differential equations, a distance measure of the form $(\sum_{i=0}^{d} |x_i - y_i|^p)^{\frac{1}{p}}$, where d is the dimensionality of the problem domain and p defines the type of the *norm*, is often very useful. The most common norm is the 2-norm, which is equal to the Euclidean distance between the two points x and y. Other common norms are the 1-norm and the ∞-norm. Different norms may be relevant in different representations of the problem. The 2-norm is equal to the Euclidean distance and is often used in the physical domain, but if x and y are processor addresses in a network configured as a Boolean n-cube, then if x_i and y_i are the coordinates (0 or 1) in the n-dimensional space, the 1-norm is equal to the *Hamming* distance between the two points. The Hamming distance is equal to the minimum number of communication links a data item must traverse to move from processor x to processor y in a Boolean cube network. The 1-norm is not ideal for all networks. In a completely interconnected network, all points are at unit distance from each other, and the 0-norm is a relevant distance measure. This chapter will focus on Boolean cube networks and use the 1-norm as a distance measure when data items are represented in the address space of the architecture. In the physical domain we assume the distance to be defined by the 2-norm, unless stated otherwise.

Mathematical models of physical phenomena are often based on local rules, whether the models are continuum models resulting in (partial) differential equations or discrete models for particle interaction, as for instance in the "N-body" problem and in cellular automata models of macroscopic phenomena. Continuum models need to be discretized before they can be used on a conventional computer. Techniques such as finite differences, finite element, and finite volume are all based on a set of local interdependencies. In so-called *explicit methods* for the solution of the discretized equations, these interdependencies completely define the spatial data interactions in a step of the algorithm, i.e., the communication requirements in the discretized physical domain. The computations exhibit locality in the physical domain. *Implicit methods*, on the other hand, require that a system of equations be solved in each step of the computation.

Iterative methods for the solution of linear systems of equations, such as Jacobi's method, successive overrelaxation (SOR), and symmetric successive overrelaxation (SSOR) all require only local communication in each step of the computation. The conjugate gradient method requires a global reduction operation for the computation of a scaling factor and a global copy, or broadcasting, operation for the distribution of this scaling factor in addition to the same local communication required in the Jacobi iteration. Though each step in the iterative methods involves only local communication in the physical domain, most problems require global communication to attain a correct solution. Elliptic problems are of this type [32].

The requirement for nonlocal, or global, communication is more apparent in direct methods. Factoring matrices by Gaussian elimination, or Householder transformations, can be performed as a sequence of rank-1 updates of the submatrix that remains to be factored. For a dense matrix, the pivot row is distributed to all the rows of the remaining submatrix and the pivot column to all the remaining columns. Global communication is required. For sparse matrices, the communication may be local. Most elements are zero in a sparse matrix. The number of non-zero elements is typically less than 10%, or even less than 1% of the total number of matrix elements. Sparse matrices often occur when a set of equilibrium equations is defined for a network or a discretized physical domain. The non-zero structure of the matrix reflects the topology of the network. For a sparse matrix, the pivot row needs to be distributed only to the rows that have non-zero entries in the pivot column. Similarly, the pivot column needs to be distributed only to the columns for which the pivot row has non-zero entries. The rank-1 update for a sparse matrix affects the nodes adjacent to the node being eliminated [90] and hence is local. In Gaussian elimination, one variable at a time is eliminated from the system of equations, and, depending on the topology of the graph, it may require only local communication throughout the elimination process, which then require a time proportional to the diameter of the graph. *Perfect elimination graphs* [26] have this property. Divide-and-conquer methods for solving linear systems of equations, such as odd-even cyclic reduction, nested dissection, and multi-grid methods perform a recursive subdivision of the physical domain. For some of the steps, these methods require communication between subdomains that are not adjacent.

In determining the address map for optimum performance, it is necessary to consider the communication implied by various functions on distributed data sets. The communication implies a selection of *paths* for the data and the specification of appropriate *schedules* for moving data along the paths. We refer to path selection and scheduling as *routing*. The traditional quantities considered in choosing the address map are the *dilation* and the *expansion* of embedding the *guest graph* in the *host graph*. In our case, the guest graph is the computation graph, and the host graph is a model of the network of processors with attached local storage units. The dilation is the maximum path length of an edge in the guest graph when embedded in the host graph. The expansion is the ratio between the number of nodes in the host and guest graphs. With path lengths greater than one, many host graphs allow for several possible paths between source and destination.

Finding the *shortest paths* for all messages is one criterion for a routing algorithm, but such a criterion does not necessarily imply completion of the communication in minimum time. Many paths may share the same network edge, and minimizing the *congestion* may be a more important criterion. For bit-serial, pipelined communication systems, the congestion is more important than path length. In many networks each node has a number of incoming and outgoing edges. Supporting concurrent communication on all the channels of a node requires many independent data paths in a node and may therefore be too expensive to realize. The *active degree* of a node defines how many of its channels are required to operate concurrently by a given routing algorithm. The *node load* determines the total number of messages a node is required to serve for a given routing algorithm. Optimizing the routing algorithm implies a trade-off among the various criteria. We will give estimates of the congestion, active-degree, and node load for some of the communication primitives we describe, and determine the expansion and dilation associated with the graph embeddings we present.

Frequently occurring operations on sets of data that involve communication are *reduction* and *copy*. The computation of inner-products and the application of a convolution kernel or difference stencil to a data set are examples of reduction operations. We have already mentioned rank-1 updates in connection with Gaussian elimination. The rank-1 update implies a copy operation. The reduction and copy operations can be modeled by spanning trees on the data/processor set they encompass. Another very useful communication primitive is shifting of data a given distance along some direction. The shifting may be periodic, as in Cannon's algorithm for matrix multiplication [17] and for some of the operations in our implementations of quantum electrodynamics and quantum chromodynamics computations [4]. The shifting operation is a special type of *permutation*: Each node sends and receives one data item. Permutations may be required between different phases of a computation, since different address maps may be optimum for the different phases. Examples are the conventional binary encoding of addresses and Gray code addresses frequently used in Boolean cubes. Conversion between different allocation schemes, described in more detail later, often consists of multiple permutations that we call *personalized communication*. Each processor sends a unique set of data to a set of other processors. Examples of permutations are matrix transposition and bit-reversal.

A convolution kernel is applied to a number of data points, and so are difference molecules. Tree-like communication for identical trees (one kernel) rooted at every processor is required. Each processor belongs to as many trees as there are nodes in the convolution kernel. A slightly different situation occurs if every processor broadcasts data to every other processor in a given set: *all-to-all broadcasting* [66]. Such communication is required in many types of finite element codes and also in N-body codes. (It is also required in the divide-and-conquer algorithm for computing eigenvalues of tridiagonal systems by Dongarra and Sorensen [27].) Personalized communication may also require concurrent communication in multiple trees involving the same set of processors, but rooted in different nodes. We will present algorithms for *all-to-all* communication in Boolean cubes.

When the number of elementary objects is much larger than the number of processors, several objects are mapped to the same processor. It is important to use the full bandwidth of the communication system. If the network provides several edge-disjoint paths between pairs of nodes, it is desirable to make effective use of this property. Communication schemes that make full use of the communication bandwith can be devised by finding multiple embeddings of the same communication structure between the same set of processors such that the set of communications links used by the different instances are disjoint. For instance, in a Boolean n-cube there exist n edge-disjoint spanning trees rooted at the same node, and hence the data set passed over any link is reduced by a factor of n compared to a single spanning tree used for broadcasting.

The concurrent embedding of different data sets in disjoint sets of processors can increase the utilization of the system. Working on independent data sets concurrently, in effect, is a software partitioning of the set of processors. We will give some examples of multiple, concurrent embeddings of similar communication structures.

For irregular and dynamic communication graphs, identifying and taking advantage of locality is a very difficult problem. One strategy is to take an approach that guarantees an acceptable worst-case behavior. Randomization of the routing or the address map may be used to reduce the likelihood of severe congestion in the communication system.

The efficient utilization of the communication system is often the dominating issue for problems with a high degree of regularity. For irregular problems, *load balance* is equally important. In solving small problems on highly parallel architectures, it is sometimes feasible to instantiate several copies of the problem and have each instance compute part of the solution. This technique is most efficient if the computational complexity is proportional to some power greater than 1 in the number of unknowns, such as matrix multiplication and $O(N^2)$ algorithms for the so called N-body problem. For large problems the number of virtual processors typically exceeds the number of physical processors. Two obvious choices for allocating virtual processors to physical processors are *cyclic* and *consecutive* assignment. In cyclic assignment, an object is assigned to a processor with the address computed as the object index modulo the number of processors. If there are 2^n processors, the n lowest order bits of the index of the elementary object determines the processor address. In a *consecutive* assignment, objects with successive indices are allocated to the same processor. The total number is determined by the ratio of the number of objects to the number of processors. If both the number of objects and processors is a power of 2, the highest order n bits of the object index determines the processor address. More complicated address allocations may be required for load balance [63]. The two forms of data allocation are illustrated below.

Cyclic assignment:

$$\underbrace{\left(w_{p-1}w_{p-2}\ldots w_k\right.}_{vp}\underbrace{\left.w_{k-1}w_{k-2}\ldots w_0\right)}_{rp}.$$

Consecutive assignment:

$$\underbrace{\left(w_{p-1}w_{p-2}\ldots w_{p-k}\right.}_{rp}\underbrace{\left.w_{p-k-1}w_{p-k-2}\ldots w_0\right)}_{vp}.$$

The outline of this chapter is as follows. In the next section we justify some of the communication functions discussed in detail in the main parts of the chapter by describing the communication in some typical applications in scientific computation. The importance of preserving locality is shown in section two. Interconnection networks are discussed briefly in section three. Section four presents the model of the address space used throughout this chapter, and defines the address maps that are being considered. Section five introduces some necessary graph theoretic notations and results. Lattice emulation is treated in section six, butterfly network emulation in section seven, and tree embeddings in section eight. The emulation of pyramid networks, and hyper-pyramids are treated in section nine. Lower bounds and algorithms for a restricted class of permutations, dimension permutations, are presented in section ten, and performance issues in network emulation in wafer-scale integration is treated in section eleven. The chapter is concluded with a consideration of the emulation of shared memory on distributed memory architectures.

4.1 Communication Requirements in Scientific Computation

Many scientific and engineering computations require global communication. For instance, in elliptic problems the solution in the interior depends on the boundary conditions and the forcing function at any point in the domain. The final value of any variable representing the state at an internal point depends on all other points [32]. Similar arguments can be made for other types of partial differential equations and problems in linear algebra. The global dependence can be resolved through a sequence of local interactions, through long-range communications, or through a mixture of the two. Note, however, that the distance measure in the problem domain is not necessarily the same as the distance measure in the network. We will now briefly describe a few applications in scientific computing, the algorithms used in these applications, and the type of data interaction required in the physical space. The problems we consider are from fluid and solid mechanics, acoustics, lattice-gauge physics, and linear algebra. We consider explicit and implicit methods for solving the problems, direct and iterative equation solvers for dense and sparse problems, and Fourier transforms. In each case we focus on the type of data motion required and the operation for which the data motion is needed. The purpose is to justify and define a set of communication primitives similar to the well-known computational primitives.

4.1.1 An Explicit Navier-Stokes Compressible Flow Solver

The Navier-Stokes equation describes the balance of mass, linear momentum and energy, and it models the turbulent phenomena that occur in viscous flow. In three dimensions the equations are of the form

$$\frac{\partial \mathbf{q}}{\partial \tau} = \frac{\partial \mathbf{F} + \mathbf{F}_\nu}{\partial \xi} + \frac{\partial \mathbf{G} + \mathbf{G}_\nu}{\partial \eta} + \frac{\partial \mathbf{H} + \mathbf{H}_\nu}{\partial \zeta}, \tag{4.1}$$

where the variable vector $\mathbf{q}(\xi, \eta, \zeta, \tau)$ has five components: one for density, three for the linear momentum in the three coordinate directions x, y and z, and one component for the total energy. The coordinates of the physical domain are x, y and z, whereas ξ, η and ζ are coordinates in the computational domain. $\mathbf{F}, \mathbf{G}$ and $\mathbf{H}$ are the flux vectors, and $\mathbf{F}_\nu, \mathbf{G}_\nu$ and $\mathbf{H}_\nu$ are the viscous flux vectors. The exact form of these functions is beyond the scope of this chapter. It is sufficient to say here that they are functions of the vector $\mathbf{q}$, the transformation between the physical and computational domains, and the derivatives of this transformation. For details see [89].

For regular computational domains, the solution to the Navier-Stokes equation can be approximated by discretizing the domain by a three-dimensional lattice. The spacing between nodes along the normal to boundaries in the form of solid walls often needs to be much smaller and closer to the boundary in order to compute the flow in the boundary layer with sufficient accuracy. A stretched grid is one way to accomplish this task. Such a grid is topologically equivalent to a regular grid, and any efficient embedding of such grids can be used advantageously. In solving the Navier-Stokes equations by a finite difference method, the partial derivatives are approximated by differences between computed values in neighboring lattice points. For a first order accurate approximation, values from two neighboring lattice points suffice. The higher the accuracy of the approximation, the larger the number of points involved in the approximation. A few typical stencils in three dimensions are shown in Figure 4.1. In the interior, symmetric stencils are often the correct approximation. However, on the boundary the stencils cannot be symmetric. If the same order of accuracy is required for the approximation of the boundary derivatives as in the interior, then the width of the boundary stencils is the same as the width of the stencils in the interior. In some cases, the model problem in [89], for example, the order of approximation on the boundary can be lower than in the interior without loss of accuracy. This property may allow the boundary stencils to be subgraphs of the stencils in the interior. The values associated with the edges of the subgraph may be different from the values associated with the corresponding edge of the graph for the complete stencil, however.

In addition to a need for difference stencils to be location-dependent for a given term of the partial differential equation, there is also a need for different stencils at the same location for different terms of the equation. In the case of the Navier-Stokes equation, artificial viscosity needs to be introduced to stabilize the explicit numeric method used in [89]. The artificial viscosity is introduced through a fourth order derivative, and the difference stencil for this term includes five points in each dimension, centered at the interior point for which the evaluation is desired. For the other terms of the Navier-Stokes equations, the stencils include only three points in each dimension. In the case of the Navier-Stokes equation, therefore, under very simple conditions two difference stencils are being used in each lattice point, and the stencils vary for interior points and points on or close to a boundary surface, edge, and corner. Given the directional dependence, there are one interior stencil, six face stencils, twelve edge stencils,

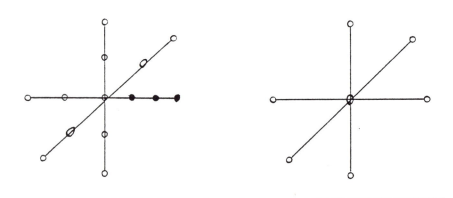

Figure 4.1   Samples of difference stencils.

and eight corner stencils of second order accuracy. The number of different types of stencils increases with the order of approximation. A central difference stencil with $2N + 1$ points in each of three dimensions gives rise to a total of $8N^3 + 12N^2 + 6N + 1$ stencils because of the boundaries. For the derivates $N = 1$ in our case, and for the artificial viscosity $N = 2$, and the total numbers of stencils are 27 and 125, respectively. Note that each difference stencil defines a combining operation on a set of points and that the stencils for all the points can be applied concurrently.

If an explicit method is used for the solution of the partial differential equation, then the derivatives are evaluated by applying the difference stencils associated with each of the points in the lattice; and the integration in time performed, by an explicit method, such as a Runge-Kutta method.

4.1.2   Stress Analysis by the Finite Element Method

In the finite element method [126], the solution to a set of partial differential equations in a domain is approximated by polynomial solutions over subdomains. The subdomains are called elements. The most typical shapes of the elements in two dimensions are triangles and rectangles. In three dimensions, brick, prism, and tetrahedral elements are common. The order of the element determines the order of the polynomial approximation. In order to specify the polynomial, it is necessary to specify either its value or the values of its derivatives in a number of points consistent with its order. These points are the nodal points of the elements. The solution is computed for the nodal points, each of which has a polynomial associated with it. The value of a polynomial is 1 at the nodal point with which it is associated and 0 at all other nodal points. The desired solution is expressed as a linear combination of the polynomials. The equations for the displacements of the nodal points as a function of the applied forces require that a set of equations involving a *stiffness matrix* be solved. It relates the displacements to the forces. Evaluation of the elements of the stiffness matrix requires that products of the polynomials be integrated over the element. The numerical integration, or *quadrature*, is performed by evaluating the product at a number of locations on the element and computing a weighted sum of these values. The locations and weighting factors depend on the particular quadrature formula being used. The computations can be organized such that the computations involve an *all-to-all reduction*, or they may be completely local [69].

The nodal points on the elements are often determined by Lagrange's formula in each dimension, resulting in a regular sublattice for brick elements. The elemental stiffness matrices are dense. For the solution of the equilibrium equations, the elemental stiffness matrices are often *assembled* into a global stiffness matrix by introducing a global node ordering. For the elemental stiffness matrices, an ordering local to an element will suffice.

If the matrix is assembled, then every node couples to every node on every element that shares the node. Interior nodes couple to all other nodes on the element, face nodes to all nodes on two elements, edge nodes to every node on four elements, and corner nodes to every node on eight elements. A pth order Lagrange element has $p+1$ nodes in each dimension. In three dimensions, four types of stencils describe the coupling between nodes. The stencils involve $(p+1)^3$, $(p+1)^2(2p+1)$, $(p+1)(2p+1)^2$, and $(2p+1)^3$ points, respectively. For second order elements, the numbers are 27, 45, 75, and 125. The stencils are considerably more complex than in a finite difference method of the same order. Figure 4.2 illustrates the stencils in two dimensions.

Solving the equilibrium equations is traditionally done by a direct method, but the use of iterative solvers is increasing, in particular as three dimensional problems are being addressed. A direct method requires a global ordering of the nodal points. It also requires that an elimination order be established, i.e., an order in which variables are eliminated from the system of equations. The ordering normally results in a sparse stiffness matrix. Elimination in most cases implies fill-in of the matrix. If Gaussian elimination is used, communication in the form of a mesh is required. If a divide-and-conquer scheme like nested dissection is used, then communication as in odd-even cyclic reduction is required. For an iterative method, communication as defined by the finite element discretization and the order of the elements is required. For the conjugate gradient method, a global reduction operation is also required.

If finite elements are chosen as the elementary objects in a data parallel implementation, then the concurrency is limited to the number of finite elements. With nodal points selected as the elementary objects, the degree of concurrency is typically one to two orders of magnitude higher than if finite elements are the elementary objects. In the latter case, interprocessor communication is equivalent to finite element interactions. All operations on an element are local storage references. With nodal points being elementary objects, intra-element interactions also become interprocessor communication.

In conclusion, lattice-like communication may be required for finite element discretization of regular domains. The communication may be as defined by the elements and their order for an iterative method, or by a two-dimensional mesh as defined by a total order and Gaussian elimination (even for three-dimensional problems), or by a combination of tree and mesh structures as for a nested dissection solver. Some of the communication may take the form of *all-to-all* communication within disjoint subsets of processors.

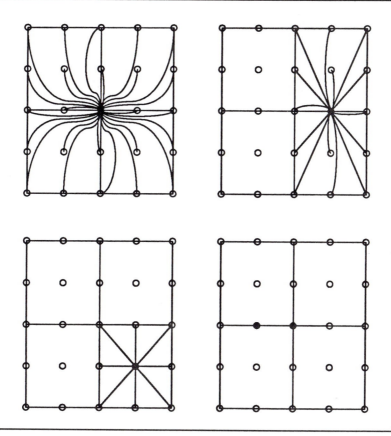

Figure 4.2   Second order stencils for rectangular finite elements.

4.1.3   Acoustic Field Computation by an Alternating Direction Method

The forward propagation of acoustic waves by the so-called Wide Angle Wave Equation [75] implies the solution of an equation of the form

$$(1+\frac{1}{4}(1-\delta)X)(1-\frac{1}{4}Y)u(r+\Delta r) = (1+\frac{1}{4}(1+\delta)X)(1+\frac{1}{4}Y)u(r), \quad (4.2)$$

where k_0 is a reference wave number, and $n(r,\theta,z) = k(r,\theta,z)/k_0$ $\delta = ik_0\Delta r$, and

$$X = \frac{1}{k_0^2}\frac{\partial^2}{\partial z^2} + (n^2(r,\theta,z)-1), \quad \text{and} \quad Y = \frac{1}{k_0^2 r^2}\frac{\partial^2}{\partial\theta^2}.$$

This equation is a parabolic approximation of Helmholtz equation. The solution to equation (4.2) can be marched out in the range (r) direction with an Alternating Direction Method [101, 70]. Tridiagonal matrix-vector multiplications are performed in the θ and z directions, followed by the solution of tridiagonal systems in the same directions. Both operations consist of a number of one-dimensional problems that can be solved independently and concurrently. In addition, each system can be solved concurrently by substructuring, by pipelined Gaussian elimination, by partial or complete transposition of equations, by odd-even cyclic reduction, or by any combination of these methods [61] (which for multiple systems may be performed as balanced cyclic reduction). The communication pattern (in one dimension) of odd-even cyclic reduction is given in Figure 4.3, and of balanced cyclic reduction in Figure 4.4. The communication topology of balanced cyclic reduction is known as a *data manipulator* network or a *PM2I* [108] network. Hence, in the case of the underwater acoustics problem, communication as defined by the difference stencil is required for matrix-vector multiplication, but for the solution of the systems of tridiagonal systems of equations, the communication depends on the selected algorithm: For pipelined Gaussian elimination, communication in the form of a Hamiltonian path is required; for equation transposition the communication is equivalent to *all-to-all personalized communication* (or *all-to-some some-to-all* personal communication), which can be performed through butterfly network communication [63, 66]; for balanced cyclic reduction, communication is required in the form of a data manipulator network. The communication requirements for odd-even cyclic reduction is a subtree of the data manipulator graph with the root at the top center and the leaf nodes being all nodes at the bottom level.

4.1.4   Lattice-Gauge Physics

A popular technique for investigating gauge theories is to discretize them onto a lattice and simulate numerically by a computer, yielding so-called lattice-gauge theories. Currently the most important gauge theory to be solved is that describing the subnuclear world of high energy physics: Quantum Chromodynamics (QCD). A simple and well-known example of a gauge theory is Quantum Electrodynamics (QED), the theory that describes the interaction of electrons and photons. Both quantum electrodynamics and quantum chromodynamics computations are based on four-dimensional lattices. In constructing a lattice-gauge theory, the gauge symmetry must be kept explicit in the lattice formulation so that in the continuum limit (when the lattice spacing tends to zero) the original gauge theory is recovered. An elegant formulation is the original one by Wilson [125], in which the action S for the gauge fields U is local, involving only the product of the gauge

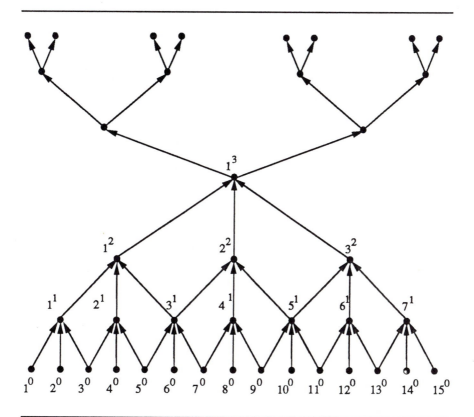

Figure 4.3   The communication topology for odd-even cyclic reduction.

fields around elementary squares, called plaquettes, on the lattice. The action S is expressed as

$$S(U) = \beta E(U) = \beta \sum_p \left(1 - \frac{1}{N} ReTrU_p \right)$$

with

$$U_p = U_\mu(n)U_\nu(n + \hat{\mu})U_\mu^\dagger(n + \hat{\nu})U_\nu^\dagger(n),$$

and is illustrated in Figure 4.5.

The gauge fields $U_\mu(n)$ are elements of the gauge group of the theory being studied, $U(1)$ for QED (the abelian group), $SU(3)$ for QCD. $U(1)$ is represented by a scalar and $SU(3)$ by a 3×3 complex matrix. $U(1)$ and $SU(3)$ are associated with links on the four-dimensional lattice joining sites n and $n + \hat{\mu}$, where $\hat{\mu}$ is a unit lattice vector in the μ-direction. $U_\mu(n)$ is a

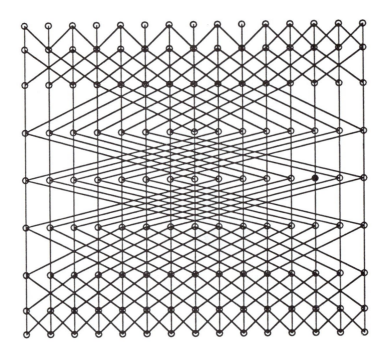

Figure 4.4   The communication topology of balanced cyclic reduction.

directed variable: In going from $n + \hat{\mu}$ to n, we use $U_{\mu}^{\dagger}(n)$. The parameter β determines the interaction strength or "inverse temperature" of the theory. The constant N is the dimensionality of the group elements (1 for QED, 3 for QCD).

Two classes of algorithms dominate the simulation of the lattice-gauge theories: stochastic and deterministic. The most popular stochastic algorithms are based on Metropolis algorithm [86]. The Monte Carlo algorithm changes the energy of a system while keeping its temperature constant, whereas the microcanonical algorithm conserves the total energy while allowing temperature to vary. The two approaches may be combined, as in [4], where a Monte Carlo method is used to bring the lattice-gauge theory into equilibrium at a specified temperature (coupling), and then a microcanonical algorithm is used to evolve the system for measurements of its properties. The microcanonical algorithm is computationally less demanding. During the latter phase it may be required to switch back periodically to a Monte Carlo algorithm in order to obtain ergodicity [28].

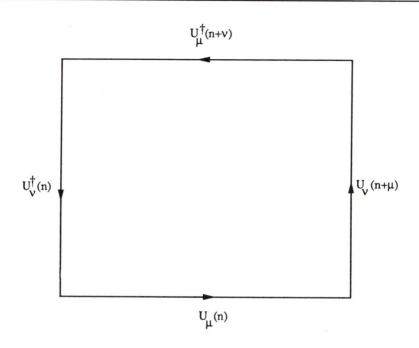

Figure 4.5   The plaquette calculations.

Monte Carlo algorithms [86] cycle through all the gauge field links of the lattice, changing their values by a random procedure until they settle down into physically correct configurations, C. These are such that, when statistical equilibrium is reached, the probability of finding any one of them is proportional to its Boltzmann factor $e^{-S(C)}$, where S is the action of the gauge theory. A sufficient condition for statistical equilibrium to be attained is that, at each step of the Monte Carlo algorithm, the probability of changing a configuration C into a new one C' is the same as the probability of changing C' back to C. This state is called *detailed balance*. It has important consequences for parallel computer implementations, namely, that in order to preserve detailed balance, one cannot simultaneously update gauge field links that interact. As the action involves interactions around plaquettes, one can therefore update only half the links in any one dimension simultaneously and preserve detailed balance, as shown in Figure 4.6, in two dimensions for simplicity.

On a parallel computer, full processor utilization is obtained by observing that there are two plaquettes to be calculated for each dimension and link update and scheduling half of the processors to calculate the "positive plaquettes" and half to calculate the "negative plaquettes".

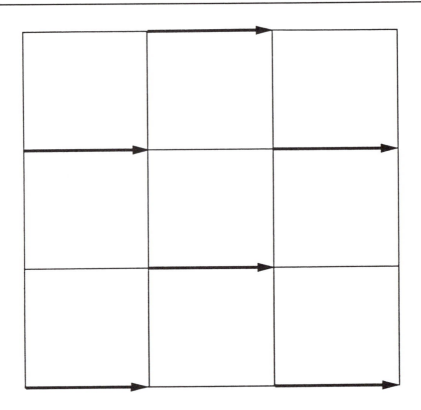

Figure 4.6   Link updates in parallel Monte Carlo lattice gauge theories.

The Metropolis algorithm proceeds computationally as follows. In order to change a configuration C into another C', the change in action (β times the change in energy) is computed: $\Delta S = S(C') - S(C)$. If $\Delta S \leq 0$, the change is accepted and C replaced with C'; otherwise, if $\Delta S > 0$, the new configuration is accepted with the probability $e^{-\Delta S}$. In practice this is done by generating a pseudorandom number r in the interval $(0, 1]$ with uniform distribution. If $r < e^{-\Delta S}$, the change is accepted; otherwise it is rejected.

Monte Carlo algorithms move through configuration space stochastically, whereas the microcanonical algorithm moves deterministically. Here, one enlarges the configuration space by adding fictitious momenta canonical to the degrees of freedom one wants to study, and then solves Newton's equations [16, 94]. Computationally this means solving coupled partial differential equations using difference approximations. The gauge theory action $S = \beta E$ is interpreted as β times the potential energy for a classical dynamics governed by Newton's law: $\ddot{U}(t) = -E(U)$. Then the canonical momentum $\Pi = \dot{U}$ and the Hamiltonian $H(\Pi, U) = \Pi^2/2 + E(U)$ are introduced. From the resulting kinetic energy term T, one obtains the temperature of the system as $\beta^{-1} = 2T/N_{indep}$, where N_{indep} is the number of linearly independent variables among the N gauge field links U. $N_{indep} < N$ because of gauge invariance in the theory.

The edges of the four-dimensional lattice are directed, and the values associated with the edges, or bonds, can be stored at, for instance, the node at the tail end of the edge. The main quantity being computed for each bond in a step of the computation is its contribution to the total *action*, which involves all bond values of the plaquettes of which the given bond is a part. A bond is part of six plaquettes in four dimensions. With the bond values stored at the tail end of the directed edges, the stencil defining the communication in any plane is given in Figure 4.7.

4.1.5   Linear Algebra

We have already mentioned several techniques in numerical linear algebra, such as Gaussian elimination, nested dissection, odd-even cyclic reduction, matrix transposition, balanced cyclic reduction, and iterative techniques like the conjugate gradient method. We have also mentioned that two-dimensional mesh structures often fit direct methods well and that part of the communication in iterative methods is defined by the difference stencils, or the finite element (local) approximation. Before summarizing the set of communication primitives introduced through the above examples, we discuss two more basic methods of linear algebra: the multigrid method and a divide-and-conquer method for computing eigenvalues of tridiagonal systems.

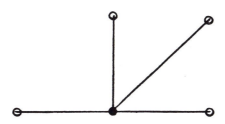

Figure 4.7   Communication stencil in a plane for quantum electrodynamics and chromodynamics computations.

The multigrid method [14] is a technique for solving partial differential equations. It is based on a sequence of grid refinements, with interpolation from a grid to the points of the one-level refined grid, and a smoothing operation in going from a grid to the next level coarser grid. The idea is that slow variations are resolved on coarse grids and fast variations on fine grids, and hence the communication in each dimension is similar to the case of odd-even cyclic reduction in that every other grid point is excluded in moving to the next coarser grid (and the communication distance doubles in the physical domain). With smoothing by relaxation and a five-point stencil, the communication required for this operation in the multigrid computation is indeed the same as for odd-even cyclic reduction. It is also possible to formulate the multigrid algorithm so that instead of a single coarser grid, multiple coarse grids are considered with disjoint points, and all grid points are considered at every level, but the points are part of different subproblems. The communication in each dimension is similar to the communication in balanced cyclic reduction.

Dongarra and Sorensen [27] have suggested a parallel algorithm for computing eigenvalues of tridiagonal systems by a divide-and-conquer method that is fully parallel. The algorithm proceeds by tearing the tridiagonal system into two smaller tridiagonal systems of approximately equal size, recursively. The computations starts from the bottom level of the recursion by computing eigenvalues for a large number of small tridiagonal systems. Then the tridiagonal systems are paired in the next step in such a way that eigenvalues are computed on half as many systems of approximately twice the size. The eigenvalues are computed as the roots of equations of the type

$$1 + \rho \sum_{j=1}^{n} \frac{\zeta_j^2}{\delta_j - \lambda} = 0,$$

where n is the number of equations in a system. With the components of ζ and δ distributed, the computations can be organized either with reduction and copy operations within segments representing the independent systems or by an all-to-all broadcasting and concurrent computation in every processor at every step of the computation. The data structure for the problem is preferably organized as a dynamic two-dimensional array of elementary objects. During the course of computation, the array is reshaped from an array with few rows and many columns to an array with a single column [109].

4.1.6 Permutations

Familiar permutations in addition to cyclic shifts are matrix transposition and bit-reversal. Bit-reversal and matrix transposition for matrices with the rows and columns being powers of two are examples of a class of permutations called dimension permutations. They can be defined as permutations on the bits of the address field rather than as a permutation on the addresses themselves. Conversion between the cyclic and consecutive storage schemes and many other storage schemes are also dimension permutations, and so are shuffle permutations. We will give lower bounds and optimal algorithms for dimension permutations on Boolean cube networks.

4.1.7   Summary

In summary, in the above examples for the computation of approximations to solutions of partial differential equations describing the fluid flow, the stresses and displacements in a solid under pressure, the acoustic field, or the internals of matter, the discretization of the domain was defined by a two-dimensional lattice for the acoustics problem, by a three-dimensional stretched grid for the Navier-Stokes equations, by a deformed three-dimensional lattice for the stress analysis problem, and by a four-dimensional lattice for the lattice-gauge physics problem. The required communication for each processor was defined by the difference stencils in the Navier-Stokes equations solved by an explicit method, and for the right-hand side (matrix-vector multiplication) in the acoustics problem. Solving the equilibrium equations by a finite element method and an iterative solver also implies communication according to a local stencil (defined by the elements). The lattice-gauge physics examples again used only local communication that could be modeled by a fairly simple stencil (11-points). In this example a single stencil suffices for the entire domain, in part because of periodic boundary conditions. Other boundary conditions may cause a proliferation in the number of stencils required, as was the case in the Navier-Stokes example, and not all interior points may be equivalent, as is the case in the finite element method with assembled elementary stiffness matrices. The stencils define combining operations in the form of +-reduction on weighted variables.

The direct equation solvers require long-range communication in the physical space. If the problem is of full rank, then global communication is required for the solution [32]. The iterative solvers accomplish this in the iterative process. The examples we gave all used a hierarchy of grids. The balanced cyclic reduction algorithm resulted in a *PM2I* network, and the odd-even cyclic reduction in a subtree (for the reduction) of this network. Conventional multigrid algorithms have a structure similar to odd-even cyclic reduction, and the super-convergent multigrid algorithm has a structure similar to the balanced cyclic reduction algorithm.

Algorithms making use of transposition of systems of tridiagonal equations, as might be used in the acoustics problem, may use communication in the form of butterfly networks, the ideal communications network for

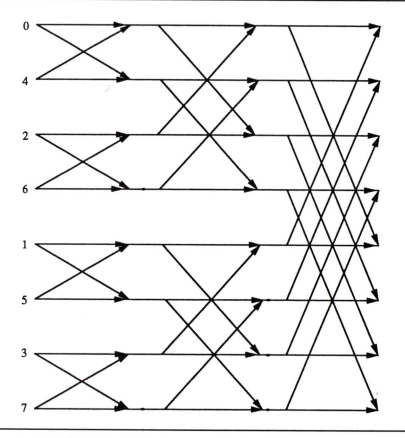

Figure 4.8   A butterfly network on 8 nodes.

the Fast Fourier Transform. Many other functions, such as sorting and permutations, can also be performed well on butterfly networks. Figure 4.8 shows a butterfly network.

In the following we consider communication in the form of

- Emulation of lattices of arbitrary dimensionality and shape
- Emulation of butterfly networks
- Emulation of *PM2I* networks, and subtrees thereof
- Emulation of pyramid networks
- Spanning trees, spanning graphs, and multi-graphs thereof
- Dimension permutation

4.2   The Value of Locality in Computation

The previous section presented a few applications for which simple discretizations of the physical domain were appropriate and defined the communication needs of the algorithms used to find the solution of a set of discretized equations. In this section we first give some design parameters for a 1 trillion operations per second supercomputer based on MOS technology. Then show the importance of exploiting locality for a sustained performance close to the peak performance.

Assuming a 25 MHz clock rate and one operation per cycle for each functional unit, 1 trillion operations per second requires 40,000 units. With 64-bit RISC-like processors, the area requirement for each processor is estimated to 30 Mλ^2 [37, 38, 72, 81, 106, 107]. For the floating-point unit, the area estimate is 100 Mλ^2, based on the area requirements of the single-chip units that have appeared during the late 1980s. It is highly desirable to have separate communication circuitry for routing and management of buffers. A preliminary design of the routing circuitry for the Fluent supercomputer [99] suggests that 30 Mλ^2 is a realistic estimate. The area estimates are expressed in terms of λ units. The minimum feature size of the technology is 2λ [83]. With a system with a performance of about 1 Tflop/sec, we assume a primary storage in the range of 64–128 GB. To address every location in the primary storage requires up to 40 bits for bit addressing and 37 bits for byte addressing. With 64-bit operands, a three operand instruction requires 300–400 bits for operands, addresses, and the operation code. With one instruction per cycle it is necessary that a functional unit and a substantial amount of storage reside on the same chip such that most of the required information resides on-chip. Otherwise, the limitations of the current packaging technologies will degrade the performance. With 16 Mbit/chip and 40,000 chips, the total primary storage becomes 80 GB. Assuming 100 λ^2 per bit [106], the estimate for the total chip area is 1800 Mλ^2, which in 0.5 μm technology means a chip of size 10 × 10 mm. The total processing capacity and memory capacity can be furnished by a number of chips consistent with reliably working systems.

With 10^3–10^4 channels per side of a chip, the total data motion capacity of the 40,000 chips is 100–1,000 TB/sec without sharing of on-chip channels between different data paths. But, assuming current standard packaging technologies of 100–300 pins per chip, the data motion capacity at the chip boundary is about 10 TB/sec. At the board boundary, assuming about 500 pins, the data motion capacity for a 200 board system is about 0.16 TB/sec. The data transfer rate at the chip boundary is at least two orders of magnitude less than on the chip, and the transfer rate at the board boundary yet another two orders of magnitude less than at the chip boundary. As pointed out earlier, the data delivery rate required to sustain a Tflop/sec processing rate is about 40 TB/sec. With complete locality of reference, therefore, the technology has the capacity to support the processing rate, but with no locality of reference the data motion capacity falls short by about three orders of magnitude.

In order to assess the value of exploiting locality, we consider three often used computations: matrix multiplication, nearest neighbor communication in three-dimensional grids, as in 3-D relaxation, and butterfly based computations, as in computing Fast Fourier Transforms and bitonic sort. In 3-D relaxation on a regular lattice with k variables per lattice point and 2 operations per variable, the ratio of the number of operations per remote reference is $\frac{1}{2d}(\frac{M}{k})^{\frac{1}{d}}$. For $d = 3$ the number of operations per remote reference is $\frac{1}{6}(\frac{M}{k})^{\frac{1}{3}}$. In Tables 4.1 and 4.3, $k = 8$. For a Navier-Stokes code, a more realistic value of k is 100–150 [89]. If the variables form matrices, then the number of arithmetic operations per variable is higher. Several linear algebra operations, including finite difference operators and iterative equation solvers for partial differential equations, have a ratio of operations to remote references that follows the rule $\frac{1}{\alpha}(\frac{M}{\beta})^{\frac{1}{\gamma}}$ for suitable values of α, β and γ. For butterfly based algorithms such as FFT and sorting, the dependence is of the form $\alpha \log(\frac{M}{\beta})$. For the FFT, the ratio is $1.25 log_2(M/2)$ real operations per remote reference using a radix-M algorithm, which is optimum [49].

Table 4.2 gives the number of bits that have to cross the chip, board, and system boundaries during a single cycle for the computations in Table 4.3, assuming either the optimum locality and no locality of reference. It is assumed that each chip has one processing unit, and that a board has 256 processing units.

Exploiting locality reduces the required communication bandwidth by a factor of up to 300 at the chip boundary for these computations, a factor of up to $\sim$ 7500 at the board level, and a factor of at least 160 at the I/O interface. The value of exploiting locality is apparent, but the techniques for accomplishing this task are not.

Table 4.1   Average number of operations per remote reference of a single variable.

Computation	Registers only	4 Mbit chips	256 4 Mbit chips (board)	256 boards
Mtx mpy	0.5	104	1600	26000
3-D Relaxation	0.17	4.27	26.7	170.7
FFT	1	18.8	28.8	38.8

Table 4.2   Number of bits across the chip/board/system boundary per cycle.

comp	4 Mbit, 1 proc. 1 chip	256 procs. Board	256 boards Machine
Mtx mpy	1	10	160
3-D Relaxation	32	480	24600
FFT	3	1140	160000
no locality	300	76800	19660800

4.3   Networks

The choice of network for interconnecting processors is a trade-off between several factors, such as the ability to accommodate different communication needs, area or volume, maximum wire length, pin requirements at technology boundaries such as chip and board boundaries, and fault tolerance. Long wires require a large amount of power to drive the signals from one end to the other at a high speed. The on-chip wires often determine the clock rate of many designs in MOS technology. Wire lengths are becomming more critical as the minimum feature sizes of the technology continue to decrease. Wires also affect the performance adversely by requiring a large area. Most chip designs are wire limited. More devices could fit on many chips, but there is not enough space to interconnect them. Devices typically occupy less than 5% of the total area and functional blocks (including internal wiring) less than 50% of the total area.

For MOS technologies two models for wire delays are proposed. The most frequently used model is the capacitive model, but it is likely that the wire resistance must be accounted for as feature sizes move into the submicron range. In a purely capacitive model, the wire delay is reduced in proportion to the scaling factor with which device features are reduced, if all dimensions are scaled, i.e., width, length, and thickness. However, if the length of the wire is not scaled, as for instance might be the case for a bus interconnecting, more devices as the feature sizes are reduced. The wire delay remains constant. Reducing wire width and thickness does not affect the delay in the capacitive model. Optimizing the driver for minimum delay, assuming minimum feature size logic on the input side, yields a delay τ_w proportional to the switching time τ of the technology. The wire delay is also proportional to the logarithm of the length L of the wire [83]. The effect of scaling the design by a factor α is shown in Table 4.3. Accounting

Table 4.3   Scaling of wire delays in MOS technology with optimized drivers.

Model	Opt. Delay	$L \to \dfrac{L}{\alpha}$	
	τ_w	τ_w	$\dfrac{\tau_w}{\tau}$
Capacitive	$const \cdot \tau \cdot log_e(const \cdot L)$	$\tau_w \to \tau_w/\alpha$	$\dfrac{\tau_w}{\tau} \to \dfrac{\tau_w}{\tau}$
Resistive	$const \cdot \sqrt{\tau L}$	$\tau_w \to \tau_w/\sqrt{\alpha}$	$\dfrac{\tau_w}{\tau} \to \dfrac{\tau_w}{\tau}\sqrt{\alpha}$

Model	Opt. Delay	$L \to L$	
	τ_w	τ_w	$\dfrac{\tau_w}{\tau}$
Capacitive	$const \cdot \tau \cdot log_e(const \cdot L)$	$\tau_w \to \tau_w/\alpha\left(1 + \dfrac{log_e\alpha}{log_e(const \cdot L)}\right)$	$\dfrac{\tau_w}{\tau} \to \dfrac{\tau_w}{\tau}\left(1 + \dfrac{log_e\alpha}{log_e(const \cdot L)}\right)$
Resistive	$const \cdot \sqrt{\tau L}$	$\tau_w \to \tau_w\sqrt{\alpha}$	$\dfrac{\tau_w}{\tau} \to \dfrac{\tau_w}{\tau}\alpha^{\frac{3}{2}}$

for wire resistance increases the wire delay and degrades its behavior under scaling. The length of the clock cycle may have to be increased.

Several networks have been proposed for multiprocessor systems, with two-dimensional meshes, butterfly, Boolean, and Cube Connected Cycles networks being the most common. The area requirements, maximum wire lengths, and partitioning properties of some networks are indicated in Tables 4.4 and 4.5.

Configuring processors as linear arrays and complete binary trees requires a total number of interconnections equal to the number of processors. Both configurations scale in an excellent way with increasing ensemble size and decreasing feature size with respect to area, wire length, and partitioning. With several processors on a single chip, the required bandwidth at the chip boundary only grows at the rate of the clock frequency, regardless of the number of processors per chip and the size of the machine being built [80]. The tree has the advantage over the linear array that its diameter (the distance between the processors that are farthest apart) is $2(log_2 N - 1)$ compared to N (or $\frac{1}{2}N$ for an array with end-around connections). The diameter of the network topology defines a lower bound for the speed of many types of computations [32].

The area requirements for networks in VLSI technology are mostly based on the Thompson grid model [115, 116] for (planar) layout. In this model, processing elements can be located only at the intersections of vertical and horizontal tracks, i.e., at the grid points. Tracks are spaced at unit distance.

Table 4.4   Topological properties of some common networks.

Configuration	Nodes	Diam	Fan-out	Edges
Linear Array	2^k	2^{k-1}	2	$2^k - 1$
2-d mesh	2^k	$2(2^{k/2} - 1)$	4	$2(2^k - 2^{k/2})$
Binary tree	$2^k - 1$	$2(k-1)$	3(1)	$2^k - 2$
Boolean cube	2^k	k	k	$k \cdot 2^{k-1}$
CCC	$k2^k$	$2k - 1$	3	$3k \cdot 2^{k-1}$
Shuffle-exchange	2^k	$2k - 1$	≤ 3	$\approx (1.5)2^k$

Table 4.5   Layout properties of some common networks.

Configuration	Nodes	Edge len.	Area	Pin Count
Linear Array	2^k	$O(1)$	$O(2^k)$	2
2-d mesh	2^k	$O(1)$	$O(2^k)$	$4\sqrt{M}$
Binary tree	$2^k - 1$	$O(2^{k/2}/k)$	$O(2^k)$ $-O(2^k \cdot k)$	4
Boolean cube	2^k	$O(2^{\frac{k}{2}})$	$O(2^{2k})$	$M(k - \log M)$
CCC	$k2^k$	$O(2^{\frac{k}{2}})$	$O(k^2 2^{2k})$	$M - (\frac{M}{k} \log_2 \frac{M}{k})$
Shuffle-exchange	2^k	$O(2^k/k)$	$O(2^{2k}/k^2)$	

Processor interconnections run along the tracks, with the restriction that only one processor interconnection can be allocated to any track between a pair of grid points. Processor interconnections may cross at grid points. The required area for the complete binary tree is of order $O(N)$ [83] if the nodes can be placed arbitrarily in the plane, in which case the maximum wire length is $\frac{\sqrt{N}}{\log N}$ [91, 9]. Placing all the leaf nodes of the complete binary tree along the boundary yields an area requirement of order $O(N \log_2 N)$ [15] and a maximum wire length of order $O(\frac{N}{\log N})$.

Linear arrays and complete binary trees have small bandwidth and present communication bottlenecks for many important computations. The two-dimensional mesh and mesh of trees [77] offer higher bandwidth and are preferable for many matrix computations. The first two networks can be realized in a small area on a wafer ($O(N)$ for the N node mesh and $O(\frac{N}{\log^2 N})$ for the N node mesh of trees) with wire lengths $O(1)$ for the mesh and $O(\frac{\sqrt{N} \log N}{\log \log N})$ for the mesh of trees. The advantage of the mesh of trees over the mesh is its logarithmic diameter ($2\log N$ compared with $2\sqrt{N}$ for the mesh).

Other networks with a diameter of $O(\log_2 N)$ are the shuffle-exchange, the Ω-network, the Cube Connected Cycles [95] network, and Boolean n-cube networks. All these networks require layout area almost quadratic in the number of nodes and wire lengths that grow almost linearly with the number of nodes. Correspondingly, the cost per communication is high. Currently, a 64-input 1-bit wide Ω-network with simple switching elements, or two 32-input Batcher sorting networks, fits on a single chip [73]. Large systems must be partitioned across many chips and boards. Table 4.5 gives the number of channels crossing a unit boundary for the different networks, assuming M processors per unit.

We will focus on Boolean cube networks, in part because of their versatility and their current popularity as interconnection networks in parallel systems with more than 100 processors, such as the Intel iPSC, the NCUBE, and the Connection Machine system.

4.4   Address Maps

We model the communication problem as a graph embedding problem. The task of emulating the communication specified by one graph, the *guest graph*, on the Boolean n-cube, the *host graph*, is equivalent to embedding the guest graph in the n-cube. The nodes of the guest graph are assumed to represent a unit of computation, and the edges correspond to a unit of communication. We associate a fixed cost to communication across an edge in the Boolean n-cube. We assume that the time for a local memory reference is negligible compared to the interprocessor communication time. Clearly the communication time is minimized if all the nodes of the guest graph are assigned to the same node in the host graph. But the computation is sequential. To balance the load we assume that each node of the guest graph is assigned to a unique node in the host graph. To perform the communication as defined by the guest graph, it is desirable that nodes that are adjacent in the guest graph also are adjacent when embedded in the host graph. Such adjacency cannot always be preserved, however. For instance, a Boolean 10-cube cannot be embedded in a two-dimensional mesh with adjacency preserved.

The *edge-dilation* is the length of the path that represents a guest graph edge in the host graph. The path is usually chosen to be of minimum length (a longer path may reduce congestion) or the load on certain nodes. The *dilation* of the embedding is the maximum of all edge-dilations. The ratio between the number of nodes in the host graph and the guest graph is the *expansion* of the embedding. A trade-off often exists between the expansion and the dilation of an embedding, as will be shown in the embedding of lattices in Boolean cubes. In addition to dilation and expansion, it is often important to consider *edge-congestion*, i.e., the number of guest graph edges that share a host graph edge when mapped to the host graph, and the maximum congestion for any edge in the host graph. We refer to the maximum edge-congestion as the *congestion* of the embedding. Designing processors so that they can support concurrent communication on multiple channels implies many independent data paths in each node, which requires significant hardware resources. The *active-node-degree* is a measure of the number of communication paths going through a given node. The *active degree* of an embedding is simply the maximum active-node-degree. Finally, we also consider the maximum number of messages any node in the host graph has to service for a given embedding, the *node load*. These measures of an embedding are precisely defined in the next section.

If several paths exist between a pair of processors, as is the case for the Boolean n-cube, then this feature can be used to increase the effective bandwidth between processors by considering different embeddings of the guest graph in the host graph. The different embeddings share the node set, but should have disjoint edge sets, ideally. Embeddings using disjoint sets of host graph edges provide a mechanism for fault tolerance with respect to communication links and a means for fully using the communications bandwidth at a high virtual processor ratio.

Also important is the simultaneous embedding of several guest graphs into disjoint subgraphs of the host graph, a packing problem that corresponds to a partitioning of the machine.

One desirable feature of a highly parallel computing system is that all functions can be distributed and the amount of centralized control kept minimal. All communication algorithms presented in this chapter require only local control.

The task of finding communication-efficient embeddings of the guest graph in the host graph consists of three parts: finding an address map, selecting routing paths, and scheduling messages along the paths.

There are 2^n nodes in a Boolean n-cube. Each dimension has two coordinate points. The nodes can be given addresses such that the addresses of adjacent nodes differ in precisely 1 bit. The Boolean cube is a recursive structure. An n-cube can be extended to an $n + 1$-cube by taking two n-cubes and connecting corresponding vertices of the two cubes. All nodes of one of the cubes have the highest order bit equal to 0, and all nodes of the other cube have the highest order bit equal to 1. The recursive nature of the Boolean cube is illustrated in Figure 4.9.

Each node has n neighbors, and the maximum distance between an arbitrary pair of nodes is n. The total number of internode connections is $n2^{n-1}$. Between any pair of nodes there exist n edge-disjoint paths. For nodes at a distance k there are k paths of length k, and $n - k$ paths of length $k + 2$. There are $\binom{n}{k}$ nodes at distance k from any node. Note that the number of nodes at distance k follows a binomial distribution. The Boolean n-cube can be laid out in area $O(2^{2n})$. The maximum wire length is of order $O(2^n)$ [77], i.e., proportional to the number of nodes.

The *machine address space* is $\mathcal{A}$ and the *logic address space* is $\mathcal{L}$. The *machine address space* $\mathcal{A} = \{(a_{q-1}a_{q-2}\ldots a_0)|\ a_i = 0, 1;\ 0 \le i < q\}$ is the Cartesian product of the *processor address space* and *local storage address*

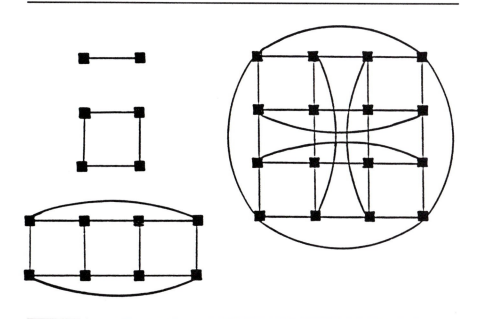

Figure 4.9   Boolean n-cubes for $n = 1 - 4$.

space. The processor address space requires n bits, or *dimensions.* The storage per node is 2^{q-n} elements. The n low-order dimensions of the machine address space are used for processor addresses, and the $q - n$ high-order dimensions are used for local storage addresses:

$$\big(\underbrace{a_{q-1}a_{q-2}\ldots a_n}_{s}\,\underbrace{a_{n-1}a_{n-2}\ldots a_0}_{p}\big).$$

The set of *machine dimensions* is $Q = \{0, 1, \ldots, q-1\}$, the set of *processor dimensions* is $Q_p = \{0, 1, \ldots, n-1\}$, and the set of *local storage dimensions* is $Q_s = \{n, n+1, \ldots, q-1\}$.

The *logic address space* $\mathcal{L} = \{(w_{m-1}w_{m-2}\ldots w_0)|w_i = 0, 1; 0 \le i < m\}$ encodes a set of $|\mathcal{L}| = 2^m$ elements. The set of *logic dimensions* is $\mathcal{M} = \{0, 1, \ldots, m-1\}$. The relationship between the number of processor dimensions, the local storage dimensions, and the number of logic dimensions is arbitrary. For instance, if $m \le q - n$ and $m \le n$, then the entire data set can be allocated either to the local storage of a single processor or across processors with one element per processor using 2^m processors.

We model multiple elementary objects per node by virtual processors. For simplicity we assume that the number of elementary objects is 2^m

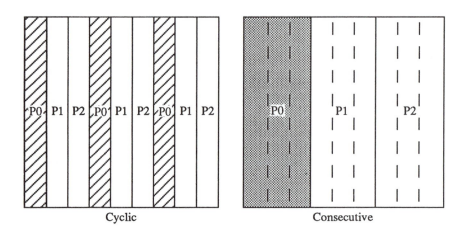

Figure 4.10   Cyclic and consecutive one-dimensional partitioning. [63]

and always allocate virtual processors such that the virtual processor ratio is 2^k for some $k < m$. We have already mentioned two frequently used schemes for the allocation of virtual processors: *cyclic* and *consecutive*. In the cyclic assignment, processor j contains elementary objects with indices congruent to j modulo N. In the consecutive assignment, 2^k consecutive elements are assigned to the same processor, with successive sets being allocated to successive processors. The n lowest order bits of the binary encoded index determines the processor to which an object is assigned in the cyclic partitioning. In the consecutive assignment, the n highest order bits determines the processor assignment.

Different languages have different rules for the order in which axes are assigned bits of the address space. Since we have made the simplifying assumption that the axes' lengths are powers of two, the segment of the address space defined by the appropriate number of bits is fully utilized. No waste of storage results from considering bits only. The partitioning of the address space for matrices (or arrays with two axes) for a few frequently used allocations is illustrated in Figures 4.10 and 4.11. In Figure 4.10, one axis is entirely in storage, or virtual. Having one axis entirely in storage is beneficial if most of the communication is along that axis, given the assumption of significantly faster local storage references than references to the storage of a different processor. With u denoting the part of the logic address space encoding rows and v the part that encodes columns, some sample partitionings of the logic address space are

One-dimensional cyclic column partitioning

$$\big(\underbrace{u_{p-1}u_{p-2}\ldots u_0\ \ v_{q-1}v_{q-2}\ldots v_{n_0}}_{vp}\ \underbrace{v_{n_0-1}\ldots v_0}_{rp}\big).$$

One-dimensional consecutive column partitioning

$$\big(\underbrace{u_{p-1}u_{p-2}\ldots u_0}_{vp}\ \underbrace{v_{q-1}v_{q-2}\ldots v_{q-n_0}}_{rp}\ \underbrace{v_{q-n_0-1}\ldots v_0}_{vp}\big).$$

For the cyclic two-dimensional assignment the address field is partitioned as

$$\big(\underbrace{u_{p-1}u_{p-2}\ldots u_{n_1}}_{vp}\ \underbrace{u_{n_1-1}\ldots u_0}_{rp}\ \underbrace{v_{q-1}v_{q-2}\ldots v_{n_0}}_{vp}\ \underbrace{v_{n_0-1}\ldots v_0}_{rp}\big),$$

and for the consecutive assignment the address field is partitioned as

$$\big(\underbrace{u_{p-1}u_{p-2}\ldots u_{p-n_1}}_{rp}\ \underbrace{u_{p-n_1-1}\ldots u_0}_{vp}\ \underbrace{v_{q-1}v_{q-2}\ldots v_{q-n_0}}_{rp}\ \underbrace{v_{q-n_0-1}\ldots v_0}_{vp}\big).$$

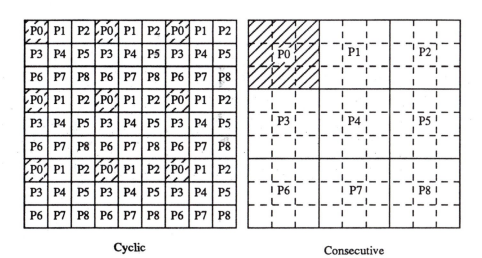

Cyclic   Consecutive

Figure 4.11   Cyclic and consecutive two-dimensional partitioning. [63]

With $m = p + q$ the number of dimensions used for both the consecutive and cyclic mapping is $m - n_0$ (or $m - n_1$) in the one-dimensional case and $m - n$ in the two-dimensional case. For column partitioning, $n_1 = 0$, $0 \leq n_0 \leq n$. For row partitioning, $n_0 = 0$, $0 \leq n_1 \leq n$. In the following we label the array axes from right to left with the rightmost axis being axis 0.

Definition   1   A *dimension allocation function*, π, is a one-to-one mapping from the set of logic dimensions, $\mathcal{M}$, to the set of machine dimensions, $\mathcal{Q}$; $\pi : \mathcal{M} \to \mathcal{Q}$. ∎

Let $\mathcal{R} = \{r_{m_p-1}, r_{m_p-2}, \ldots, r_0\}$ be the set of logic dimensions mapped to *processor dimensions*, i.e., $\pi(i) \in \mathcal{Q}_p$, $\forall i \in \mathcal{R}$ and $\mathcal{V} = \{v_{m_s-1}, v_{m_s-2}, \ldots, v_0\}$ be the set of logic dimensions mapped to *local storage dimensions*, i.e., $\pi(i) \in \mathcal{Q}_s$, $\forall i \in \mathcal{V}$. Then, $|\mathcal{R}| = m_p \leq n$, $|\mathcal{V}| = m_s \leq q - n$, $\mathcal{R} \cup \mathcal{V} = \mathcal{M}$, $\mathcal{R} \cap \mathcal{V} = \phi$, $m_p + m_s = m$. $\Gamma_p = \{\pi(i) | \forall i \in \mathcal{R}\}$ and $\Gamma_s = \{\pi(i) | \forall i \in \mathcal{V}\}$ are the sets of processor and local storage dimensions used for the allocation of elements: $\Gamma = \Gamma_p \cup \Gamma_s$. The inverse of the dimension allocation function π^{-1} is a mapping: $\Gamma \to \mathcal{M}$, such that $\pi^{-1} \circ \pi = I$, where I is the identity function. Figure 4.12 illustrates the relationships between the different sets and the allocation function. We will refer to the dimensions in Γ_p

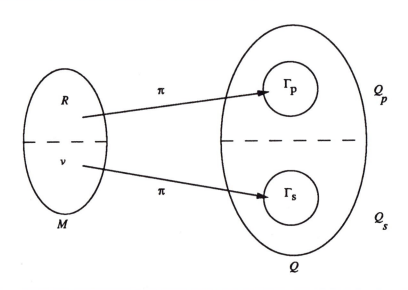

Figure 4.12   The relationship between the sets of address dimensions.

(processor dimensions) as *real dimensions* and the dimensions in Γ_s (local storage dimensions) as *virtual dimensions*.

Definition   2   The *real distance* between two locations with addresses a and a', $a, a' \in \mathcal{A}$, is $Hamming_r(a, a') = \sum_{i=0}^{n-1}(a_i \oplus a_i')$ and the *virtual distance* between a and a' is $Hamming_v(a, a') = \sum_{i=n}^{q-1}(a_i \oplus a_i')$.   ∎

Lemma   1 $Hamming(a, a') = Hamming_r(a, a') + Hamming_v(a, a')$.   ☐

The cyclic and consecutive assignments have different properties with respect to load balance and communication requirements. Let the axes be numbered $0, 1, \ldots, d-1$ and the number of real and virtual processor dimensions in dimension i be rp_i and vp_i.

Lemma   2   *The number of objects on a face with normal along a given axis is equal to the product of the virtual processors per axis for all axes but the one orthogonal to the face. If the total number of virtual processor dimensions that is greater than the highest real processor dimension for an axis is x, then the number of faces held by a processor that are orthogonal to the axis is 2^x.*

☐

Lemma 2 provides an easy way of computing the data volume that needs to be communicated between a pair of processors. The number of objects on a face orthogonal to axis i is $2^{\sum_{j \neq i} vp_j}$. If there is only one axis, then the number of objects is one, which is the size of the face if a path is cut into pieces. When a consecutive allocation is applied to an axis, then there is clearly only one face per processor (Figure 4.11). In the cyclic allocation scheme a processor holds several faces with normals along the same axis (Figure 4.10). The number of faces is equal to the number of virtual processors along the axis only if all virtual processor dimensions are of higher order than the real processor dimensions along the axis. The size of each fraction of a face that a processor holds is independent of the number of fractions.

Corollary   1   *The number of objects that must be communicated between a pair of processors is minimized if the number of faces in a processor that are orthogonal to a given axis is minimized and if the size of each face is minimized.*   ☐

Corollary 1 is a generalization of the often used surface to volume argument. If communication takes place only along one axis, then that axis should have all virtual processors, and a single face with normals along that axis, i.e., the consecutive allocation, should be used for that axis. With the same frequency of communication along each axis, and the same cost of communicating along each axis, the sum of the sizes of the faces with normals along the different axes should be minimized. The number of virtual processor dimensions for the different axes should be as equal as possible. The number of faces should also be minimized. The consecutive allocation scheme should be used for each axis.

The consecutive allocation scheme minimizes the data that needs to be communicated along an axis, but the cyclic partitioning may yield a better load balance [56] or allow for more efficient utilization of the communication system [65]. A simple example of cyclic allocation of rows and columns yielding a better load balance is Gaussian elimination without pivoting on a dense matrix. In this case the first row and column is used only once, the second row and column twice, etc. With a consecutive allocation, one row and one column of processors becomes idle after the first 2^{vp_0} elimination steps, assuming for simplicity that $vp_0 = vp_1$. After each set of 2^{vp_0} elimination steps another row and column become idle. As the computations progress, an increasing number of the values of the higher order row and column dimensions encode groups of indices that represent rows and columns that are no longer used in the elimination process. By using a cyclic assignment, those logic dimensions are assigned to local storage dimensions instead of physical processor dimensions. The work a processor must do decreases as a function of the elimination step, but not the number of active processors, as long as there are active virtual processors for each physical processor.

In practice, communication and arithmetic or logic operations are often associated with a start-up time. In order to reduce the impact of such overheads it is desirable to amortize it over as long a sequence of operations as possible. Sometimes this optimization favors consecutive allocation, or a more complicated partitioning of the address space, as in the following example for the factorization of banded matrices by a combination of block operations and parallelism through a divide-and-conquer scheme. The algorithm is described in detail in [54, 52, 55]. The non-zero elements of the matrix, the right-hand sides, and the solution vectors can be stored in a rectangular array by conventional row/column storage of the matrix, or by a row/diagonal organization. We do not discuss here the techniques for band matrix storage and their consequences for the solution procedure. For the illustration here we simply assume that the relevant elements are stored in an array of $P = 2^p$ rows and $Q = 2^q$ columns. Then, for a two-dimensional partitioning with 2^{n_0} processors in both the row and column directions, blocks of size $2^{q-n_0} \times 2^{q-n_0}$ elements may be stored in the same processor, and blocks assigned cyclically with respect to the row addresses, i.e., the address field is partitioned as

$$\underbrace{\big(u_{p-1}u_{p-2}\ldots u_q}_{vp}\ \underbrace{u_{q-1}\ldots u_{q-n_0}}_{rp}\ \underbrace{u_{q-n_0-1}\ldots u_0}_{vp}\ \underbrace{v_{q-1}\ldots v_{q-n_0}}_{rp}\ \underbrace{v_{q-n_0-1}\ldots v_0\big).}_{vp}$$

The total number of real processor dimensions is $2n_0$. For the row assignment the n_0 contiguous dimensions of the address field used for real processor addresses divides the address space into two parts: one part of $q - n_0$ dimensions used for consecutive assignment and $p - q$ dimensions used for cyclic assignment. For the concurrent elimination of multiple vertices, the matrix is partitioned into S block rows. With $S = 2^s$, the s highest order bits of the matrix row addresses are used for real processor addresses. With the previous assignment for each such block, the address field is partitioned as

$$\underbrace{\big(u_{p-1}u_{p-2}\ldots u_{p-s}}_{rp}\ \underbrace{u_{p-s-1}u_{p-2}\ldots u_q}_{vp}\ \underbrace{u_{q-1}\ldots u_{q-n_0}}_{rp}$$

$$\underbrace{u_{q-n_0-1}\ldots u_0}_{vp}\ \underbrace{v_{q-1}\ldots v_{q-n_0}}_{rp}\ \underbrace{v_{q-n_0-1}\ldots v_0\big),}_{vp}$$

and hence in this case the dimensions used for real processor addresses for the rows form two fields. The number of dimensions for real processors in the row direction is $s + n_0$, and the total number of real processor dimensions is $s + 2n_0$. The notions of cyclic and consecutive partitioning are now conditioned on the part of the real processor address fields.

As yet another example of the partitioning of the address space, the one used in the quantum chromodynamics computations is given below:

$$\underbrace{\big(w_{k_3+n_0-1}\ldots w_{k_3}}_{rp_3}\ \underbrace{w_{k_3-1}\ldots w_{k_2+n_0}}_{vp_3}\ \underbrace{w_{k_2+n_0-1}\ldots w_{k_2}}_{rp_2}\ \underbrace{w_{k_2-1}\ldots w_{k_1+n_0}}_{vp_2}$$

$$\underbrace{w_{k_1+n_0-1}\ldots w_{k_1}}_{rp_1}\ \underbrace{w_{k_1-1}\ldots w_{k_0+n_0}}_{vp_1}\ \underbrace{w_{k_0+n_0-1}\ldots w_{k_0}}_{rp_0}\ \underbrace{w_{k_0-1}\ldots w_0\big).}_{vp_0}$$

For a complete application, and sometimes also for an indiviual algorithm, different phases of the computation have different optimal data allocation functions. Changing one data allocation into another is treated in the section on permutations.

4.5   Graph Notions

In the following, node i has address $(i_{n-1}i_{n-2}\ldots i_0)$, and node s has address $(s_{n-1}s_{n-2}\ldots s_0)$. The bit-wise *exclusive-or* operation is denoted $\oplus$, and

$i \oplus s = c = (c_{n-1}c_{n-2}\ldots c_0)$, where $c_m = i_m \oplus s_m$. c is the *relative address* of node i with respect to node s. Address bits are numbered from 0 through $n - 1$ with the lowest order bit being the 0th bit. The mth bit corresponds to the mth dimension in a Boolean space. $||x||$ is the number of 1-bits in the binary representation of x, i.e., $||x|| = Hamming(x,0)$. 0^m is a string of m 0-bits, and 1^m a string of m 1-bits. Calligraphic letters are used for sets. The set of *node* addresses is $\mathcal{N} \equiv \{0, 1, \ldots, N - 1\}$, and the set of dimensions is $\mathcal{D} \equiv \{0, 1, \ldots, n - 1\}$. $|\mathcal{S}|$ is used to denote the cardinality of a set $\mathcal{S}$. d or m is used to denote an arbitrary dimension, $d, m \in \mathcal{D}$. $\mathcal{A}_G(i)$ is the set of nodes adjacent to node i: $\mathcal{A}_G(i) = \{j | (i, j) \in \mathcal{E}(G)\}$. $x^{\{m\}} = x \oplus 2^m$.

Definition 3 A *Boolean n-cube* is a graph $B_n = (\mathcal{V}, \mathcal{E}^{B_n})$ such that $\mathcal{V} = \mathcal{N}$ and $\mathcal{E}^{B_n} = \{(i, j) | i \oplus j = 2^m, \forall m \in \mathcal{D}, \forall i, j \in \mathcal{N}\}$. An edge (i, j) such that $i \oplus j = 2^m$ is in dimension m, and nodes i and j are connected through the mth port. ∎

An *edge* (i, j) is *directed* from node i to node j, and of unit length, i.e., $Hamming(i, j) = 1$. It is a $0 \rightarrow 1$ edge if the bit that differs in i and j is zero in i; otherwise, it is a $1 \rightarrow 0$ edge. In a directed graph all edges are directed. A node with no edges directed to it is a *root (source)* node, and a node with no edges directed away from it is a *leaf (sink)* node. A node that is neither a leaf node nor a root node is an *internal* node. If (i, j) is a directed edge, then node i is the *parent* of node j, and j is the *child* of i.

Definition 4 With *Rotation of a node* we mean a right cyclic rotation of its address, $Ro(i) = (i_0 i_{n-1} \ldots i_2 i_1)$. With *Rotation of a graph* $G(\mathcal{V}, \mathcal{E})$ we mean a graph $Ro(G(\mathcal{V}, \mathcal{E})) = G(Ro(\mathcal{V}), Ro(\mathcal{E}))$, where $Ro(\mathcal{V}) = \{Ro(i) | \forall i \in \mathcal{V}\}$ and $Ro(\mathcal{E}) = \{(Ro(i), Ro(j)) | \forall (i, j) \in \mathcal{E}\}$. Moreover, Ro^{-1} is a left rotation and $Ro^k = Ro \circ Ro^{k-1}$ for all k. ∎

The right rotation operation is also known as an *unshuffle* operation and the left rotation as a *shuffle* operation.

Definition 5 With *Reflection of a node* we mean a bit-reversal of its address, $Re(i) = (i_0 i_1 \ldots i_{n-2} \, i_{n-1})$. With *Reflection of a graph* $G(\mathcal{V}, \mathcal{E})$ we mean a graph $Re(G(\mathcal{V}, \mathcal{E})) = G(Re(\mathcal{V}), Re(\mathcal{E}))$, where $Re(\mathcal{V}) = \{Re(i) | \forall i \in \mathcal{V}\}$ and $Re(\mathcal{E}) = \{(Re(i), Re(j)) | \forall (i, j) \in \mathcal{E}\}$. ∎

Definition 6 With *Translation of a node* i by s we mean a bit-wise exclusive-or of the addresses, $Tr(s, i) = c$. With *Translation of a graph*

$G(\mathcal{V},\mathcal{E})$ with respect to node s, we mean a graph $Tr(s,G(\mathcal{V},\mathcal{E})) = G(Tr(s,\mathcal{V}),Tr(s,\mathcal{E}))$, where $Tr(s,\mathcal{V}) = \{Tr(s,i)|\forall i \in \mathcal{V}\}$ and $Tr(s,\mathcal{E}) = \{(Tr(s,i),Tr(s,j))|\forall(i,j) \in \mathcal{E}\}$. ∎

Lemma   3   Rotations, Reflections, *and* Translations *of a graph preserve the Hamming distance between nodes. The Rotation operation* Ro^k *maps every edge in dimension* d *to dimension* $(d-k)$ mod n, *the* Reflection *operation maps every edge in dimension* d *to dimension* $n-1-d$, *and the* Translation *operation preserves the dimension of every edge.* Rotation *and* Reflection *preserve the direction of every edge.* Translation *reverses the direction of all edges in the dimensions for which* $s_m = 1$, $m \in \mathcal{D}$. ☐

Corollary   2   *The topology of a graph remains unchanged under* Rotation, Reflection, *and* Translation. ☐

Definition   7   For a binary number i the *Period* of i, $P(i) = \min_{m>0} Ro^m(i) = i$. A binary number i is *cyclic* if $P(i) < n$; and *noncyclic* otherwise. A *cyclic node* is a node with cyclic *relative* address. ∎

The period of the number (011011) is 3. A cyclic node is defined only when the source node is given. Node (001000) is cyclic with respect to the source node (000001).

The *embedding function* f maps each vertex in the *guest graph* $G = (V_G, E_G)$ into a unique vertex in the *host graph* $H = (V_H, E_H)$. V_G and V_H denote the node sets of the guest and host graphs respectively, and E_G and E_H the edge sets. $|V_G|$ denotes the cardinality of the set V_G.

Definition   8   An embedding f of a *guest* graph, G, into a *host* graph, H, is a one-to-one mapping from $\mathcal{V}(G)$ to $\mathcal{V}(H)$. The *expansion* of the embedding f is

$$exp_f = \frac{|\mathcal{V}(H)|}{|\mathcal{V}(G)|}. \quad ∎$$

Under the mapping function f node $i \in V_G$ is mapped to node $f(i) \in V_H$. In order to consider *dilation* and *congestion*, we specify the path from $f(i)$ to $f(j)$ in H for every edge $e_G = (i,j) \in \mathcal{E}(G)$. Let $path_H(e_G) = p_0, p_1, \ldots, p_k$, where $p_m \in \mathcal{V}(H)$, for all $0 \le m \le k$, $p_0 = f(i)$ and $p_k = f(j)$. Moreover, let $\mathcal{E}(path_H(e_G)) = \{(p_m, p_{m+1})|0 \le m < k\}$, i.e., the set of edges along the path.

Definition 9   The dilation of an edge $e_G \in \mathcal{E}(G)$ is the length of the path $path_H(e_G)$:

$$dil_f(e_G) = |\mathcal{E}(path_H(e_G))|.$$

The *dilation* of the embedding f is

$$dil_f = \max_{\forall e_G \in \mathcal{E}(G)} dil_f(e_G).$$

The *average edge dilation* is

$$\frac{1}{|E_G|} \sum_{\forall e_G \in \mathcal{E}_G} dil_f(e_G)$$

We will sometimes also consider dilation of a set of edges S as

$$dil_f(S) = \max_{\forall e_G \in S} dil_f(e_G). \quad \blacksquare$$

Definition 10   The *congestion* of an edge $e_H \in \mathcal{E}(H)$, $cong_f(e_H)$, is the number of edges in G with images including e_H,

$$cong_f(e_H) = \sum_{\forall e_G \in \mathcal{E}(G)} |\{e_H\} \cap \mathcal{E}(path_H(e_G))|.$$

The *congestion* of the mapping f is

$$cong_f = \max_{\forall e_H \in \mathcal{E}(H)} cong_f(e_H). \quad \blacksquare$$

Congestion is sometimes referred to as *load-factor* [7].

Definition 11   The *active degree* of a node i, α_i, is the number of edges of host graph node i being part of any $path_H$.

$$\alpha_i = \sum_{\forall j \in \mathcal{A}_H(i)} |\{(i, j)\} \cap (\cup \mathcal{E}(path_H(e_G)), \forall e_G \in \mathcal{E}(G))|.$$

The *active degree* of the mapping f is

$$active\ degree_f = \max_{\forall i \in \mathcal{V}(H)} \alpha_i. \quad \blacksquare$$

The *active degree* is a measure of the number of ports that need to be serviced concurrently in case the communication is pipelined. The *node load* measures the total load on any node, which is of particular importance in case a node can service only one port at a time.

Definition   12   The *node load* of node i, β_i, is the number of messages that node i needs to service

$$\beta_i = \sum_{\forall j \in \mathcal{A}_H(i)} cong_f((i,j)).$$

The *node load* of the mapping f is

$$node\ load_f = \max_{i \in \mathcal{V}(H)} \beta_i. \quad \blacksquare$$

In the following, subcube 0 refers to a set of Boolean cube nodes that all have the same address bit equal to 0. The other half of the nodes form subcube 1.

Definition   13   In *one-port* communication a processor can send *and* receive on only one of its ports at any given time. Different processors communicate on different ports. In *n-port* communication a processor can communicate on all its ports concurrently. $\blacksquare$

We also assume that there is an overhead, start-up time τ, associated with each communication of B elements, each of which requires a transfer time t_c. B_{opt} denotes an optimal packet size. For the analysis it is convenient to assume that communication takes place during distinct time intervals. The duration of a *Routing Cycle* is $\tau + Bt_c$. Routing cycles are labeled from 0.

4.6   Lattices

Boolean cube networks can emulate multidimensional lattices very effectively, but two-dimensional lattices cannot emulate higher dimensional lattices well in a topological sense [102]. However, with the characteristics of VLSI technology it is not necessarily true that a two-dimensional lattice is slower than a higher dimensional lattice for high-dimensional lattice emulation [97]. The same characteristics apply to the emulation of butterfly networks, the communication structure for many efficient algorithms, such as the Fast Fourier Transform. Before considering the emulation of lattices of an arbitrary number of dimensions and shapes, we consider the embedding of loops.

4.6.1 One-Dimensional Arrays

Binary numbers have the property that $Hamming(i,j)$ in general is not equal to one if $i \neq j$. For instance, $\frac{N}{2} = 2^{n-1}$ and $\frac{N}{2} - 1$ differ in n bits. Hence embedding a loop of length $|L| = 2^n$ in an n-cube by a binary encoding of the indices of the nodes in the loop, assuming that they are numbered $0 - L - 1$, does not preserve proximity. Gray codes have the property that successive integers differ only in 1 bit. One such code that is frequently used for Boolean cubes [68] is the *binary-reflected* Gray code [100]. Let the n-bit code of 2^n integers be $G(n)$.

$$G(n) = \begin{pmatrix} G_0 \\ G_1 \\ \vdots \\ G_{2^n-2} \\ G_{2^n-1} \end{pmatrix}$$

Then $G(n+1) = \begin{pmatrix} 0G_0 \\ 0G_1 \\ \vdots \\ 0G_{2^n-2} \\ 0G_{2^n-1} \\ 1G_{2^n-1} \\ 1G_{2^n-2} \\ \vdots \\ 1G_1 \\ 1G_0 \end{pmatrix}$, or alternatively, $G(n+1) = \begin{pmatrix} G_0 0 \\ G_0 1 \\ G_1 1 \\ G_1 0 \\ G_2 0 \\ G_2 1 \\ \vdots \\ G_{2^n-1} 1 \\ G_{2^n-1} 0 \end{pmatrix}$

The Gray codes of successive integers differ in precisely 1 bit. It is also easy to show that any loop of length $2k, k \in \mathcal{N}$ can be embedded in a Boolean cube with dilation 1 and minimum expansion, and that any loop of length $2k + 1$ must have dilation 2 [56].

Algorithms for which the data interaction is described by the butterfly network require communication between nodes with indices of the form i and $i \oplus 2^k$. Data manipulator networks require communication between nodes i and $i \pm 2^k$ for properly assigned addresses. In the binary representation $i \oplus 2^k$ differs in 1 bit, but $i \pm 2^k$ may differ in arbitrarily many bits.

Lemma 4   *The binary-reflected Gray code encoding of i and $j = (i + 2^k) mod 2^n, k > 0$ differs in precisely 2 bits.*

Proof [56] Let the binary encoding of i be $(i_n i_{n-1} \ldots i_0)$ and that of j be $(j_n j_{n-1} \ldots j_0)$. Furthermore, let the Gray code of i be $G(n) = (g_n g_{n-1} \ldots g_1)$ and that of j be $H(n) = (h_n h_{n-1} \ldots h_1)$. Then $i_m = j_m, m = \{0, 1, \ldots, k-1\}$ and $j_m = \bar{i}_m, m = \{k, k+1, \ldots, s\}$, where $s \geq k$ is the bit where the carry stops propagating. It follows from the encoding formula that $h_m = g_m, m = \{1, 2, \ldots, k-1, k+1, \ldots, s\}$ and $h_k = \bar{g}_k, h_{s+1} = \bar{g}_{s+1}$. □

The property of binary-reflected Gray codes stated in lemma 4 is important for algorithms such as the FFT, bitonic sort, recursive doubling, cyclic reduction, and nested dissection. Two interprocessor communications are required for all but one of the steps of an FFT computation with a Gray code embedding of loops and arrays.

The conversion between the binary encoding of $i = (i_n i_{n-1} i_{n-2} \ldots i_1 i_0)$ and the Gray code encoding $G_i = (g_n g_{n-1} g_{n-2} \ldots g_1)$ of an integer i is defined by

$$g_j = (i_j + i_{j-1}) \bmod 2 \quad \text{and conversely} \quad i_j = (\sum_{k=j+1}^{n} g_k) \bmod 2.$$

4.6.2 Multidimensional Arrays

Definition   14   A $l_1 \times l_2 \times \cdots \times l_d$ array $M(l_1, l_2, \ldots, l_d)$ is a graph with vertex set

$$
\begin{aligned}
&\mathcal{V}(M(l_1, l_2, \ldots, l_d)) \\
&\quad = \{(x_1, x_2, \ldots, x_d) | 0 \leq x_1 < l_1, 0 \leq x_2 < l_2, \ldots, 0 \leq x_d < l_d\}
\end{aligned}
$$

and edge set

$$\mathcal{E}(M(l_1, l_2, \ldots, l_d)) = \{(v, v') | v, v' \in \mathcal{V}(M(l_1, l_2, \ldots, l_d)), |v - v'|_1 = 1\}. \ \blacksquare$$

For the embedding of multidimensional arrays by a binary-reflected Gray code, we partition the address into as many segments as there are dimensions of the lattice. For dimension i of the lattice we assign $\lceil \log_2 N_i \rceil$ address bits. The number of dimensions of the cube is $n = \lceil \log_2 N_1 \rceil + \lceil \log_2 N_2 \rceil + \cdots \log_2 N_d \rceil$ for a $N_1 \times N_2 \times \cdots \times N_d$ mesh. The expansion is

$$\frac{2^{\lceil \log_2 N_1 \rceil + \lceil \log_2 N_2 \rceil + \cdots + \lceil \log_2 N_d \rceil}}{(N_1 N_2 \cdots N_d)},$$

which is in the range of 1 to 2^d. Figure 4.13 shows the embedding of a 4×4 mesh by a binary-reflected Gray code in a 4-cube. With the ordering of dimensions used in Figure 4.13, dimensions 0 and 2 are assigned to the encoding of column indices, and dimensions 1 and 3 to row indices.

Theorem 1   [36] *If an $N_1 \times N_2 \times \cdots \times N_d$ mesh is embedded in an n-cube with dilation one, then $n \geq \lceil log_2 N_1 \rceil + \lceil log_2 N_2 \rceil + \cdots + \lceil log_2 N_d \rceil$.*

Theorem 1 was first proved by Havel and Móravek. It was later independently rediscovered [10, 16, 37, 28]. Each node of the $N_1 \times N_2$ mesh is represented by an address (i, j), where $0 \leq i \leq N_1 - 1$, $0 \leq j \leq N_2 - 1$. There is a one-to-one mapping from edges of the mesh to edges of the cube in a dilation 1 embedding. An edge (i, j) is in dimension k, if i and j differ in bit k. The label on an edge is its dimension.

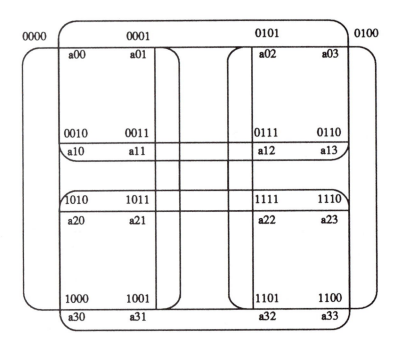

Figure 4.13   Embedding of a 4 × 4 mesh in a 4-cube. [56]

Proof [43] Consider a two-dimensional mesh. Label each edge in the mesh with a number that represents the dimension of the corresponding edge in the cube. A legal labeling should satisfy the following two properties:

(1) Any cycle in the mesh must contain any label an even number of times.

(2) Any path in the mesh must contain some label an odd number of times. ☐

For any cycle of length 4 in the mesh, the edge labels form a sequence (p, q, p, q), where $p \neq q$, Figure 4.14 (a). By extending this argument, it follows that any vertical or horizontal cut will only cut edges having the same label, Figure 4.14 (b). Now consider the labels of any horizontal path connecting all the N_2 nodes of a row. Since the corresponding cube nodes are distinct and the longest node-disjoint path in an n-cube is $2^n - 1$, the number of distinct labels on the edges forming a row is at least $\lceil \log_2 N_2 \rceil$. The same argument applies to any vertical path. Moreover, the set of labels on the horizontal paths must be disjoint from the set of labels on the vertical paths; otherwise there exist two adjacent edges with the same label, Figure 4.14 (c). Hence, the minimum number of dimensions required is $\lceil \log_2 N_1 \rceil + \lceil \log_2 N_2 \rceil$.

Since the set of cube dimensions used as labels for mesh edges in mesh dimension i is disjoint from the set of cube dimensions used as labels for mesh edges in dimension j, $i \neq j$, the proof is complete.

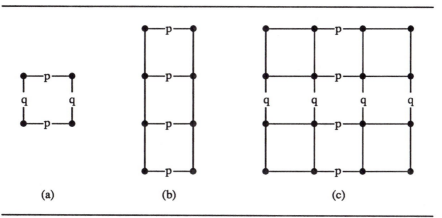

(a)   (b)   (c)

Figure 4.14   (a) Dimensions of a cycle of length 4. (b) Vertical or horizontal cuts. (c) The set of dimensions of vertical edges and the set of dimensions of horizontal edges are disjoint.

The percentage of lattices for which the binary-reflected Gray code embedding offers minimal expansion is a function of the dimensionality of the lattice. Let $a_i \in [\frac{1}{2}, 1], i \geq 1$ be uniformly distributed over the interval, and let $\alpha \in [\frac{1}{2}, 1]$. Then, with $N_i = a_i 2^{\lceil \log_2 N_i \rceil}$ the number of lattice shapes for which the binary-reflected Gray code provides minimum expansion is equal to the number of lattices for which $\Pi_{i=1}^d a_i \cdot 2^{\sum_{i=1}^d \lceil \log_2 N_i \rceil} \geq \frac{1}{2} 2^{\sum_{i=1}^d \lceil \log_2 N_i \rceil}$. The probability that $\frac{1}{2} \leq \prod_{i=1}^k a_i \leq 1$ can be shown to be $2^d (1 - \frac{1}{2} \sum_{i=0}^{d-1} \frac{(-1)^i \ln^i \frac{1}{2}}{i!})$ [47], asymptotically. Figure 4.15 shows the value of this expression as a function of the number of dimensions.

4.6.3   Arbitrary Lattices

Recently it has been shown that any two-dimensional lattice can be embedded in a Boolean cube with minimum expansion and dilation two [19]. This embedding is based on a modified version of the so-called line-compression technique [2]. A block partitioning technique yielding dilation two minimum expansion embeddings for a large class of two-dimensional lattices was described earlier in [43]. Techniques for the embedding of meshes of one shape into meshes of another shape that have been considered for the embedding of lattices in Boolean cubes [56] are the break-and-fold technique [79] and the step-embedding and the modified step-embedding techniques [2]. Lattices of higher dimensionality can be approached by pairing dimensions, but such a treatment does not necessarily yield optimal embeddings.

Using meshes as intermediate graphs
Given an $N_1 \times N_2 (a_1 2^{n_1} \times a_2 2^{n_2})$ mesh, $1 < a_1, a_2 < 2$ and $1 < a_1 \cdot a_2 < 2$, we seek a small dilation embedding into an $M_1 \times M_2 (2^{m_1} \times 2^{m_2})$ mesh such that $m_1 + m_2 = n_1 + n_2 + 1$. The dilation is 3 for *step embedding* and *modified step embedding*, and 2 for *folding* and *line compression*. For convenience, we assume that $N_1 \leq N_2$.

Step embedding
Figure 4.16 (a) shows the embedding of a 3×9 mesh into a 4×8 mesh by step embedding. Row i of the guest mesh will occupy part of rows i and $i + N_2 - M_2$ of the host mesh. The dilation is 3. The number of edges with edge dilation 3 is $(N_2 - M_2)(N_1 - 1)$. The average edge dilation is $1 + \frac{2(N_2 - M_2)(N_1 - 1)}{|E_G|} \approx 1 + \frac{N_2 - M_2}{N_2}$, which is in the range 1–1.5 for $M_2 = 2^{\lfloor \log_2 N_2 \rfloor}$. A savings of one cube dimension occurs only

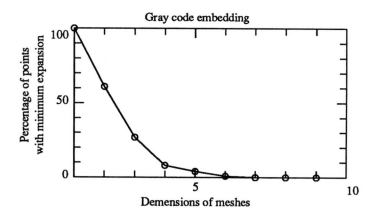

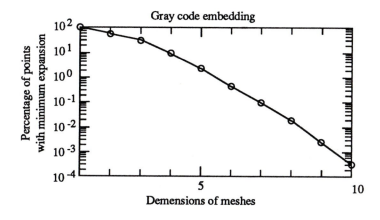

Figure 4.15   The fraction of the domain $2^{n_i} \leq N_i < 2^{n_i+1}$ for which minimum expansion is attained by Gray code embedding. The right plot has a logarithmic scale for the y-axis. [47]

if $M_1 = 2^{\lceil \log_2 N_1 \rceil} \geq\geq N_2 - M_2 + N_1$. Note that $M_2 \geq N_1$. Moreover, for $N_2 = M_2 + 1$, the dilation can be reduced to 2 as shown in 4.16 (b). The step embedding provides minimal expansion, dilation 3 embeddings for $\approx \frac{3}{8}$ of the two-dimensional meshes for which the Gray code embedding does not yield minimum expansion.

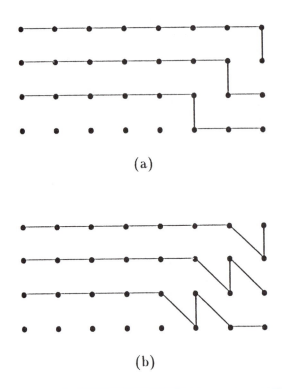

Figure 4.16 Embedding of a 3 × 9 mesh into a 4 × 8 mesh by step embedding: (a) dilation 3, (b) dilation 2. The three solid lines show the three rows of the guest mesh. [47]

Modified step embedding

In step embedding the rows of the guest graph traverse only one column of the host mesh. In modified step embedding, a row may traverse many columns. Each traversal can save at most $M_1 - N_1$ columns from N_2. With $k > 0$ column traversals, $M_1 \geq \lceil \frac{N_2 - M_2}{k} \rceil + N_1$. $k = 1$ corresponds to step embedding. Figure 4.17 gives an example of modified step embedding. The pairs for which the modified step embedding yields expansion one, but the step embedding does not satisfy the condition $|n_1 - n_2| \geq 2$. For example, a 3 × 10 mesh will be mapped to a 5 × 8 mesh using step embedding, but to a 4 × 8 mesh using modified step embedding (see Figure 4.17). The average edge dilation is the same as in step embedding. If $\lfloor \frac{N_2 - M_2}{k} \rfloor = 1$, then the dilation can be reduced to 2, but the average edge dilation will increase.

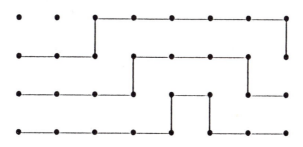

Figure 4.17   Embedding of a 3×10 mesh into a 4×8 mesh by modified step embedding. [47]

Folding

Folding is used in [3] in squaring up an $N_1 \times N_2$ mesh with $N_1 \ll N_2$. If there exists values M_1 and M_2 such that $m_1 + m_2 = \lceil \log_2 N_1 \rceil + \lfloor \log_2 N_2 \rfloor$ and $\lfloor \frac{M_1}{N_1} \rfloor \geq \lceil \frac{N_2}{M_2} \rceil$, then one cube dimension can be saved. Figure 4.18 shows the embedding of a 3×20 mesh into a 4×16 mesh. The dilation of embedding by folding is 2. The average edge dilation is $\approx 1 + \frac{N_1}{M_2}$. The set of pairs for which folding yields expansion one, but Gray coding does not is largely disjoint with the set of pairs for which step embedding yields minimum expansion. However, folding and the modified step embedding are mostly useful for the same set of meshes.

Line compression

In line compression [3] a "tile" of size $a \times b$ is compressed into a tile of size $b \times a$. Let $b = a + 1$, then $\lfloor \frac{M_2}{a} \rfloor \geq N_2 - M_2$ and $M_1 \geq N_1 + \lceil \frac{N_1}{a} \rceil$. In order to satisfy these two constraints, $a > 1$ and $N_2 \leq \frac{3}{2} M_2$ must be satisfied.

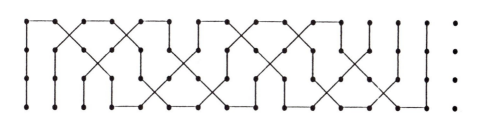

Figure 4.18   Embedding of a 3 × 20 mesh into a 4 × 16 mesh by folding. [47]

The dilation is 2 and the average edge dilation is $\approx 1 + \frac{a+2b}{2ab} \approx 1 + \frac{3}{2a}$. Minimizing the average edge dilation is equivalent to maximizing a. Figure 4.19 shows the embedding of a 5×12 mesh into a 4×16 mesh by the line compression method. The set of pairs for which line compression offers expansion one embedding largely includes the set of pairs for which step embedding, modified step embedding, and folding are useful.

Embeddings based on partitioning

In using this technique, the mesh to be embedded in the Boolean cube is partitioned into blocks, with widths and lengths being powers of two, except for the boundary blocks. All blocks with the exception of the boundary blocks are of the same shape. Since each block has sides being powers of two, they can be embedded in a Boolean cube with dilation one, expansion one by using a binary-reflected Gray code embedding [100, 68, 56]. The difficulty in finding an embedding with small dilation is reduced to

- Finding a good partitioning of the lattice into blocks with sides being powers of two;

- Finding an assignment of blocks to subcubes with a small dilation;

- Finding an orientation of the blocks within each subcube that minimizes the dilation.

Theorem 2 *[47] If an $\ell_1 \times \ell_2 \times \cdots \times \ell_k$ mesh M can be embedded in an n-cube with dilation d, then an $\ell'_1 \times \ell'_2 \times \cdots \times \ell'_k$ mesh M' where $\ell'_i \leq \ell_i 2^{n_i}, 1 \leq i \leq k$ can be embedded in an $(n + \sum_{i=1}^{k} n_i)$-dimensional cube with dilation d.*

Proof   Let M'' be a k-dimensional mesh of form $\ell_1 2^{n_1} \times \ell_2 2^{n_2} \times \cdots \times \ell_k 2^{n_k}$. Since the mesh M' is a subgraph of mesh M'', we prove instead

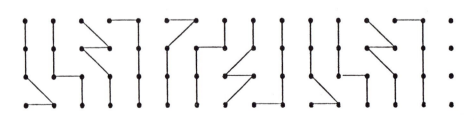

Figure 4.19   Embedding of a 5×12 mesh into a 4×16 mesh by line compression. [47]

that the mesh M'' can be embedded into an $(n + \sum_{i=1}^{k} n_i)$-dimensional cube with dilation d. Let f be an embedding function that maps mesh M into an n-cube with dilation d. So, for any two adjacent nodes x and x' of mesh M, $Hamming(f(x), f(x')) \le d$. Let $G(i)$ be the binary-reflected Gray code of i. Also let y_i be a binary string of length n_i. Define a new embedding function $\tilde{f}$ for mesh M'' as follows:

$$\tilde{f}((x_1||y_1, x_2||y_2, \ldots, x_k||y_k))$$
$$= f((x_1, x_2, \ldots, x_k))||\tilde{G}(x_1, y_1)||\tilde{G}(x_2, y_2)|| \cdots ||\tilde{G}(x_k, y_k),$$
$$\text{where } \tilde{G}(x_i, y_i) = \begin{cases} G(y_i), & \text{if } x_i \text{ is even,} \\ G(2^{n_i} - 1 - y_i), & \text{if } x_i \text{ is odd.} \end{cases}$$

Here, $\tilde{G}(x_i, y_i)$ is either $G(y_i)$ or the complement of $G(y_i)$, depending on x_i being even or odd. Note that since y_i is of length n_i, we have $0 \le x_i < \ell_i$, i.e., $f((x_1, x_2, \ldots, x_k))$, which is well-defined. It can be shown easily that $\tilde{f}$ is a one-to-one mapping. We now show that if z and z' are adjacent in mesh M'', then $Hamming(\tilde{f}(z), \tilde{f}(z')) \le d$. Let $z = (x_1||y_1, x_2||y_2, \ldots, x_k||y_k)$ be a node in mesh M'', and z' be adjacent to node z in mesh M'' with its address differing on the ith coordinate. Assume that the ith coordinate of z' is greater than that of z without loss of generality. Consider the ith coordinate of node z, i.e., $x_i||y_i$, for the following two cases:

- $y_i < 2^{n_i} - 1$, which means that the ith coordinate of z' is $x_i||(y_i + 1)$. So,

$$Hamming(\tilde{f}(z), \tilde{f}(z')) = Hamming(\tilde{G}(x_i, y_i), \tilde{G}(x_i, y_i + 1)) = 1.$$

- $y_i = 2^{n_i} - 1$, which means that the ith coordinate of z and z' are $x_i||11\ldots1$ and $(x_i + 1)||00\ldots0$, respectively. Hence,

$$Hamming(\tilde{f}(z), \tilde{f}(z'))$$
$$= Hamming(f((x_1, x_2, \ldots, x_i, \ldots, x_k)), f((x_1, x_2, \ldots, x_i + 1, \ldots, x_k))$$
$$+ Hamming(\tilde{G}(x_i, 11\ldots1), \tilde{G}(x_i + 1, 00\ldots0)).$$

The *Hamming* function of the first term (to the right of the equal sign) is less than or equal to d. The *Hamming* function of the second term is equal to 0. This follows from the fact that if x_i is even, then $\tilde{G}(x_i, 11\ldots1) = G(11\ldots1) = \tilde{G}(x_i + 1, 00\ldots0)$, and if x_i is odd, then $\tilde{G}(x_i, 11\ldots1) = G(00\ldots0) = \tilde{G}(x_i + 1, 00\ldots0)$. $\square$

Note that in Theorem 2, the expansion of mapping the mesh M'' to an $(n + \sum_{i=1}^{k} n_i)$-cube is the same as for mapping the mesh M with function φ, but the expansion for mapping the mesh M' may be higher.

Theorem 2 shows that any lattice embedding can serve as a generating lattice for a large class of lattices, all having the same expansion and dilation. We present dilation 2 minimum expansion embeddings of 5×3, 9×7, and 11×11 meshes into 16, 64, and 128 node Boolean cubes [43]. These three meshes cover 70% of all two-dimensional meshes for which the binary-reflected Gray code does not provide minimum expansion. The dilation of the block embedding is 2. All three embeddings have the property that the dilation of the edges in the upper-left submesh defined by the first $2^{\lfloor \log_2 \ell_1 \rfloor}$ rows and first $2^{\lfloor \log_2 \ell_2 \rfloor}$ columns is one.

Embedding a 5×3 mesh into a 4-cube

The mapping used for embedding a 5×3 mesh into a 4-cube is given in Figure 4.20. The numbers represent the cube addresses of the mesh nodes. $\bar{x}$ represents the node in subcube one that corresponds to node x in subcube zero. Subcubes are defined with respect to the most significant bit. The '•' sign on a dashed line means that the Hamming distance is two between two nodes when embedded in the cube. The dilation is apparent from the figure.

In order to determine the congestion of the embedding, it is necessary to specify all length-two paths. A '•' sign above a doubled arrow identifies a cube edge that is used for both a dilation one and two path.

$$\bar{0} \leftrightarrow 0 \overset{\bullet}{\leftrightarrow} 1, \qquad \bar{1} \leftrightarrow 1 \overset{\bullet}{\leftrightarrow} 3, \qquad \bar{3} \leftrightarrow 3 \overset{\bullet}{\leftrightarrow} 7,$$
$$\bar{7} \leftrightarrow 7 \overset{\bullet}{\leftrightarrow} 5, \qquad \bar{4} \leftrightarrow 4 \overset{\bullet}{\leftrightarrow} 5, \qquad \bar{6} \leftrightarrow 6 \overset{\bullet}{\leftrightarrow} 4.$$

By inspection the congestion is two; the number of cube edges with congestion two is 6; the active degree is 4; and the node load is 6.

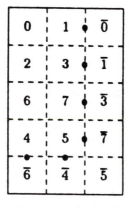

Figure 4.20   Embedding a 5×3 mesh into a 4-cube. [47]

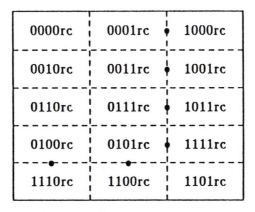

0000rc	0001rc	1000rc
0010rc	0011rc	1001rc
0110rc	0111rc	1011rc
0100rc	0101rc	1111rc
1110rc	1100rc	1101rc

Figure 4.21   Block addresses of embedding a $5 \cdot 2^{n_1} \times 2 \cdot 2^{n_2}$ mesh into a $(4 + n_1 + n_2)$-cube. [47]

Figures 4.21 and 4.22 show the effect of applying Theorem 2 to the embedding. In Figure 4.21, 'r' and 'c' denote the local addresses within a block (subcube); it is defined by the $\tilde{G}(x_i, y_i)$ function in Theorem 2. Figure 4.22 shows the embedding of a 10×6 mesh into a 6-cube.

The number of edges with dilation 2 is $N_1 + N_2$, and the total number of edges is $\frac{1}{4}(15 N_1 N_2 - 5 N_1 - 6 N_2)$. As both N_1 and N_2 increase, the average edge dilation approaches 1. Table 4.6 gives the expression for the average dilation and its maximum.

Embedding a 9×7 mesh into a 6-cube

Figure 4.23 shows the embedding of a 9×7 mesh in a 6-cube. The length-two paths in the cube are specified as follows:

$$\overline{4} \leftrightarrow \overline{5} \leftrightarrow \overline{21}. \qquad \overline{20} \leftrightarrow \overline{21} \overset{\bullet}{\leftrightarrow} \overline{23}. \qquad \overline{16} \leftrightarrow \overline{18} \leftrightarrow \overline{22}.$$
$$\overline{16} \leftrightarrow 16 \leftrightarrow 18. \qquad \overline{24} \leftrightarrow \overline{26} \leftrightarrow \overline{18}. \qquad \overline{0} \leftrightarrow \overline{16} \leftrightarrow \overline{17}.$$
$$\overline{0} \leftrightarrow 0 \leftrightarrow 16.$$

By inspection the dilation is two; the number of edges with dilation two is 7; the congestion is two; the number of edges with congestion two is one; the active degree is 6 (i.e., node $\overline{16}$); and the node load is 6 (since all the intermediate nodes of the length-two paths are different).

Embedding an 11×11 mesh into a 7-cube

Figure 4.24 shows the embedding of an 11×11 mesh in a 7-cube. The length-two paths are specified as follows:

000000	000001	000101	000100	100000	100001
000010	000011	000111	000110	100010	100011
001010	001011	001111	001110	100110	100111
001000	001001	001101	0011)0	100100	100101
011000	011001	011101	011100	101100	101101
011010	011011	011111	011110	101110	101111
010010	010011	010111	010110	111110	111111
010000	010001	010101	010100	111100	111101
111000	111001	110001	110000	110100	110101
111010	111011	110011	110010	110110	110111

Figure 4.22   Embedding of a 10 × 6 mesh into a 6-cube.

$\overline{51} \leftrightarrow \overline{55} \leftrightarrow \overline{39},$	$\overline{49} \leftrightarrow \overline{53} \leftrightarrow \overline{37},$	$\overline{48} \leftrightarrow \overline{52} \leftrightarrow \overline{36}$
$\overline{48} \leftrightarrow \overline{16} \leftrightarrow \overline{0},$	$\overline{0} \overset{\bullet}{\leftrightarrow} \overline{1} \leftrightarrow \overline{17},$	$8 \leftrightarrow \overline{24} \leftrightarrow \overline{16}$
$\overline{16} \leftrightarrow \overline{20} \leftrightarrow \overline{22},$	$\overline{22} \leftrightarrow \overline{23} \leftrightarrow \overline{19},$	$32 \leftrightarrow 32 \leftrightarrow 36$
$\overline{32} \overset{\bullet}{\leftrightarrow} \overline{33} \leftrightarrow \overline{1},$	$\overline{33} \leftrightarrow \overline{37} \leftrightarrow 37,$	$\overline{33} \overset{\bullet}{\leftrightarrow} \overline{35} \leftrightarrow \overline{3}$
$\overline{35} \leftrightarrow \overline{39} \leftrightarrow 39,$	$\overline{35} \overset{\bullet}{\leftrightarrow} \overline{34} \leftrightarrow \overline{2},$	$\overline{50} \overset{\bullet}{\leftrightarrow} \overline{18} \leftrightarrow \overline{2}$
$\overline{18} \overset{\bullet}{\leftrightarrow} \overline{26} \leftrightarrow \overline{10},$	$\overline{34} \leftrightarrow \overline{38} \leftrightarrow 38,$	$\overline{42} \leftrightarrow 42 \overset{\bullet}{\leftrightarrow} 34$
$\overline{43} \leftrightarrow \overline{35} \leftrightarrow 35,$	$\overline{41} \leftrightarrow \overline{33} \leftrightarrow 33,$	$\overline{40} \leftrightarrow 40 \overset{\bullet}{\leftrightarrow} 32$

Note that the edges specified by all the paths are unique, and thus the congestion is at most two. By inspection, the dilation is 2; the number of edges with dilation 2 is 21; the congestion is 2; the number of edges with congestion 2 is 8 (as marked by the '•' signs).

4.7   Butterfly Network Emulation

Assume that the nodes of the butterfly network are labeled $(i,j), 0 \leq i < 2^n - 1, 0 \leq j \leq n$. Then node (i,j) is connected to nodes $(i, j+1)$ and $(i \oplus 2^j, j+1)$ for $0 \leq j < n$. Clearly, if we let $i, 0 \leq i \leq 2^n - 1$ be a processor address and let j denote time, then the required interconnection over the course of the computation corresponds to a Boolean cube. Algorithms that require butterfly network communication and proceed from rank to rank in a monotonically increasing order are known as normal algorithms

Table 4.6   Summary of embeddings for various aspect ratios.

Embedding	# of Edges w. Dil. 2	Average Dilation	Max Avg Dilation
$(\frac{5}{4}, \frac{3}{2})$	$N_1 + N_2$	$1 + \frac{4(N_1+N_2)}{15N_1N_2 - 5N_1 - 6N_2}$	1.3
$(\frac{9}{8}, \frac{7}{4})$	$\frac{3}{8}N_1 + N_2$	$1 + \frac{12N_1+32N_2}{63N_1N_2 - 36N_1 - 56N_2}$	1.2
$(\frac{11}{8}, \frac{11}{8})$	$\frac{5}{8}N_1 + 2N_2$	$1 + \frac{20N_1+64N_2}{121N_1N_2 - 44N_1 - 44N_2}$	1.1

[120]. With this type of emulation the processor utilization is one. In step j all processors communicate in dimension j. Only $\frac{1}{n}$th of the total communication bandwidth is used in any step.

It is often the case that many butterfly emulations need to be performed, for instance, when performing Fast Fourier Transforms along columns (or rows) of a multidimensional array (as in so-called fast Poisson solvers). A similar need occurs if a butterfly emulation is to be carried out for a virtual processor ratio greater than one. If the cyclic assignment is used, then the first $p - n$ ranks of butterfly computations are local to a processor. The last n ranks require interprocessor communication. For the consecutive assignment, the first n steps require interprocessor communication, and the last $p - n$ steps are local to a processor, assuming normal order input. With a single large butterfly computation as well as many small computations, the communication efficiency can be improved by the following techniques:

- Pipelining of successive butterfly computations.

- With the first several steps being local, all the computations on the local data set are performed first followed by a data reallocation such that a new butterfly computation on the same size data set, or a smaller set if $p - n > n$, is performed locally. With 2^{p-n} local elements this procedure is equivalent to emulating the butterfly as a radix-2^{p-n} butterfly.

- Perform a set of radix-Q_{vp} butterflies locally, $Q_{vp} \leq 2^{p-n}$, and then perform a data reallocation requiring interprocessor communication for the computation of radix-Q_c butterflies, $Q_c \leq 2^n$, and pipeline these computations and reallocations.

- Divide the *virtual processors* into n sets and perform a reallocation of

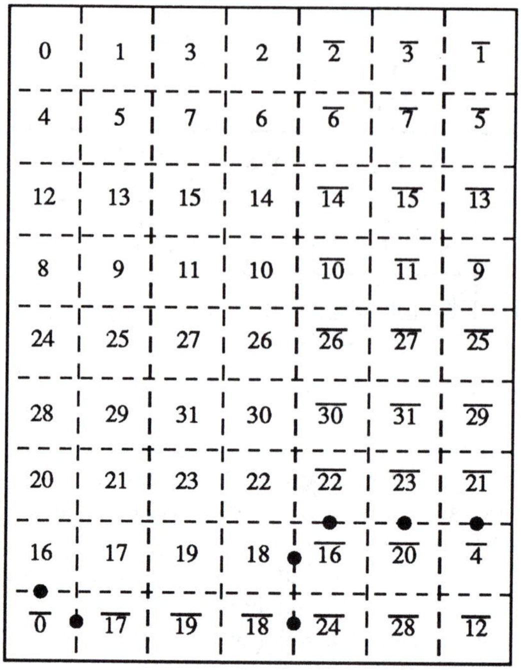

0	1	3	2	$\overline{2}$	$\overline{3}$	$\overline{1}$
4	5	7	6	$\overline{6}$	$\overline{7}$	$\overline{5}$
12	13	15	14	$\overline{14}$	$\overline{15}$	$\overline{13}$
8	9	11	10	$\overline{10}$	$\overline{11}$	$\overline{9}$
24	25	27	26	$\overline{26}$	$\overline{27}$	$\overline{25}$
28	29	31	30	$\overline{30}$	$\overline{31}$	$\overline{29}$
20	21	23	22	$\overline{22}$	$\overline{23}$	$\overline{21}$
16	17	19	18	$\overline{16}$	$\overline{20}$	$\overline{4}$
$\overline{0}$	$\overline{17}$	$\overline{19}$	$\overline{18}$	$\overline{24}$	$\overline{28}$	$\overline{12}$

Figure 4.23   Embedding of a 9 × 7 mesh in a 6-cube. [47]

the sets such that the first rank of each set requires communication in a distinct dimension. The same requirement holds for the second and all following ranks. The reallocation is obtained by m shuffle operations on the *virtual processor* set with index m (with sets labeled consecutively from zero).

For pipelining of computations we consider cyclic assignment of data elements to processors. After the first $p - n$ ranks of butterfly computations there are 2^{p-n} independent butterflies of size 2^n to be computed. This fact was used in [50] for devising sorting algorithms on Boolean cubes. Each butterfly has one element per processor, and each requires only one communication in each of the n cube dimensions. The computations of the 2^{p-n} butterflies each of size 2^n can be pipelined in such a way that a total of $n + 2^{p-n} - 1$ communications is required with concurrent communication on all ports. The communication efficiency, measured as (the sum over time of the communication resources used)/((total number of available communication resources)*(time)), for the stages requiring communication

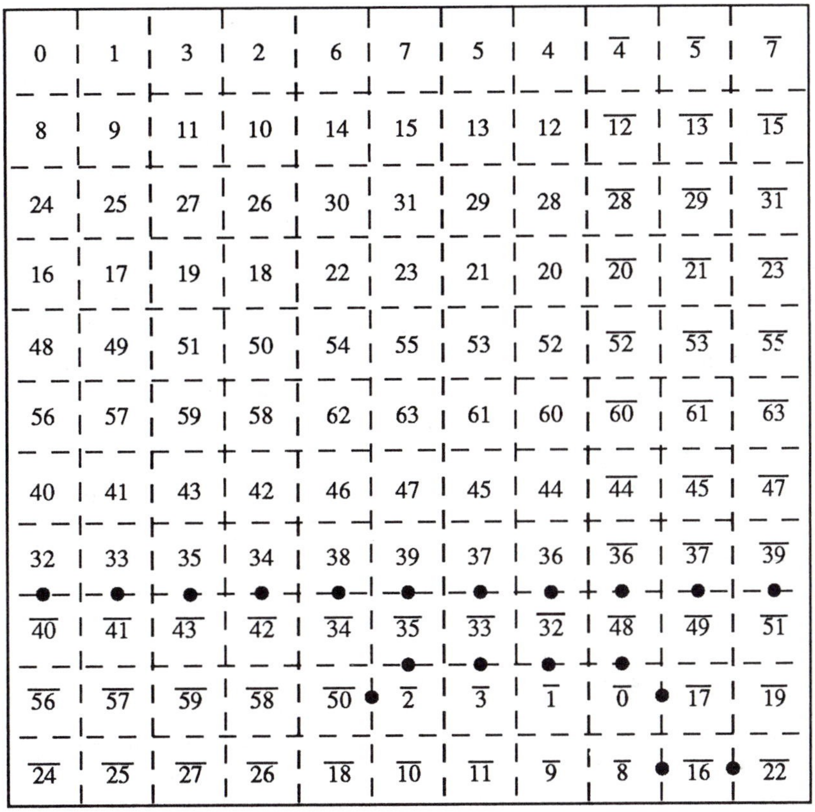

0	1	3	2	6	7	5	4	$\overline{4}$	$\overline{5}$	$\overline{7}$
8	9	11	10	14	15	13	12	$\overline{12}$	$\overline{13}$	$\overline{15}$
24	25	27	26	30	31	29	28	$\overline{28}$	$\overline{29}$	$\overline{31}$
16	17	19	18	22	23	21	20	$\overline{20}$	$\overline{21}$	$\overline{23}$
48	49	51	50	54	55	53	52	$\overline{52}$	$\overline{53}$	$\overline{55}$
56	57	59	58	62	63	61	60	$\overline{60}$	$\overline{61}$	$\overline{63}$
40	41	43	42	46	47	45	44	$\overline{44}$	$\overline{45}$	$\overline{47}$
32	33	35	34	38	39	37	36	$\overline{36}$	$\overline{37}$	$\overline{39}$
$\overline{40}$	$\overline{41}$	$\overline{43}$	$\overline{42}$	$\overline{34}$	$\overline{35}$	$\overline{33}$	$\overline{32}$	48	49	51
$\overline{56}$	$\overline{57}$	$\overline{59}$	$\overline{58}$	$\overline{50}$	2	3	1	0	17	19
$\overline{24}$	$\overline{25}$	$\overline{27}$	$\overline{26}$	$\overline{18}$	$\overline{10}$	$\overline{11}$	9	8	16	22

Figure 4.24   Embedding of an 11 × 11 mesh in a 7-cube. [47]

is $(\frac{n+1}{2} + 2^{p-n})/(n + 2^{p-n})$, if $2^{p-n} > n$. The efficiency is approximately one for $2^{p-n} \gg n$. The total communication efficiency measured as (the sum of the communication resources used over time)/((available communication resources)*(total time)) is $(\frac{n+1}{2} + 2^{p-n})/(n + 2^{p-n} + (p - n)2^{p-n})$, which for $2^{p-n} \gg n$ is approximately equal to $\frac{1}{p-n}$. The reason for this low efficiency is simply that most of the computations are local to a processor.

An alternative to this straightforward pipelined algorithm is obtained by recursively partitioning the set of 2^{p-n} butterflies such that one half of them is computed in a half-sized Boolean cube and the other half of the butterflies in the other half of the cube. The number of elements per processor of a given butterfly doubles for every step. The communication time for this pipelined, recursive partitioning of butterflies is $2^{p-n-1} + n - 1$, which is approximately half of that of the straightforward pipelined algorithm. The recursive partitioning technique as a means of improving the load balance was also used in [58, 61] for *Balanced Cyclic Reduction* on Boolean cubes.

Performing the butterfly computation as a sequence of radix-2^{p-n} butterflies requires a data reallocation of the form *all-to-all personalized communication* [66] in cubes of dimension $p - n$. There exist 2^{2n-p} distinct such cubes in an n-cube, assuming $2n - p > 0$. The total number of reallocations required is $\lceil \frac{n}{p-n} \rceil$, each requiring a total of 2^{p-n-1} element transfer times, with concurrent communication on all $p - n$ ports.

Pipelining and high radix butterfly computations can be combined. After the butterfly computations on the local data set are carried out, radix-Q_c butterflies involving communication are performed, and the different butterflies pipelined with respect to communication, assuming $Q_c < 2^n$. The total number of element transfers in sequence for this algorithm is $(2^{p-n-q_c} + \lceil \frac{n}{q_c} \rceil - 1)2^{q_c-1}$, where $Q_c = 2^{q_c}$. This expression specializes to the pipelined, recursive partitioning radix-2 algorithm for $q_c = 1$, and the non-pipelined radix-2^{p-n} algorithm for $q_c = p - n$. The value of the expression is minimized for $q_c = 2$ if $q_c < n$. However, the improvement of such a radix-4 algorithm over a radix-2 algorithm is insignificant: one element transfer.

We conclude that with respect to communication efficiency a pipelined, partitioning, radix-2 or radix-4 algorithm shall be used if $p - n < n$. Otherwise, a radix-q_c algorithm with $q_c \geq n$ shall be chosen. In such a case the pipeline filling time of $n - 1$ is avoided. For a large number of *virtual processors* this time is only a small improvement.

A fourth alternative is to divide the *virtual processors* into n sets and perform a reallocation by cyclic rotation of the addresses m steps for set m with the sets numbered consecutively from 0. After such a reallocation, the communication required for any rank is in different dimensions for the different sets of *virtual processors*. The communication system is fully utilized. A final reallocation is required after the completion of the computations to realign the data. A one step rotation is a shuffle operation; $sh(u) = (u_{n-2}u_{n-3}\ldots u_0 u_{n-1})$, where $u = (u_{n-1}u_{n-2}\ldots u_0)$. A rotation of j steps is defined by $sh^j = sh \cdot sh^{j-1}$. The communication time required for the two sets of shuffle operations and the butterfly computation is at least $2 \cdot 2^{p-n-1}\frac{n-1}{n} + 2^{p-n-1}$. This bound is obtained by dividing the local data set into n subsets and noticing that the average communication distance is $\frac{n}{2}$. Butterfly computations based on shuffle operations are clearly inferior to the pipelined or high radix algorithms.

4.8   Tree Embeddings

This section addresses four different communication problems in Boolean n-cube configured multiprocessors: (1) one-to-all broadcasting: distribution of common data from a single source to all other nodes; (2) one-to-all personalized communication: a single node sending unique data to all other nodes; (3) all-to-all broadcasting: distribution of common data from each node to all other nodes; and (4) all-to-all personalized communication: each node sending a unique piece of information to every other node. We will present optimum communication algorithms for each of these four cases. In *broadcasting*, a data set is copied from one node to all other nodes, or a subset thereof. In *personalized communication*, a node sends a unique data set to all other nodes, or a subset thereof. We consider broadcasting from a single source to all other nodes, *one-to-all broadcasting*, and concurrent broadcasting from all nodes to all other nodes, *all-to-all broadcasting*. For personalized communication we consider *one-to-all personalized communication* and *all-to-all personalized communication*. The difference between broadcasting and personalized communication is that in the latter no replication/reduction of data takes place. The bandwidth requirement is highest at the root and is reduced monotonically toward the leaves.

For single-source broadcasting and personalized communication a *one-to-all communication graph* is required. Graphs of minimum height have minimum propagation time, which is the overriding concern for small data volumes, or a high overhead for each communication action. For large data volumes it is important to use the bandwidth of a Boolean cube effectively. Ho and Johnsson [66] present three spanning graphs that make full use of the bandwidth for some of the communication primitives considered here. The three graphs are n edge-disjoint binomial trees (nESBT), n rotated binomial trees (nRSBT), and *balanced trees* (SBnT). These graphs allow lower bound communication as shown in Table 4.7.

A *Spanning Tree* rooted at node s is denoted $T^{id}(s)$, where id is used to identify the type of the spanning tree. A spanning tree rooted at node s is a translation of the same type spanning tree rooted at node 0, $T^{id}(s) = Tr(s, T^{id}(0))$. The root of a spanning tree has *level* 0. The node i in a spanning tree has a level that is one more than the level of its parent. The height h of a tree is the largest level of all the nodes.

Definition 15   A *one-to-all communication graph*, henceforth called an *o-graph*, with source node s is a connected, directed graph $G^{id}(s) = Tr(s, G^{id}(0))$ where $G^{id}(0)$ is either a spanning tree, $G^{id}(0) = T^{id}(0)$, or a composition of n distinctly rotated spanning trees, $G^{id}(0) = \cup_{d \in \mathcal{D}} Ro^d(T^{id'}(0))$. The weight of every edge in an *o-graph* is 1 if the graph is a tree and $\frac{1}{n}$ otherwise. id' identifies the generating spanning tree for the *o-graph*. ∎

Table 4.7 Lower bounds for some Boolean cube communications. [66]

Comm. model	Communication task	Lower bound
one-port	One-to-all broadcasting	$\max((M + n - 1)t_c, n\tau)$
	One-to-all personalized comm.	$\max((N - 1)Mt_c, n\tau)$
	All-to-all broadcasting	$\max((N - 1)Mt_c, n\tau)$
	All-to-all personalized comm.	$\max(\frac{nNM}{2}t_c, n\tau)$
n-port	One-to-all broadcasting	$\max((\frac{M}{n} + n - 1)t_c, n\tau)$
	One-to-all personalized comm.	$\max(\frac{(N-1)M}{n}t_c, n\tau)$
	All-to-all broadcasting	$\max(\frac{(N-1)M}{n}t_c, n\tau)$
	All-to-all personalized comm.	$\max(\frac{NM}{2}t_c, n\tau)$

The weight $\frac{1}{n}$ is introduced to account for splitting the data to be communicated into n pieces for an *o-graph* composed of n spanning trees. There exist n paths from the root to any other node. If no two edges of $T^{id_x}(s)$ and $T^{id_y}(s)$ are mapped to the same cube edge $\forall x, y \in \mathcal{D}$, $x \neq y$, then the paths of the *o-graph* are *edge-disjoint*, and there is no contention problem.

Definition   16   For an *o-graph* G^{id}, the total weight of the edges in dimension d between levels l and $l + 1$ are denoted $e^{id}(d, l)$, and the total weight of edges in dimension d is $E^{id}(d)$. ∎

Lemma   5   *Given an o-graph* $G^{id}(s) = Tr(s, G^{id}(0))$ *where* $G^{id}(0) = \cup_{d \in \mathcal{D}} Ro^d(T^{id'}(0))$, *then,* $E^{id}(d) = \frac{N-1}{n}$ *and* $e^{id}(d, l) = \frac{1}{n}\sum_{d \in \mathcal{D}} e^{id'}(d, l)$.

Proof   The lemma follows from definition 16 and the rotation property in lemma 3. □

We define three basic spanning trees: a *Hamiltonian* path, T^H, a *Spanning Binomial Tree*, T^{SBT}, and a *Spanning Balanced n-Tree*, T^{SBnT}. These trees are used to form composition graphs that provide multiple paths (not necessarily edge-disjoint) to all other nodes. The $children^{id}(i,s,k)$ function generates the set of children addresses of node i in the kth spanning tree of $G^{id}(s)$. The $parent^{id}(i,s,k)$ function generates the address of the parent of node i in the kth spanning tree of $G^{id}(s)$. For G^{id} being a spanning tree we omit the last parameter.

Definition   17   A *greedy spanning tree* rooted at node s of a Boolean n-cube is a spanning tree such that $\forall i \in \mathcal{N}$; the level of node i is $Hamming(i,s)$. A *greedy o-graph* of a Boolean n-cube is a composition of greedy spanning trees.   ■

Lemma   6   *A spanning tree is greedy iff* $|\{i \mid Hamming(i,s) = l\}| = \binom{n}{l}$. *A greedy* o-graph *is acyclic and of minimal height.*   □

Corollary   3   *If* $G^{id}(0) = \cup_{d \in \mathcal{D}} Ro^d(T^{id}(0))$, *and* T^{id} *is greedy, then* $e^{id}(d,l) = \frac{1}{n}\binom{n}{l+1}$.   □

A greedy *o-graph* contains only $0 \rightarrow 1$ edges "relative" to the root. Lemma 6 follows from definition 17. Corollary 3 follows from lemmas 5 and 6. Note that an *o-graph* of minimum height is not necessarily a greedy *o-graph*. In all-to-all personalized communication only greedy *o-graphs* have scheduling disciplines that accomplish minimal data transfer time. For *broadcasting* the data is replicated $|children^{id}(i,s,k)|$ times in node i for spanning tree k of the *o-graph* $G^{id}(s)$. With *n-port* communication, and negligible time for replication, all ports are scheduled concurrently. With *one-port* communication the order of communications on different ports is important. In *personalized communication* the source node sends a unique message to every other node. An internal node needs to receive and forward all the data for every node of the subtree of which it is a root. The ordering of data for a port is important for the communication time both for *one-port* and *n-port* communication. The *scheduling discipline* defines the communication order for each port and the order between ports for every non-leaf node. We assume the same data-independent scheduling discipline for every node. In a *reverse-breadth-first* scheduling discipline for *one-to-all personalized communication* based on an *o-graph* of height h, the root sends out the data for the nodes at level $h-p$ during the pth cycle, $0 \le p < h$. The data received by an internal node is propagated to the next level during the next cycle, if the data is not destined for the node itself. In a *postorder* [1] scheduling discipline, each node sends out the entire data set to each of its children nodes before accepting its own data.

The analysis of the complexity of the communication algorithms is considerably simplified for *n-port* communication if only the subtree with the maximum number of nodes needs to be considered. The following lemma guarantees that this is the case if the subtree with the maximum number of nodes also has the maximum height.

Lemma 7 *Given a spanning tree, let $\phi^{id}(i, x)$ be the number of nodes at distance x from node i in the subtree rooted at node i. If $\phi^{id}(i, x) \geq \phi^{id}(j, x)$ for any child node j of node i, then the data transfer time of n-port one-to-all personalized communication based on reverse-breadth-first ordering is dominated by the data transfer over the edges from the root.*

Proof The lemma follows from the fact that with *reverse-breadth-first* scheduling, the propagation time for the internal nodes is at most the same as the transmission time for the root during each routing cycle. □

Definition 18 An *all-to-all communication graph*, henceforth called an *a-graph*, is a graph $G^{id}(*) = \cup_{s \in \mathcal{N}} G^{id}(s)$.

The quantities $v^{id}(d, l)$ and $u^{id}(d, l)$ defined next are useful in deriving the time complexity of *all-to-all personalized communication* for the *a-graph*. ∎

Definition 19 Define $v^{id}(d, l)$ of an *o-graph* as the total weight of all edges within all subtrees rooted at level $l + 1$ of all spanning trees with subtree roots connected to a parent node through an edge in dimension d, inclusive of the edge to the parent node. Let $\mathcal{S}_d^{id} = \{j | j \in \mathcal{V}, (i, j) \in \mathcal{E}^{id}$ and $i \oplus j = 2^d\}$. Define $u^{id}(d, l)$ to be the total weight of edges terminating on all nodes k such that k is a descendent of j at distance l, $\forall j \in \mathcal{S}_d^{id}$. ∎

Lemma 8 $v^{id}(d, l) = u^{id}(d, l) = \sum_{x=l}^{h-1} e^{id}(d, x) = \frac{1}{n} \sum_{x=l}^{h-1} \sum_{d \in \mathcal{D}} e^{id'}(d, x)$, $\forall l \in [0, h-1]$, $d \in \mathcal{D}$ for an o-graph $G^{id}(0) = \cup_{d \in \mathcal{D}} Ro^d(T^{id'}(0))$, where h is the height of $T^{id'}$.

Proof From the n distinct rotations and lemma 3, it follows that $v^{id}(d, l) = \frac{1}{n} \times$ (the sum of the number of nodes at levels x, $l + 1 \leq x \leq h$) of $T^{id'}$. Similarly, $u^{id}(d, l) = \frac{1}{n} \times$ (the sum of the number of nodes at levels x, $l + 1 \leq x \leq h$) of $T^{id'}$. By lemma 5, $e^{id}(d, l) = \frac{1}{n} \times$ (the number of edges between levels l and $l + 1$) of $T^{id'} = \frac{1}{n} \sum_{d \in \mathcal{D}} e^{id'}(d, l)$. □

Corollary 4 *If an o-graph $G^{id}(0) = \cup_{d \in \mathcal{D}} Ro^d(T^{id'}(0))$ and $T^{id'}$ is greedy, then $v^{id}(d, l) = u^{id}(d, l) = \frac{1}{n} \sum_{i=l+1}^{n} \binom{n}{i}$.*

Proof   The corollary follows from lemma 8 and corollary 3.   ▯

Lemma   9   *For an a-graph $G^{id}(*)$, the total weight of communication graph edges mapped to every cube edge in dimension d is $E^{id}(d)$. The total weight of communication graph edges between levels l and l+1 in dimension d is $e^{id}(d,l)$.*

Proof   Since $G^{id}(*) = \cup_{s \in \mathcal{N}} G^{id}(s) = \cup_{s \in \mathcal{N}} Tr(s, G^{id}(0))$, every o-graph edge is mapped to a distinct cube edge in the same dimension through N distinct exclusive-or operations. Hence, the total weight of the *a-graph* edges mapped to every cube edge in dimension d is $E^{id}(d)$. The bit-wise exclusive-or operation preserves the topology of a spanning tree, and hence the number of edges at a given distance from the source node.   ▯

4.8.1   Spanning Graphs

A spanning binomial tree
A 0-level binomial tree has 1 node. An n-level binomial tree is constructed out of two $(n-1)$-level binomial trees by adding one edge between the roots of the two trees and by making either root the new root, [1, 29]. The familiar spanning tree rooted in node 0 of a Boolean n-cube generated by complementing leading zeroes of the binary encoding of a processor address i [31, 56, 82, 103, 104] is a *spanning binomial tree* (SBT). Let p be such that $c_p = 1$ and $c_m = 0, \forall m \in \{p+1, p+2, \ldots, n-1\} \equiv \mathcal{M}^{SBT}(c)$ and let $p = -1$ if $c = 0$. The set $\mathcal{M}^{SBT}(c)$ is the set of leading zeroes of c. Then the spanning binomial tree rooted at node s, $T^{SBT}(s)$, is defined as follows:

$$children^{SBT}(i,s) = \{(i_{n-1}i_{n-2}\ldots\bar{i}_m\ldots i_0)\}, \forall m \in \mathcal{M}^{SBT}(c),$$
$$parent^{SBT}(i,s) = \begin{cases} \phi, & i = s; \\ (i_{n-1}i_{n-2}\ldots\bar{i}_p\ldots i_0), & i \neq s. \end{cases}$$

It is easy to verify that the parent and children functions are consistent, i.e., that node j is a child of node i iff node i is the parent of node j. Figure 4.25 shows the $T^{SBT}(0)$ for a 4-cube.

Subtree k of the $T^{SBT}(s)$ consists of all nodes such that $c_k = 1$ and $c_m = 0, \forall m \in \{0, 1, \ldots, k-1\}$.

Lemma   10   *There are 2^{n-k-1} nodes in subtree k, and the maximum degree of any node at level l in subtree k is $n - k - l$, $0 < l \leq n - k$.*   ▯

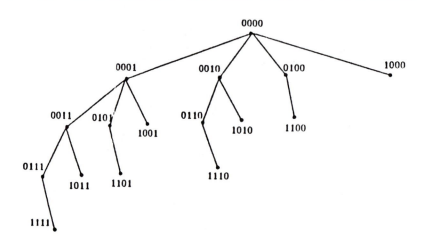

Figure 4.25   A spanning binomial tree in a 4-cube. [66]

Lemma 11   *Let* $\phi^{SBT}(i,s,x)$ *be the number of nodes at distance* x *from node* i *in the subtree rooted at node* i *of the SBT rooted at node* s. *Then,* $\phi^{SBT}(i,s,x) \geq \phi^{SBT}(j,s,x)$, *if* $j \in children^{SBT}(i,s)$.

Proof   From the definition of the SBT, the subtree rooted at node j is a connected subgraph of the subtree rooted at node i. $\Box$

A spanning balanced n-tree and a spanning balanced graph
In the *spanning balanced n-tree* [48] (SBMT) the node set is divided into n sets of nodes with approximately an equal number of nodes. Each such set forms a subtree of the source node. The maximum number of elements that need to traverse any edge directed away from the source node is minimized for personalized communication.

Definition 20   *Let* $\mathcal{K}(i,s) = \{k_1, k_2, \ldots, k_m | 0 \leq k_1 < k_2 < \cdots < k_m < n\}$, *be such that* $Ro^\alpha(c) = Ro^\beta(c)$, $\forall \alpha, \beta \in \mathcal{K}(i,s)$, *and* $Ro^\alpha(c) < Ro^\gamma(c)$, $\forall \alpha \in \mathcal{K}(i,s)$, $\gamma \notin \mathcal{K}(i,s)$. *Then,* $base^{SBnT}(i,s) = k_1$. ∎

For example, $base^{SBnT}((011100), 0) = 2$ and $base^{SBnT}((110100), (000010)) = 1$. Note that $|\mathcal{K}(i,s)| = n/P(c)$ where $P(c)$ is the period of c. The value of the base equals the minimum number of right rotations that minimize the value of c. For noncyclic nodes $|\mathcal{K}(i,s)| = 1$, but for a cyclic node c, $P(c) < n$, and $|\mathcal{K}(i,s)| > 1$. The notion of $base^{SBnT}$ is similar to the notion of the distinguished node used in [77] in that $base^{SBnT} = 0$ distinguishes a node from a generator set (necklace). To simplify the notation we omit the subscript on k in the following. Subtree k of the *Spanning Balanced n-Tree* rooted at node s consists of all nodes $i \neq s$ such that $base^{SBnT}(i,s) = k$. Note that all nodes in subtree k have $c_k = 1$, but not all nodes with $c_k = 1$ are in the kth subtree. Let $base^{SBnT}(i,s) = k$. For $c = 0$ let $p = -1$, else if $c_k = 1$ then $p = k$, else let p be the first bit cyclically to the right of bit k that is equal to 1 in c, i.e., $c_p = 1$, and $c_m = 0, \forall m \in \{(p+1) \bmod n, (p+2) \bmod n, \ldots, (k-1) \bmod n\} \equiv \mathcal{M}^{SBnT}(i,s)$ with $k = n$ if $c = 0$. The spanning tree $T^{SBnT}(s)$ is defined through

$$children^{SBnT}(i,s) =$$
$$\begin{cases} \{(i_{n-1}i_{n-2}\ldots \bar{i}_m \ldots i_0)\}, \forall m \in \mathcal{M}^{SBnT}(i,s), & \text{if } c = 0; \\ \{q_m = (i_{n-1}i_{n-2}\ldots \bar{i}_m \ldots i_0)\}, & \\ \quad \forall m \in \mathcal{M}^{SBnT}(i,s) \text{ and } base^{SBnT}(q_m,s) = base^{SBnT}(i,s), & \text{if } c \neq 0. \end{cases}$$

$$parent^{SBnT}(i,s) = \begin{cases} \phi, & \text{if } c = 0; \\ (i_{n-1}i_{n-2}\ldots \bar{i}_p \ldots a_0), & \text{otherwise.} \end{cases}$$

The $parent^{SBnT}$ function preserves the base, since for any node i with base k, c_p is the highest order bit of $Ro^k(c)$. Complementing this bit cannot change the base. It is also readily seen that the $parent^{SBnT}$ and $children^{SBnT}$ functions are consistent.

Theorem 3   *The $parent^{SBnT}(i,s)$ function defines a spanning tree rooted at node s.*

Proof   For every node i the $parent^{SBnT}(i,s)$ function generates a path to node s, and hence the graph is connected. Moreover, the parent node of a node at distance l from node s is at distance $l-1$ from node s, and each node only has one parent node. Hence, the graph is a spanning tree. □

Figure 4.26 shows a spanning balanced 5-tree in a 5-cube.

Lemma 12   *The SBnT is a greedy spanning tree.*

Proof   From the definition of the $parent^{SBnT}(i,s)$ function, it follows that the distance from node i to node s is $Hamming(i,s)$. □

Lemma 13   *[48] Let $\phi^{SBnT}(i,s,x)$ be the number of nodes at distance x from node i in the subtree rooted at node i of the SBnT rooted at node s. Then, $\phi^{SBnT}(i,s,x) \geq \phi^{SBnT}(j,s,x)$, if $j \in children^{SBnT}(i,s)$.* □

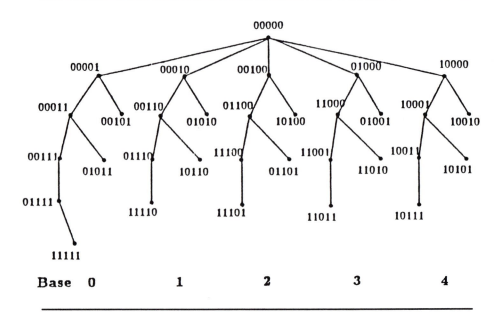

Figure 4.26   A spanning balanced 5-tree in a 5-cube. [66]

Theorem   4   [66] Excluding node $(\bar{s}_{n-1}\bar{s}_{n-2}\ldots\bar{s}_0)$, *all the subtrees of the root of the SBnT are isomorphic if n is a prime number. Furthermore, the kth subtree can be derived by* $(k-j)$ *mod n left rotation steps of each node of the jth subtree.*

Proof   For n, a prime number, there are no cyclic nodes, except nodes with relative addresses $(00\ldots0)$ and $(11\ldots1)$. For all other nodes $|\mathcal{K}(i,s)| = 1$, and each generator set has n members. Any subtree has the same number of nodes as any other subtree at every level. It follows from the definition of $base^{SBnT}$ that if $j \in children^{SBnT}(i,s)$, then $Ro^k(j) \in children^{SBnT}(Ro^k(i),s), \forall k \in \mathcal{D}$. $\square$

If n is not a prime number, then some subtrees of the root of the SBnT will contain more nodes than others. It can be shown that the number of nodes in a subtree is $O(\frac{N}{n})$ [48]. The imbalance is illustrated in Table 4.8. This imbalance is important for personalized communication. The number of elements transferred over the edges can be perfectly balanced in the sense that the maximum at any level is minimized by allowing multiple paths to cyclic nodes. With multiple paths to cyclic nodes the graph is no longer a tree. We call the SBnT so modified a *spanning balanced graph* (SBG).

Table 4.8. A comparison of subtree sizes of SBT and SBnT. The last column contains the ratio of SBnT(max) to $\frac{N-1}{n}$. [66]

n	SBT(max)	SBnT(max)	SBnT(min)	$(N-1)/n$	factor
2	2	2	1	1.50	1.33
3	4	3	2	2.33	1.29
4	8	5	3	3.75	1.33
5	16	7	6	6.20	1.13
6	32	13	9	10.50	1.24
7	64	19	18	18.14	1.05
8	128	35	30	31.88	1.10
9	256	59	56	56.78	1.04
10	512	107	99	102.30	1.05
11	1024	187	186	186.09	1.00
12	2048	351	335	341.25	1.03
13	4096	631	630	630.08	1.00
14	8192	1181	1161	1170.21	1.01
15	16384	2191	2182	2184.47	1.00
16	32768	4115	4080	4095.94	1.00
17	65536	7711	7710	7710.06	1.00
18	131072	14601	14532	14563.50	1.00
19	262144	27595	27594	27594.05	1.00
20	524288	52487	52377	52428.75	1.00

The SBG can be defined as a composition of n distinctly rotated SBnT's as follows.

Define $G^{SBG}(0) = \cup_{d \in \mathcal{D}} Ro^d(T^{SBnT}(0))$ and $G^{SBG}(s) = Tr(s, G^{SBG}(0))$. The number of different paths to node i in an SBG rooted at node s is $\frac{n}{P(i \oplus s)}$ where $P(i)$ is the period of i.

4.8.2   Spanning Graphs Composed of n Spanning Trees

n rotated Hamiltonian paths

A spanning graph rooted at node s and consisting of n rotated Hamiltonian paths is $G^{nRH}(s) = Tr(s, G^{nRH}(0))$, where $G^{nRH}(0) = \cup_{d \in \mathcal{D}} Ro^d(T^H(0))$. The paths generated through n distinct rotations of the $G^H(s)$ path are not edge-disjoint for $n > 2$. For instance, the edge $(2,6)$ is used in two paths of the graph $G^{3RH}(0)$, and these paths are part of every graph $G^{nRH}(0)$ for $n > 3$. Path $Ro^k(T^H(s))$ and path $Ro^m(T^H(s))$ share $2^{n-\alpha-1} + 2^{n-\beta-1} - 2$ edges, where $\alpha = (k - m) \bmod n$ and $\beta = (m - k) \bmod n$. Each of the n cube edges $((00\ldots01_d0\ldots0), (00\ldots01_{(d+1)\bmod n}1_d0\ldots0))$, $\forall d \in \mathcal{D}$ are shared by $n - 1$ paths for $s = 0$. Figure 4.27 shows the graph $G^{3H}(0)$. The

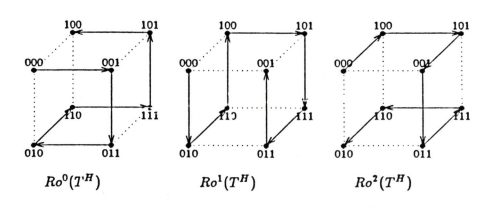

Figure 4.27   Three rotated binary-reflected Gray code paths in a 3-cube. [66]

fact that the paths are not edge-disjoint limits the potential for pipelining. Note that even though the n rotated Hamiltonian paths are not edge-disjoint, it is possible to have edge-disjoint embeddings of several Hamiltonian paths generated in other ways. Figure 4.28 shows four edge-disjoint paths in a 4-cube.

n rotated spanning binomial trees
The graph consisting of n rotated SBTs is defined by $G^{nRSBT}(s) = Tr(s, G^{nRSBT}(0))$, and $G^{nRSBT}(0) = \cup_{d \in \mathcal{D}} Ro^d(T^{SBT}(0))$. Figure 4.29 shows the nRSBT graph of a 3-cube. In general, a cube edge (i, j) is part of several rotated SBT's. The number on each cube edge in the figure shows the sum of the weights of the graph edges mapped onto it.

Lemma   14   *For any node in the graph $G^{nRSBT}(0)$, except the root, the weight of the incoming cube edge in dimension d is $\frac{p+1}{n}$, where p is the number of consecutive 0-bits immediately to the left of bit d.*

Proof   A node in the SBT graph has an incoming edge in dimension d if the bit in dimension d is the highest order bit. In the nRSBT graph the number of incoming edges (i, j) to node j is equal to the number of graphs $Ro^m(G^{SBT}(0))$, $m \in \mathcal{D}$ such that there exists an edge (i^m, j^m), where $i = Ro^m i^m$, and $j = Ro^m j^m$. Such an edge occurs in $Ro^m(\mathcal{E}^{SBT}(0))$ for $m = \{d+1, d+2, \ldots, d+p+1\}$, i.e., in these $p+1$ SBT's, bit d is the last bit complemented in reaching node j. □

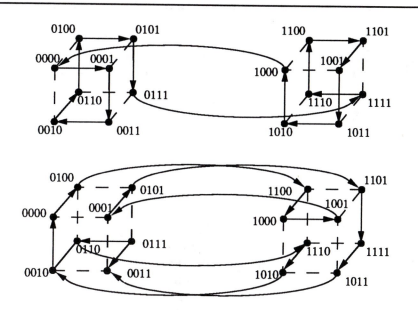

Figure 4.28   Four directed edge-disjoint Hamiltonian paths in a 4-cube. Only two paths are shown. The other two paths can be derived from the reversed paths. [66]

For instance, node (011001) has an incoming cube edge in dimension 0 with a weight of $\frac{1}{2}$, an incoming edge in dimension 3 weighted at $\frac{1}{6}$, and an incoming edge in dimension 4 weighted at $\frac{1}{3}$.

Corollary   5   *The sum of the weights of incoming edges is 1 for every node except the source node.*

Proof   From lemma 14 the weight of an incoming edge is equal to the number of dimensions between the dimension considered and the next higher dimension with a 1-bit.   ☐

Lemma   15   *In any dimension the edges of the nRSBT graph are mapped to only half of the cube edges.*

Proof   Rotation does not change the direction of edges.   ☐

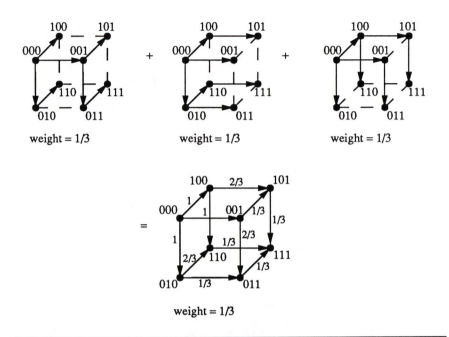

Figure 4.29   Three rotated spanning binomial trees as an *o-graph* in a 3-cube. [66]

n edge-disjoint spanning binomial trees

The nESBT (n edge-disjoint spanning binomial trees) graph is composed of n SBT's with one tree rooted at each of the nodes adjacent to the source node. The SBT's are rotated such that the source node of the nESBT graph is in the smallest subtree of each SBT. The nESBT graph is then obtained by reversing the edges from the roots of the SBT's to the source node. The nESBT graph $G^{nESBT}(s) = Tr(s, G^{nESBT}(0))$, where $G^{nESBT}(0) = \cup_{d \in D} T^{SBT_d}(0)$, and $T^{SBT_d}(0) = Tr(2^d, Ro^{n-d-1}(T^{SBT}(0)))$ (with the root being node 0 instead of node 2^d). Figure 4.30 shows an nESBT graph in a 3-cube. The nESBT graph is not a tree and contains cycles. Every node appears in every subtree of the source node. The height of the nESBT graph is $n + 1$, since the source node is adjacent to all the roots of the SBT's used in the definition of the nESBT graph. The number of distinct edges in the n SBT's is $n(N - 1)$. An alternative definition of G^{nESBT} is through the functions $children^{nESBT}$ and $parent^{nESBT}$. For a given k, let p be such that $c_p = 1$, and $c_m = 0$, $\forall m \in \{(p+1) \bmod n, (p+2) \bmod n, \ldots, (k-1) \bmod n\}$. For $c = 0$, $p = -1$, $k = n$, and for $c_k = 1$, $p = k$.

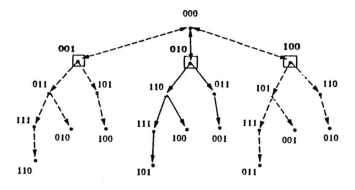

Figure 4.30   Subtrees of an nESBT viewed as SBT's. [66]

Then $\mathcal{M}^{nESBT}(i,s,k) = \{(p+1) \bmod n, (p+2) \bmod n, \ldots, (k-1) \bmod n\}$. The children and parent of node i in the kth spanning tree, $T^{SBT_k}(s)$, are

$$children^{nESBT}(i,s,k) =$$

$$\begin{cases} \{(i_{n-1}i_{n-2}\ldots \overline{i}_k \ldots i_0), & \text{if } c = 0; \\ \{(i_{n-1}i_{n-2}\ldots \overline{i}_m \ldots i_0)\}, \forall m \in \mathcal{M}^{nESBT}(i,s,k) \bigcup \{k\}, & \text{if } c_k = 1, p \neq k; \\ \{(i_{n-1}i_{n-2}\ldots \overline{i}_m \ldots i_0)\}, \forall m \in \mathcal{M}^{nESBT}(i,s,k), & \text{if } c_k = 1, p = k; \\ \phi, & \text{if } c_k = 0, c \neq 0. \end{cases}$$

$$parent^{nESBT}(i,s,k) = \begin{cases} \phi, & \text{if } c = 0; \\ (i_{n-1}i_{n-2}\ldots \overline{i}_k \ldots i_0), & \text{if } c_k = 0, c \neq 0; \\ (i_{n-1}i_{n-2}\ldots \overline{i}_p \ldots i_0), & \text{if } c_k = 1. \end{cases}$$

Dimension p is the first dimension to the right of dimension k, cyclically, which has a bit equal to 1. All nodes with $c_k = 0$, except node s, are leaf nodes of the kth subtree (or kth spanning tree). Conversely, all nodes with $c_k = 1$ are internal nodes of the kth subtree. The exceptional connection to node 0 is handled by the conditions on p. The first case defines the children for the source node, the second the set of children nodes of internal nodes of the kth subtree, except the node at level 1. The third case handles the node at level 1, and the last case handles the leaf nodes. Figure 4.31 shows that the three subtrees (spanning trees) of a 3ESBT graph are edge-disjoint. The labels on the edges will be used later. Figure 4.32 shows a 4ESBT graph with source node 0. Every 1-bit in the node address divides the outgoing edges into distinct sets. The set of 0-bits to the right of a

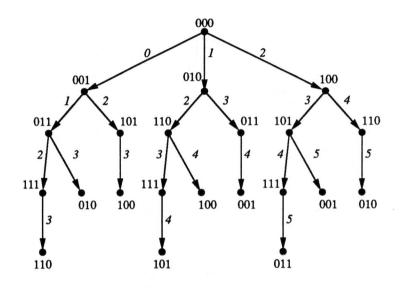

Figure 4.31   Three edge-disjoint directed spanning trees in a 3-cube. [66]

1-bit defines children nodes for a spanning subtree with the same index as the dimension of the 1-bit, Figure 4.33.

Theorem 5   *[66] The n subtrees of the nESBT graph are edge-disjoint.*

Proof   We need to prove only that for an arbitrary node the address of its parent node in each of the n subtrees is obtained by complementing a distinct bit.

From the definition of the $children^{nESBT}(i, s, k)$ (or $parent^{nESBT}$ (i, s, k)) function, it is clear that a node is a leaf node of the kth subtree iff $c_k = 0$, with the exception of node s. If a node is a leaf node in a particular subtree, then its parent address in that subtree is obtained by complementing the corresponding bit in its address (bit k for the kth subtree). If a node is an internal node of the kth subtree, then the corresponding bit is 1, and the parent address is obtained by complementing the first bit cyclically to the right of the kth address bit that is equal to 1. Hence the addresses of the parent nodes for all the subtrees of which the node is an internal node are also obtained by complementing distinct bits.

□

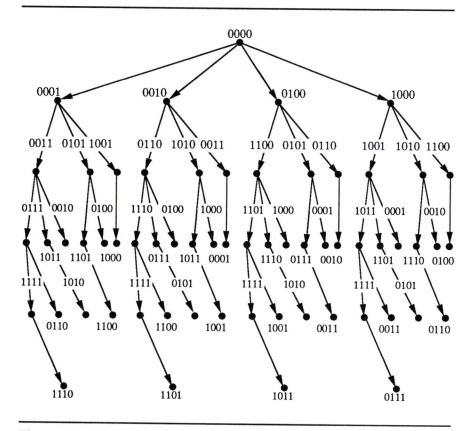

Figure 4.32   Four edge-disjoint directed spanning trees in a 4-cube. [66]

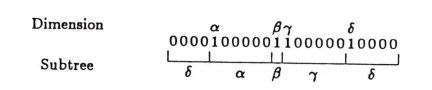

Figure 4.33   Scheduling of broadcasting operations for a node with nESBT routing. [66]

Figure 4.34 shows the parents and children of one node in a 6-cube. The numbers on the edges are the dimensions through which the node connects to its parents or children nodes in different subtrees. The labels on the nodes denote the subtree to which the parent and children nodes belong.

The nESBT graph for a Boolean n-cube is a directed graph such that all directed cube edges, except those incident on the source node, appear precisely once in the nESBT graph. The in-degree and the out-degree of any node in an nESBT graph is n, with the exception that the source has an in-degree 0 and all the neighbors of the source have an out-degree $n-1$.

Theorem 6 *[66] The height of the nESBT graph is minimal among all possible configurations of n edge-disjoint spanning trees.*

Proof To prove that with n edge-disjoint spanning trees the height $n+1$ is minimal, we show that n disjoint spanning trees with height n is impossible. The total number of directed edges in an n-cube is nN, but only the edges directed out from the source may be used. Each spanning tree has $N-1$ edges. Hence every eligible edge is used by the n edge-disjoint spanning trees. It follows that the edges directed out from the node with address $(\bar{s}_{n-1}\bar{s}_{n-2}\ldots\bar{s}_0)$ also must be used, and since this node is at distance n from the source node, the theorem follows. $\square$

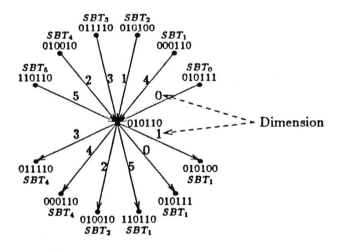

Figure 4.34   Parents and children of a node in a 6-cube. [66]

Lemma   16   *The number of nodes at level l of a subtree of the nESBT graph is*

$$\begin{cases} \binom{n-1}{l-1} + \binom{n-1}{l-2} = \binom{n}{l-1}, & \text{for } n+1 \geq l \geq 1, l \neq 2; \\ 1, & \text{for } l = 0; \\ n-1, & \text{for } l = 2. \end{cases}$$

Proof   Follows directly from the definition. □

Lemma   17   *Let $\phi^{nESBT}(i,s,x)$ be the number of nodes at distance x from node i in the subtree rooted at node i of the nESBT rooted at node s. Then, $\phi^{nESBT}(i,s,x) \geq \phi^{nESBT}(j,s,x)$, if $j \in children^{nESBT}(i,s)$.*

Proof   By the definition of the graph $G^{nESBT}(s)$, each subtree of the nESBT is a SBT with the smallest subtree removed. The lemma can be shown by lemmas 11, 16, and 10. □

4.8.3   Summary of Topological Characteristics of the Communication Graphs

The topological characteristics used for the complexity analysis are summarized in Table 4.9.

With respect to the entry for the SBT graph in Table 4.9, note that $\binom{x}{y} = 0$ if $x < y$. Moreover,

$$\max_d v^{SBT}(d,l) = v^{SBT}(\min(2l, n-1), l)$$

$$= \begin{cases} \binom{2l}{l}2^{n-2l-1}, & \text{if } l \in \{0, \ldots, \lfloor\frac{n-1}{2}\rfloor\}; \\ \binom{n-1}{l}, & \text{if } l \in \{\lfloor\frac{n+1}{2}\rfloor, \ldots, n-1\}. \end{cases}$$

The entries in Table 4.9 for the SBG graph can be proved by lemmas 5 and 12 and corollaries 3 and 4. The characteristics for the nRH graph can be proved from lemmas 5 and 8, and the entries for the nRSBT graph proved using corollaries 3, 4, and the fact that T^{SBT} is greedy. The nESBT properties can be proved using the definition of the graph $G^{nESBT}(s)$ and lemmas 5, 8, and 16.

4.8.4   One-to-All Broadcasting

Lemma   18   *A lower bound for one-to-all broadcasting with one-port communication is $\max((M+n-1)t_c, n\tau)$, and $\max((\lceil\frac{M}{n}\rceil + n-1)t_c, n\tau)$ for n-port communication.*

Table 4.9 Some topological characteristics of some *o-graphs*. [66]

Graph	$E(d)$	$e(d,l)$	$v(d,l)$	$u(d,l)$
G^H	2^{n-d-1}	$0, d \neq t_l$ or 1, o.w.	$0, d \neq t_l$ or $N-l-1$, o.w.	$\frac{N}{2^d+1} - \left\lceil \frac{l-2^d+1}{2^d+1} \right\rceil$
G^{nRH}	$\frac{N-1}{n}$	$\frac{1}{n}$	$\frac{1}{n}(N-l-1)$	$\frac{1}{n}(N-l-1)$
G^{SBT}	2^d	$\binom{d}{l}$	$\binom{d}{l}2^{n-d-1}$	$\binom{n-d-1}{l}2^d$
G^{nRSBT}	$\frac{N-1}{n}$	$\frac{1}{n}\binom{n}{l+1}$	$\frac{1}{n}\sum_{i=l+1}^n \binom{n}{i}$	$\frac{1}{n}\sum_{i=l+1}^n \binom{n}{i}$
G^{nESBT}	$\frac{N-1}{n}$	$1, l=0; n-1, l=1; \frac{1}{n}\binom{n}{l}, l \geq 2$	$N-1, l=0; N-2, l=1; \frac{1}{n}\sum_{i=l}^n \binom{n}{i}, l \geq 2$	$\frac{1}{n}\sum_{i=l+1}^n \binom{n}{i}, l \geq 2$
G^{SBG}	$\frac{N-1}{n}$	$\frac{1}{n}\binom{n}{l+1}$	$\frac{1}{n}\sum_{i=l+1}^n \binom{n}{i}$	$\frac{1}{n}\sum_{i=l+1}^n \binom{n}{i}$

Proof   The height of any *o-graph* is at least n. The root needs a time of Mt_c (or $\lceil \frac{M}{n} \rceil t_c$) to send out the data. An additional delay of at least $(n-1)t_c$ is required to reach the node at maximum distance from the root.

$\Box$

With $M = 1$, a tight lower bound is $n(t_c + \tau)$ both for *one-port* and *n-port* communication. This bound is realized by SBT routing and appropriate scheduling. In fact, it can be shown that with $M = 1$ and *one-port* communication, any routing for broadcasting that yields the lower bound is topologically equivalent to the graph G^{SBT}. With *n-port* communication any *o-graph* of height n can realize the lower bound communication for $M = 1$. The scheduling discipline we propose for the graph G^{nESBT} yields the lowest communication complexity of the broadcasting algorithms we consider, both for *one-port* and *n-port* communication, except $B_{opt} = M$ for *one-port* communication and $B_{opt} = \frac{M}{n}$ for *n-port* communication. For these two cases our scheduling for the graph G^{nESBT} is inferior only by 1 routing cycle. With $1 < M \leq n$ and *n-port* communication, our scheduling for the graph G^{nRSBT} yields the lower bound, $n(t_c + \tau)$.

The estimated communication times are summarized in Table 4.10. For an *o-graph* composed of n spanning trees the data set M is divided into n approximately equal size subsets, and each such subset communicated by one of the spanning subtrees. If the trees are not edge-disjoint, we assume no pipelining.

4.8.5   One-to-All Personalized Communication

In personalized communication no replication of information takes place during distribution, nor is there any reduction during the reverse operation; the root has M elements for every node.

Lemma   19   *The data transmission time for* one-to-all *personalized communication is bounded by* $(N-1)Mt_c$ *for* one-port *communication and by* $\frac{(N-1)M}{n}t_c$ *for* n-port *communication. A lower bound for the total time is* $\max((N-1)Mt_c, n\tau)$ *and* $\max(\frac{(N-1)M}{n}t_c, n\tau)$, *respectively.*

Proof   The root needs to send out $(N-1)M$ elements. $\Box$

In SBT routing for personalized communication, the maximum number of nodes connected to the root through one of its edges is $\frac{1}{2}NM$, and a data transmission time of the order given in the lemma is not achievable for *n-port* communication. In a SBnT graph all edges of the root connect subtrees of approximately equal sizes. For the SBG, nRSBT, and nESBT

Table 4.10   The complexity of one-to-all broadcasting. [66]

Comm.	Routing	T	B_{opt}	T_{min}
one-port	H	$(M + (N-2)B)t_c + (\lceil \frac{M}{B} \rceil + N - 2)\tau$	$\sqrt{\frac{M\tau}{(N-2)t_c}}$	$(\sqrt{Mt_c} + \sqrt{(N-2)\tau})^2$
	SBT	$Mnt_c + \lceil \frac{M}{B} \rceil n\tau$	M	$n(Mt_c + \tau)$
	nRSBT	$nMt_c + (2\sum_{i=1}^{n-1} \lceil \frac{Mi}{nB} \rceil + \lceil \frac{M}{B} \rceil)\tau$	M	$nMt_c + (2n-1)\tau$
	nESBT	$(M + nB)t_c + (\lceil \frac{M}{B} \rceil + n)\tau$	$\sqrt{\frac{M\tau}{nt_c}}$	$(\sqrt{Mt_c} + \sqrt{n\tau})^2$
n-port	SBT	$(M + (n-1)B)t_c + (\lceil \frac{M}{B} \rceil + n - 1)\tau$	$\sqrt{\frac{M\tau}{(n-1)t_c}}$	$(\sqrt{Mt_c} + \sqrt{(n-1)\tau})^2$
	nRSBT	$Mt_c + \lceil \frac{M}{nB} \rceil n\tau$	$\frac{M}{n}$	$Mt_c + n\tau$
	nESBT	$(\frac{M}{n} + nB)t_c + (\lceil \frac{M}{Bn} \rceil + n)\tau$	$\frac{1}{n}\sqrt{\frac{M\tau}{t_c}}$	$(\sqrt{\frac{Mt_c}{n}} + \sqrt{n\tau})^2$

graphs, all outgoing (cube) edges of the root transmit the same amount of data, $\frac{(N-1)M}{n}$.

Table 4.11 summarizes the communication complexities of personalized communication.

Table 4.11   The complexity of one-to-all personalized communication. [66]

Comm.	Routing	T
one-port	H	$(N-1)Mt_c + \max(\lceil \frac{(N-1)M}{B} \rceil, N-1)\tau$
	SBT	$(N-1)Mt_c + \sum_{i=0}^{n-1} \lceil \frac{2^i M}{B} \rceil \tau$
	nESBT	$\frac{n+1}{n}(N-1)Mt_c + (n\lceil \frac{(N-1)M}{nB} \rceil + \sum_{i=0}^{n-1} \lceil \frac{2^i M}{nB} \rceil)\tau$
	nRSBT	$(N-1)Mt_c + (\sum_{i=1}^{n-1} \lceil \frac{(N-2^i)M}{nB} \rceil + \sum_{i=1}^{n} \lceil \frac{(2^i-1)M}{nB} \rceil)\tau$
n-port	SBT	$\frac{NM}{2}t_c + \sum_{i=0}^{n-1} \lceil \binom{n-1}{i} \frac{M}{B} \rceil \tau$
	nESBT	$\frac{(N-1)M}{n}t_c + (\lceil \frac{M}{nB} \rceil + \lceil \frac{(n-1)M}{nB} \rceil + \sum_{i=2}^{n} \lceil \frac{M}{nB} \binom{n}{i} \rceil)\dot{\tau}$
	nRSBT	$\frac{(N-1)M}{n}t_c + \sum_{i=0}^{n-1} \lceil \binom{n}{i} \frac{M}{nB} \rceil \tau$
	SBG	$\frac{(N-1)M}{n}t_c + \sum_{i=0}^{n-1} \lceil \binom{n}{i} \frac{M}{nB} \rceil \tau$

4.8.6   All-to-All Broadcasting

Theorem 7   *A lower bound for all-to-all broadcasting is* $\max((N-1)Mt_c, n\tau)$ *for* one-port *communication and* $\max(\frac{(N-1)M}{n}t_c, n\tau)$ *for n-port communication.*

Proof   Each node receives M elements from every other node, i.e., each node receives $(N-1)M$ elements. Hence for *one-port* communication a lower bound for the data transfer time is $(N-1)Mt_c$, and with *n-port* communication the time is bounded by $\frac{(N-1)M}{n}t_c$. □

Lemma 20   *The data transfer time for* one-port *communication is minimized if one dimension is routed per cycle and all nodes use the same scheduling discipline.*

Proof   The number of elements transferred on every edge in the dimension subject to routing is the same, since the communication graphs for the different sources are translations of each other. □

Lemma   21   *A lower bound for data transmission time of all-to-all broadcasting based on the $G^{id}(*)$ and "any" scheduling discipline is*

$$\max_{\forall d \in \mathcal{D}} E^{id}(d) M t_c.$$

Proof $E^{id}(d)M$ elements need to be sent across every cube edge in dimension d. □

Theorem   8   *The communication time for all-to-all broadcasting based on $G^{id}(*)$ of height h, n-port communication, and the defined scheduling discipline requires a time of*

$$T = \sum_{l=0}^{h-1} \left(M t_c \times \max_{\forall d \in \mathcal{D}} e^{id}(d,l) + \left\lceil \frac{M}{B} \times \max_{\forall d \in \mathcal{D}} e^{id}(d,l) \right\rceil \tau \right).$$

If $B \geq \max_{0 \leq l \leq h-1, \forall d \in \mathcal{D}}(M \times e^{id}(d,l))$ then

$$T = \left(\sum_{l=0}^{h-1} \max_{\forall d \in \mathcal{D}} e^{id}(d,l) \right) M t_c + h\tau.$$

Proof   Each node broadcasts its data set M according to its own *o-graph*. During the lth routing cycle, all nodes at level $l+1$ of each *o-graph* receive messages sent out from the roots during the 0th routing cycle. By lemma 9, the amount of data contending for a communication link in dimension d is $M \times e^{id}(d,l)$. □

Corollary   6   *For a given a-graph satisfying $e^{id}(x,l) = e^{id}(y,l)$, $\forall x, y \in \mathcal{D}$, and $0 \leq l < h$, the communication time for all-to-all broadcasting based on the graph $G^{id}(*)$ of height h and n-port communication and the scheduling disciplines of Table 4.12 is*

$$T = \frac{(N-1)M}{n} t_c + h\tau, \; \text{if } B \geq \max_{0 \leq l < h}(M \times e^{id}(d,l)).$$

Proof   The corollary follows from theorem 8. □

Table 4.12 Scheduling disciplines for all-to-all broadcasting. [66]

Comm.	Scheduling discipline
one-port	1. For each spanning tree of the *a-graph*, the labels of the outgoing edges of any node are greater than the label of the incoming edge. 2. All the edges with the same label in the *o-graph* are in the same dimension.
n-port	All data sent at once, spanning trees concurrently.

From Table 4.9 and corollary 8, the nRH, nRSBT, nESBT, and the SBG routings all yield the lower bound for the data transfer time with *n-port* communication. In fact, following corollary 6, we have the following corollary.

Corollary 7   *For $G^{id}(*)$ with $G^{id}(0)$ composed of n distinctly rotated spanning trees, the data transfer time for all-to-all broadcasting is minimum with* n-port *communication and the defined scheduling.* □

Routing according to a greedy *o-graph*, while it is a necessary condition for *all-to-all personalized communication* to attain the minimum data transfer time, is not necessary for *all-to-all broadcasting*. The nRSBT and the SBG both with minimum height also attain the minimum number of start-ups.

Table 4.13 summarizes the complexity of all-to-all broadcasting.

The *one-port* $G^H(*)$ routing is employed in the matrix multiplication algorithm by Dekel [24, 60]. Messages with different source nodes are routed through different H paths. Messages are exchanged along a sequence of dimensions such as 0, 1, 0, 2, 0, 1, 0, 3, ..., etc. The SBT communication amounts to a single exchange per dimension [103].

4.8.7   All-to-All Personalized Communication

Theorem 9   *A lower bound for* one-port *all-to-all personalized communication is* $\max(\frac{nNM}{2}t_c, n\tau)$. *The packet size must be at least* $\frac{NM}{2}$ *to attain this lower bound.*

Table 4.13   The complexity of all-to-all broadcasting. [66]

Comm.	Routing	T
one-port	H	$(N-1)Mt_c + \lceil \frac{M}{B} \rceil (N-1)\tau$
	SBT	$(N-1)Mt_c + \sum_{i=0}^{n-1} \lceil \frac{2^i M}{B} \rceil \tau$
	nRSBT	$(N-1)Mt_c + (\sum_{i=0}^{n-1} \lceil \frac{(2^{i+1}-1)M}{nB} \rceil$ $+ \sum_{i=n}^{2n-2} \lceil \frac{(N-2^{i-n+1})M}{nB} \rceil)\tau$
	nESBT	$(N-1)Mt_c + (\sum_{i=0}^{n-1} \lceil \frac{2^i M}{nB} \rceil$ $+ \sum_{i=n}^{2n-1} \lceil \frac{(N-1-2^{i-n})M}{nB} \rceil)\tau$
n-port	nRH	$\frac{(N-1)M}{n}t_c + \lceil \frac{M}{nB} \rceil (N-1)\tau$
	SBT	$\frac{NM}{2}t_c + \sum_{i=0}^{n-1} \lceil \binom{n-1}{i} \frac{M}{B} \rceil \tau$
	nRSBT	$\frac{(N-1)M}{n}t_c + \sum_{i=1}^{n} \lceil \binom{n}{i} \frac{M}{nB} \rceil \tau$
	nESBT	$\frac{(N-1)M}{n}t_c + (\sum_{i=2}^{n} \lceil \binom{n}{i} \frac{M}{nB} \rceil$ $+ \lceil \frac{M}{nB} \rceil + \lceil \frac{(n-1)M}{nB} \rceil)\tau$
	SBG	$\frac{(N-1)M}{n}t_c + \sum_{i=1}^{n} \lceil \binom{n}{i} \frac{M}{nB} \rceil \tau$

Proof   The bandwidth requirement for distributing personalized data from one node is

$$\sum_{i=0}^{n} i\binom{n}{i} M = \frac{nNM}{2}.$$

The total bandwidth requirement is $\frac{nN^2M}{2}$. During each cycle only N edges of the n-cube can communicate in the case of *one-port* communication; $\frac{nNM}{2}$ is the minimum number of element transfers in sequence. The number of start-ups is at least n. The maximum packet size can be derived by dividing the total bandwidth requirement $\frac{nN^2M}{2}$ by the number of cycles n, and the number of directed edges that can be used in each routing cycle N. ☐

Theorem   10   *A lower bound for* n-port *all-to-all personalized communication is* $\max(\frac{NM}{2}t_c, n\tau)$. *The packet size must be at least* $\frac{NM}{2n}$ *to attain this lower bound.*

Proof   From theorem 9 the total bandwidth requirement is $\frac{nN^2M}{2}$. During each routing cycle nN directed edges can communicate concurrently. The maximum packet size is derived by dividing the total bandwidth requirement by the number of cycles n and the total number of links nN.

$\Box$

Theorem   11   *The time for* n-port *all-to-all personalized communication based on $G^{id}(*)$ of height h and* postorder *scheduling is*

$$T = \sum_{l=0}^{h-1} \left(Mt_c \times \max_{\forall d \in \mathcal{D}} v^{id}(d,l) + \left\lceil \frac{M}{B} \times \max_{\forall d \in \mathcal{D}} v^{id}(d,l) \right\rceil \tau \right).$$

If $B \geq \max_{0 \leq l \leq h-1,\ \forall d \in \mathcal{D}}(M \times v^{id}(d,l))$ then

$$T = \left(\sum_{l=0}^{h-1} \max_{\forall d \in \mathcal{D}} v^{id}(d,l) \right) Mt_c + h\tau.$$

Proof   For each $G^{id}(s)$, the total amount of data transmitted across all edges in dimension d during routing cycle l is $v^{id}(d,l) \times M$, $0 \leq l \leq h-1$. For $G^{id}(*) = \cup_{s \in \mathcal{N}} Tr(s, G^{id}(0))$, each *a-graph* edge is mapped to N distinct cube edges with N distinct exclusive-or operations on both endpoints. The amount of data transmitted across each cube edge in dimension d during routing cycle l is $v^{id}(d,l) \times M$. $\Box$

Theorem   12   *The time for* n-port *all-to-all personalized communication based on $G^{id}(*)$ of height h and* reverse-breadth-first *scheduling is*

$$T = \sum_{l=0}^{h-1} \left(Mt_c \times \max_{\forall d \in \mathcal{D}} u^{id}(d,l) + \left\lceil \frac{M}{B} \times \max_{\forall d \in \mathcal{D}} u^{id}(d,l) \right\rceil \tau \right).$$

If $B \geq \max_{0 \leq l \leq h-1,\ \forall d \in \mathcal{D}}(M \times u^{id}(d,l))$ then

$$T = \left(\sum_{l=0}^{h-1} \max_{\forall d \in \mathcal{D}} u^{id}(d,l) \right) Mt_c + h\tau.$$

Proof   Similar to the proof of theorem 11. $\Box$

Theorem   13   *The all-to-all personalized communication based on N translated o-graphs will attain the lower bound for the data transfer time iff the o-graph is greedy.*

Proof   The bandwidth requirement for each node is

$$\sum_{l=0}^{h-1}\sum_{d=0}^{n-1} v^{id}(d,l) = \sum_{l=0}^{h-1}\sum_{d=0}^{n-1} u^{id}(d,l) = \sum_{l=1}^{h} l \times \text{(the number of nodes at level } l).$$

Hence non-greedy *o-graphs* require more data transfer than greedy *o-graphs*.

$\square$

Theorem   14   *All-to-all personalized n-port based on $G^{id}(*)$ where $G^{id}(0) = \cup_{d \in \mathcal{D}} Ro^d(T^{id'}(0))$ and $T^{id'}(0)$ is greedy, can attain both the minimum data transmission time, $\frac{NM}{2}t_c$, and the minimum number of start-ups, n, for $B \geq \frac{(N-1)M}{n}$ both for* postorder *and* reverse-breadth-first *schedulings.*

Proof   From theorem 11 and corollary 4, the data transfer time is

$$\sum_{l=0}^{n-1}\sum_{i=l+1}^{n} \frac{M}{n}\binom{n}{i}t_c = \frac{NM}{2}t_c.$$

The packet size $\frac{(N-1)M}{n}$ occurs during routing cycle 0.   Similarly, the *reverse-breadth-first* scheduling discipline can be shown to be optimum and has the same value of the maximum packet size. It occurs during the last routing cycle. $\square$

Corollary   8   *All-to-all personalized n-port communication based on $G^{nRSBT}(*)$ and $G^{SBG}(*)$ can attain the time $\frac{NM}{2}t_c + n\tau$, which is within a factor of 2 of the lower bound.* $\square$

Notice that in *one-port* communication the data transfer time is always optimal if the routing is based on N translated greedy *o-graphs* and appropriate scheduling. But not all greedy *o-graphs* have the same number of start-ups. Only the SBT graph allows a minimum of n start-ups for sufficiently large packet sizes. The minimum number of start-ups can be decided by the same labeling rules as were used in all-to-all broadcasting. The minimum number of start-ups is the same as for all-to-all broadcasting. The difference is that the amount of data transferred during cycle i is equal to the sum of weighted subtree sizes with the root of each subtree connected through an edge labeled i to its parent.

Table 4.14 summarizes the complexity estimates.

Lemma   22   *All-to-all personalized one-port communication based on $G^{SBT}(*)$ can be accomplished in time $\frac{nNM}{2}t_c + n\tau$, which is within a factor of 2 of the lower bound.*

Table 4.14   The complexity of all-to-all personalized communication. [66]

Comm.	Routing	T
one-port	H	$\frac{(N-1)NM}{2}t_c + \sum_{i=1}^{N-1}\lceil\frac{iM}{B}\rceil\tau$
	SBT	$\frac{nNM}{2}t_c + \lceil\frac{NM}{2B}\rceil n\tau$
	nRSBT	$\frac{nNM}{2}t_c + (\sum_{i=0}^{n-1}\lceil\frac{(i+1)NM}{2nB}\rceil$ $+\sum_{i=n}^{i=2n-2}\lceil\frac{(2n-i-1)NM}{2nB}\rceil)\tau$
	nESBT	$(\frac{nN}{2}+N-2)Mt_c + (\sum_{i=0}^{n-1}\lceil(\frac{(i+2)N}{2}-1)\frac{M}{nB}\rceil$ $+\sum_{i=1}^{n}\lceil(\frac{iN}{2}-1)\frac{M}{nB}\rceil)\tau$
	SBG	$\approx \frac{nNM}{2}t_c + \max(2n-1,\frac{nNM}{2B})\tau$
n-port	nRH	$\frac{(N-1)NM}{2n}t_c + \sum_{i=1}^{N-1}\lceil\frac{iM}{nB}\rceil\tau$
	SBT	$(\sum_{l=0}^{\lfloor\frac{n-1}{2}\rfloor}\binom{2l}{l}2^{n-2l-1} + \sum_{l=\lfloor\frac{n+1}{2}\rfloor}^{n-1}\binom{n-1}{l})Mt_c$ $+(\sum_{l=0}^{\lfloor\frac{n-1}{2}\rfloor}\lceil\binom{2l}{l}\frac{2^{n-2l-1}M}{B}\rceil + \sum_{l=\lfloor\frac{n+1}{2}\rfloor}^{n-1}\lceil\binom{n-1}{l}\frac{M}{B}\rceil)\tau$
	nRSBT	$\frac{NM}{2}t_c + \lceil\frac{NM}{2nB}\rceil n\tau$
	nESBT	$(\frac{N}{2}+\frac{N-2}{n})Mt_c + (\sum_{i=2}^{n}\sum_{j=i}^{n}\lceil\binom{n}{j}\frac{M}{nB}\rceil$ $+\lceil\frac{(N-2)M}{nB}\rceil + \lceil\frac{(N-1)M}{nB}\rceil)\tau$
	SBG	$\frac{NM}{2}t_c + \sum_{i=1}^{n}\lceil\sum_{j=i}^{n}\binom{n}{j}\frac{M}{nB}\rceil\tau$

Proof   During the first routing cycle, $\frac{NM}{2}$ data is exchanged along the lowest dimension. The procedure is then applied recursively with the data set doubling for each cycle of the recursion and the dimension of the cube decreasing by 1. Let $T(i,M)$ be the time required by the stated personalized all-to-all routing algorithm with initially M data per node in an i-cube. Clearly, $T(n,M) = 2^{n-1}Mt_c + \tau + T(n-1,2M)$ and $T(1,M) = Mt_c + \tau$. Hence, $T(n,M) = \frac{nNM}{2}t_c + n\tau$. □

In the case of *n-port* communication, we can find the communication complexity from the previously derived formula. The interesting quantity for *postorder* scheduling is $\sum_{l=0}^{n-1} \max v^{id}(d,l)$.

$$\sum_{l=0}^{n-1} \max_d v^{SBT}(d,l) = \sum_{l=0}^{\lfloor \frac{n-1}{2} \rfloor} \binom{2l}{l} 2^{n-2l-1} + \sum_{l=\lfloor \frac{n+1}{2} \rfloor}^{n-1} \binom{n-1}{l},$$

which is of $O(\sqrt{n}N)$. Since $u^{SBT}(d,l) = v^{SBT}(n-d-1,l)$, Table 4.9, the *reverse-breadth-first* scheduling yields the same result.

With *one-port* communication, the SBT, nRSBT, SBG, and nESBT routings with *postorder* and *reverse-breadth-first* schedulings for all-to-all personalized communication are optimum within constant factors. With *n-port* communication, the nESBT, nRSBT, and SBG routings are optimum within a constant factor of ≈ 2 for *postorder* and *reverse-breadth-first* schedulings.

4.8.8   Summary

Table 4.15 summarizes the results. Routing for the four communication operations can also be based on Two-rooted Complete Binary Trees (TCBT) [8, 26]. Communication algorithms based on the TCBT may yield performance comparable to the algorithms presented here for *one-port* communication. For examples of such scheduling algorithms, see [66].

Packet size is very important for communication complexity. With *one-port* communication and a packet size less than or equal to the data set to be communicated to every node, all considered routings have approximately the same complexity for one-to-all personalized communication and all-to-all broadcasting.

4.9   Pyramid Embeddings

Multigrid algorithms for partial differential equations [21] and algorithms for image processing [18] are examples of algorithms that have data interactions defined by a pyramid topology [110, 87, 22, 18, 119]. This section first gives a precise definition of pyramids, and a generalization of such graphs into *hyper-pyramids*. Hyper-pyramids contain pyramids as subgraphs and can intuitively be viewed as a hierarchy of Boolean cubes interconnected in a tree-like manner. In the section about lattice embedding, we showed that arbitrary two-dimensional lattices can be embedded in Boolean cubes with dilation 2 and minimal expansion. In this section we show that hyper-pyramids can be embedded in Boolean cubes with minimal expansion and dilation d [44], where 2^d is the number of children of a node (except for the bottom level). For the conventional pyramid $d = 2$, and the embeddings we give have dilation two and congestion two. We also consider the

Table 4.15   Time complexity of communication algorithms. The last column shows the constant factors as compared to the best known lower bounds. [66]

Comm.	Data distribution	Assumption	Routing	B_{opt}	T_{min}	Factor
one-port	One-to-all	$M=1$	SBT	1	$n(t_c+\tau)$	1
	Broadcasting	$M>1$	nESBT	$\sqrt{\frac{M\tau}{nt_c}}$	$(\sqrt{Mt_c}+\sqrt{n\tau})^2$	4
	One-to-all P.C.	$M\geq 1$	SBT	$\frac{NM}{2}$	$(N-1)Mt_c+n\tau$	2
	All-to-all B.	$M\geq 1$	SBT	$\frac{NM}{2}$	$(N-1)Mt_c+n\tau$	2
	All-to-all P.C.	$M\geq 1$	SBT	$\frac{NM}{2}$	$\frac{nNM}{2}t_c+n\tau$	2
	One-to-all	$M\leq n$	nRSBT	1	$n(t_c+\tau)$	1
	Broadcasting	$M>n$	nESBT	$\frac{1}{n}\sqrt{\frac{M\tau}{t_c}}$	$(\sqrt{\frac{Mt_c}{n}}+\sqrt{n\tau})^2$	4
	One-to-all	$M\geq n$	nESBT	$\sqrt{\frac{2}{\pi}\frac{NM}{n^{3/2}}}$	$\frac{(N-1)M}{n}t_c+(n+1)\tau$	$\frac{2(n+1)}{n}$
	Personalized comm.	$M\geq 1, M\geq n$	SBG, nRSBT	$\sqrt{\frac{2}{\pi}\frac{NM}{n^{3/2}}}$	$\frac{(N-1)M}{n}t_c+n\tau$	2
n-port	All-to-all	$M\geq n$	nESBT	$\sqrt{\frac{2}{\pi}\frac{NM}{n^{3/2}}}$	$\frac{(N-1)M}{n}t_c+(n+1)\tau$	$\frac{2(n+1)}{n}$
	Broadcasting	$M\geq 1, M\geq n$	SBG, nRSBT	$\sqrt{\frac{2}{\pi}\frac{NM}{n^{3/2}}}$	$\frac{(N-1)M}{n}t_c+n\tau$	2
	All-to-all	$M\geq n$	nESBT	$\frac{(N-1)M}{n}$	$(\frac{N}{2}+\frac{N-2}{n})Mt_c+(n+1)\tau$	$\frac{2(n+2)}{n}$
	Personalized comm.	$M\geq 1$	SBG	$\frac{(N-1)M}{n}$	$\frac{NM}{2}t_c+n\tau$	2
		$M\geq n$	nRSBT	$\frac{NM}{2n}$	$\frac{NM}{2}t_c+n\tau$	2

congestion, the active degree, and the node load. Though the expansion is minimal in the embeddings presented below, a substantial number of unused nodes are still in the cube. This fact is exploited in the embedding of multiple hyper-pyramids of two different sizes in the same Boolean cube (of minimum size for the embedding of the largest hyper-pyramid).

Definition   21   A k-level pyramid $P(k, l_1, l_2)$ is a graph with vertex set

$$\mathcal{V}(P(k, l_1, l_2)) = \bigcup_{i=0}^{k} \{(i, x_1, x_2) | (x_1, x_2) \in \mathcal{V}(M(l_1^i, l_2^i))\}$$

and edge set

$$\mathcal{E}(P(k, l_1, l_2))$$
$$= \bigcup_{i=0}^{k} \{((i, x_1, x_2), (i, x_1', x_2')) | ((x_1, x_2), (x_1', x_2')) \in \mathcal{E}(M(l_1^i, l_2^i))\} \bigcup$$
$$\bigcup_{i=1}^{k} \{((i, x_1, x_2), (i-1, \left\lfloor \frac{x_1}{l_1} \right\rfloor, \left\lfloor \frac{x_2}{l_2} \right\rfloor)) | (x_1, x_2) \in \mathcal{V}(M(l_1^i, l_2^i))\}. \quad \blacksquare$$

A $P(k, l_1, l_2)$ pyramid can be viewed as consisting of a hierarchy of two-dimensional lattices $M(l_1^0, l_2^0)$ through $M(l_1^k, l_2^k)$, with each node having $l_1 \times l_2$ children, except nodes at level k, the *base* of the pyramid. Node $(i, x_1, x_2) \in V(P(k, l_1, l_2))$ is at *level i*. The node at level 0, $(0, 0, 0)$, is called the *apex*, or the *root*, of the pyramid. The nodes at level k are *leaf* nodes. Figure 4.35 shows a $P(2, 2, 2)$ pyramid. It can be viewed as a complete quad-tree with nodes at the same level being connected as a mesh.

The number of vertices in a pyramid is

$$|\mathcal{V}(P(k, l_1, l_2))| = \sum_{i=0}^{k} (l_1 l_2)^i = \frac{(l_1 l_2)^{k+1} - 1}{l_1 l_2 - 1}$$

and the number of edges is

$$|\mathcal{E}(P(k, l_1, l_2))| = \sum_{i=1}^{k} 3(l_1 l_2)^i - l_1^i - l_2^i.$$

Definition   22   [44] A *k-level hyper-pyramid* $\hat{P}(k, d)$ of *degree d* is constructed out of 2^d $\hat{P}(k-1, d)$ hyper-pyramids by interconnecting corresponding nodes in each of these hyper-pyramids as d-dimensional Boolean cubes, and connecting a new node to every root of the $\hat{P}(k-1, d)$ hyper-pyramids. $\hat{P}(0, d)$ is a root node. $\blacksquare$

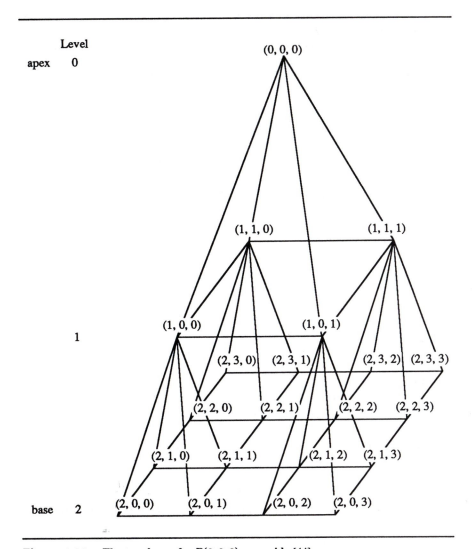

Level

apex   0   (0, 0, 0)

(1, 1, 0)   (1, 1, 1)

(1, 0, 0)   (1, 0, 1)

1   (2, 3, 0)   (2, 3, 1)   (2, 3, 2)   (2, 3, 3)

(2, 2, 0)   (2, 2, 1)   (2, 2, 2)   (2, 2, 3)

(2, 1, 0)   (2, 1, 1)   (2, 1, 2)   (2, 1, 3)

base   2   (2, 0, 0)   (2, 0, 1)   (2, 0, 2)   (2, 0, 3)

Figure 4.35   The topology of a $P(2, 2, 2)$ pyramid. [44]

A $\hat{P}(k, d)$ hyper-pyramid contains a $P(k, 2^i, 2^{d-i})$ pyramid, $0 \le i \le d$, as a subgraph. Hyper-pyramids can also be defined recursively by adding a B_{kd} cube to a $\hat{P}(k-1, d)$ hyper-pyramid. The hyper-pyramid $\hat{P}(k, d)$ is obtained by connecting each node in the B_{kd} cube to a (parent) node in the base of the $\hat{P}(k-1, d)$ hyper-pyramid. The address of a vertex is composed of the level i, $0 \le i \le k$, and an index $j, 0 \le j < 2^{id}$ identifying the node within the level. j is a binary number of length id. Node $a(i, j)$ connects to a parent node $a(i-1, (j_{id-1}j_{id-2} \ldots j_d))$, if $i \neq 0$, and to 2^d children nodes with addresses $\{(i+1, j\| *_{d-1} *_{d-2} \ldots *_0)\}$ if $i \neq k$, where $*_m = 0$ or 1 for all $0 \leq m < d$. The second argument of the parent address is obtained by removing the lowest-order d bits from j. The second argument of a root node is a null string, which is represented by ε. The second argument of the child addresses is obtained by appending a d-bit binary string to j. These edges form the "tree-edges" of the hyper-pyramid. In addition there are id "cube-edges" connecting node $a(i, j)$ to nodes $a(i, j^{\{m\}})$ where $0 \leq m < id$. Figures 4.36 and 4.37 show the topology of a $\hat{P}(2, 2)$ hyper-pyramid.

The number of nodes in a hyper-pyramid $\hat{P}(k, d)$ is

$$|\mathcal{V}(\hat{P}(k, d))| = \sum_{i=0}^{k} 2^{id} = \frac{2^{(k+1)d} - 1}{2^d - 1},$$

and the number of edges is

$$|\mathcal{E}(\hat{P}(k, d))| = \sum_{i=1}^{k} id2^{id-1} + \sum_{i=1}^{k} 2^{id}.$$

The first term in the latter formula accounts for the number of edges at each level, and the second term accounts for the edges between levels.

Lemma 23 *[44] A lower bound for the dilation of any embedding of a $\hat{P}(k, d)$ hyper-pyramid into a smallest Boolean cube B_n (i.e., $|\mathcal{V}(B_{n-1})| \leq |\mathcal{V}(\hat{P}(k, d))| < |\mathcal{V}(B_n)|$) is $\frac{d}{2}$.*

Proof   The diameter of a $\hat{P}(k, d)$ hyper-pyramid is $2k$. The smallest cube B_n that is large enough to hold a $\hat{P}(k, d)$ hyper-pyramid has $n = kd + 1$ dimensions. Since the hyper-pyramid contains more than 2^{n-1} nodes, there exist two hyper-pyramid nodes that are mapped to cube nodes at a distance of at least $n - 1$ in the B_n cube. Consider any shortest path between these two hyper-pyramid nodes. Let the length of the path be ℓ. Clearly, $\ell \leq 2k$. Each edge on the path will be stretched in the embedding such that all ℓ edges together are stretched into the path of length $\geq n - 1$ in the B_n cube. Thus, at least one of these ℓ edges is stretched into a path of length

$$\geq \frac{n-1}{\ell} \geq \frac{n-1}{2k} = \frac{d}{2}. \quad \square$$

Similarly, a lower bound of the dilation for embedding a $P(k, l_1, l_2)$ pyramid into a $B_{1+k \log_2 l_1 l_2}$ cube is $\frac{\log_2 l_1 l_2}{2}$. For the recursive construction of the embedding, we will make use of hyper-pyramids with two roots. The

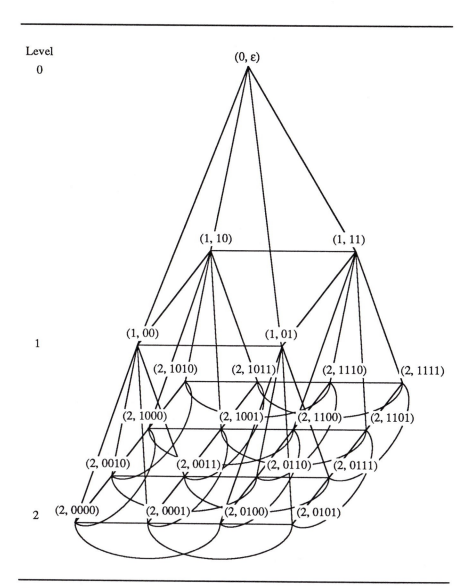

Figure 4.36   The topology of a $\hat{P}(2,2)$ hyper-pyramid.

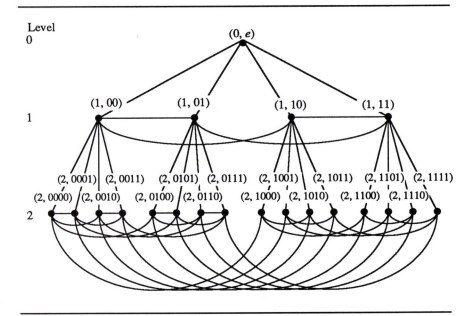

Level
0

1

2

Figure 4.37   Another view of the topology of a $\hat{P}(2,2)$ hyper-pyramid.

roots are symmetrical, and either of them can serve as the root. One of the roots will become a node at level 1 after the induction step. This node is the *real root* of the smaller hyper-pyramid out of which the larger one is constructed through the induction. The other root may either serve as one of the two roots of the hyper-pyramid created through the induction, or become unused. This root of the hyper-pyramids used in the induction step is called the *spare* root. The choice of which root is the real root and which the spare is made in the induction. The idea of using two roots for the recursive construction of tree structures has been used before, for example, by Bhatt and Leiserson [10] in constructing a complete binary tree out of chips containing smaller trees, and by Bhatt and Ipsen [8] in embedding complete binary trees into a Boolean cube.

Definition   23   [44] A *two-rooted hyper-pyramid* $\tilde{P}(k,d)$ is a $\hat{P}(k,d)$ hyper-pyramid with an additional root node and additional edges between it and all nodes at level 1. The two roots are denoted $a(0,\varepsilon)$ and $a(0',\varepsilon)$, respectively.   ∎

4.9.1   Embedding a $\hat{P}(k,d)$ Hyper-pyramid in a Boolean Cube with Minimal Expansion and Dilation d

Theorem   15   *[44] A $\hat{P}(k,d)$ hyper-pyramid, $d \geq 2$, can be embedded into a B_{kd+1} Boolean cube with dilation d.*

Proof   Instead of considering the embedding of a $\hat{P}(k,d)$ hyper-pyramid, we consider the embedding of the corresponding two-rooted hyper-pyramid $\tilde{P}(k,d)$. The dilation for the two-rooted hyper-pyramid is an upper bound on the dilation for the hyper-pyramid with a single root. Let f_k be the function that maps a two-rooted hyper-pyramid $\tilde{P}(k,d)$ into a B_{kd+1} cube with dilation d. We will define f_k (for a fixed d) by a recursive construction on k and prove the theorem by induction.

Basis: For $k = 0$, the two-rooted hyper-pyramid $\tilde{P}(0,d)$, which has only two roots, are mapped to adjacent cube nodes:

$$f_0(a(0,\varepsilon)) = 0 \text{ and } f_0(a(0',\varepsilon)) = 1.$$

Induction: Assume that there exists an embedding function f_k that embeds a $\tilde{P}(k,d)$ hyper-pyramid with dilation d in a B_{kd+1} cube such that the two roots are adjacent. Label the B_{kd+1} cubes that are needed in the induction step $0, 1, \ldots, 2^d - 1$, and assume that a hyper-pyramid $\tilde{P}(k,d)$ is embedded in each with the embedding function f_k. The hyper-pyramids $\tilde{P}(k,d)$ and their nodes are indexed by a superscript equal to the index of the B_{kd+1} cube in which they are embedded. The embedding function f_{k+1} is defined in terms of f_k by the following rules:

$$
\begin{aligned}
f_{k+1}(a(0,\varepsilon)) &= f_k(a^0(0,\varepsilon)), \\
f_{k+1}(a(0',\varepsilon)) &= f_k(a^{2^{d-1}}(0,\varepsilon)), \\
f_{k+1}(a(1,\ell)) &= \begin{cases} f_k(a^\ell(0',\varepsilon)), & \ell = 0 \text{ or } 2^{d-1}, \\ f_k(a^\ell(0,\varepsilon)), & \text{otherwise}, \end{cases} \\
f_{k+1}(a(i,\ell\|j)) &= f_k(a^\ell(i,j)), \quad i > 1.
\end{aligned}
$$

The first two equations define the two new roots. The third equation defines nodes at level 1. The last equation defines nodes at lower levels, where ℓ is a string of length d. Note that the nodes $a^l(0,\varepsilon)$ are selected as real roots in cubes B_{kd+1}^l, $l \neq 0$ and $l \neq 2^{d-1}$, but that the nodes $a^0(0',\varepsilon)$ and $a^{2^{d-1}}(0',\varepsilon)$ are selected as real roots in the cubes B_{kd+1}^0 and $B_{kd+1}^{2^{d-1}}$. Figures 4.38 and 4.39 show the induction for $d = 2$ and $d = 3$, respectively. For clarity, only the two roots of each cube are shown. In the figures, ϕ denotes unused cube nodes (after induction), and "$\rightarrow$" reads "becomes". Note that the root $a(0',\varepsilon)$ is used in cubes 0 and 2^{d-1}. Figures 4.40 and

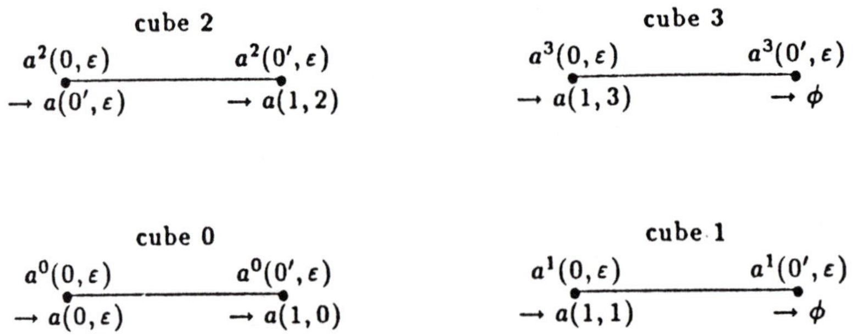

cube 2

$a^2(0,\varepsilon)$ $a^2(0',\varepsilon)$
$\rightarrow a(0',\varepsilon)$ $\rightarrow a(1,2)$

cube 3

$a^3(0,\varepsilon)$ $a^3(0',\varepsilon)$
$\rightarrow a(1,3)$ $\rightarrow \phi$

cube 0

$a^0(0,\varepsilon)$ $a^0(0',\varepsilon)$
$\rightarrow a(0,\varepsilon)$ $\rightarrow a(1,0)$

cube 1

$a^1(0,\varepsilon)$ $a^1(0',\varepsilon)$
$\rightarrow a(1,1)$ $\rightarrow \phi$

Figure 4.38 Forming a two-rooted hyper-pyramid $\tilde{P}(k+1,2)$ out of 4 copies of $\tilde{P}(k,2)$ hyper-pyramids. [44]

4.41 show the embeddings for $\tilde{P}(1,2)$ and $\tilde{P}(1,3)$, respectively. For all $0 \le j < 2^d$ and $0 \le m < d$, we have the following properties:

(1) $Hamming(f_{k+1}(a(0,\varepsilon)), f_{k+1}(a(1,j))) \le d$:

$Hamming(f_{k+1}(a(0,\varepsilon)), f_{k+1}(a(1,j))) =$

$$
\begin{cases}
Hamming(f_k(a^0(0,\varepsilon)), f_k(a^j(0,\varepsilon))) = \|j\| < d, & \text{if } j \ne 0 \\
& \text{and } j \ne 2^{d-1}, \\
Hamming(f_k(a^0(0,\varepsilon)), f_k(a^0(0',\varepsilon))) = 1, & \text{if } j = 0, \\
Hamming(f_k(a^0(0,\varepsilon)), f_k(a^{2^{d-1}}(0',\varepsilon))) = 2, & \text{if } j = 2^{d-1}.
\end{cases}
$$

(2) $Hamming(f_{k+1}(a(0',\varepsilon)), f_{k+1}(a(1,j))) \le d$: The proof follows that of 1.

(3) $Hamming(f_{k+1}(a(1,j)), f_{k+1}(a(1,j^{\{m\}}))) \le 2$: The distance is 1 except if $m \ne d-1$ *and* $j = 0$ or 2^{d-1}, for which the distance is 2.

(4) $Hamming(f_{k+1}(a(0,\varepsilon)), f_{k+1}(a(0',\varepsilon))) = 1$.

(5) The Hamming distance between corresponding nodes of adjacent cubes is 1.

(6) The dilation of an edge in $\tilde{P}(k,d)$ is preserved.

The induction hypothesis follows from these properties. $\square$

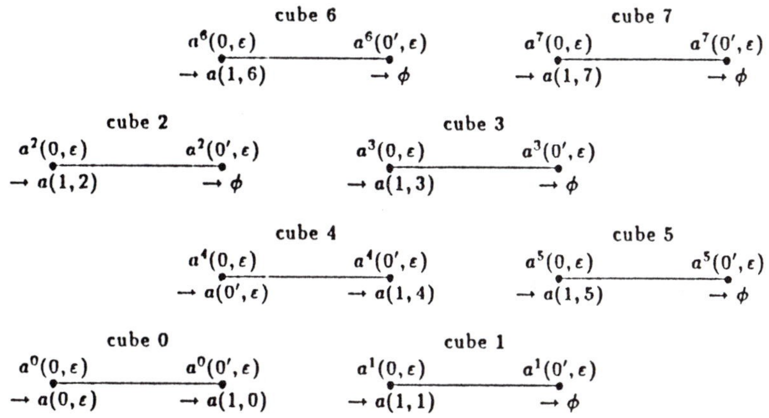

cube 6
$a^6(0,\varepsilon)$
$a^6(0',\varepsilon)$
$\to a(1,6)$
$\to \phi$
cube 7
$a^7(0,\varepsilon)$
$a^7(0',\varepsilon)$
$\to a(1,7)$
$\to \phi$
cube 2
$a^2(0,\varepsilon)$
$a^2(0',\varepsilon)$
$\to a(1,2)$
$\to \phi$
cube 3
$a^3(0,\varepsilon)$
$a^3(0',\varepsilon)$
$\to a(1,3)$
$\to \phi$
cube 4
$a^4(0,\varepsilon)$
$a^4(0',\varepsilon)$
$\to a(0',\varepsilon)$
$\to a(1,4)$
cube 5
$a^5(0,\varepsilon)$
$a^5(0',\varepsilon)$
$\to a(1,5)$
$\to \phi$
cube 0
$a^0(0,\varepsilon)$
$a^0(0',\varepsilon)$
$\to a(0,\varepsilon)$
$\to a(1,0)$
cube 1
$a^1(0,\varepsilon)$
$a^1(0',\varepsilon)$
$\to a(1,1)$
$\to \phi$

Figure 4.39 Forming a two-rooted hyper-pyramid $\tilde{P}(k+1,3)$ out of 8 copies of $\tilde{P}(k,3)$ hyper-pyramids.

By substituting f_k recursively as defined by the induction rules, an explicit expression for f_k is obtained:

$$f_k(a(i,j)) = \begin{cases} (0^{kd+1}), & i = 0, \\ (10^{kd}), & i = 0', \\ (j||x0^{(k-i)d}), & 1 \le i \le k, \end{cases}$$

where $x = 1$, if $(j_{d-2}j_{d-3}\cdots j_0) = 0$; and $x = 0$, otherwise.

$a(1,3)$
$a(1,1)$
$a(0',\varepsilon)$
$a(1,2)$
$a(0,\varepsilon)$
$a(1,0)$

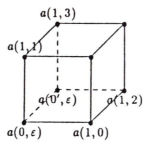

Figure 4.40 A two-rooted $\tilde{P}(1,2)$ hyper-pyramid embedded in a B_3 cube with dilation 2.

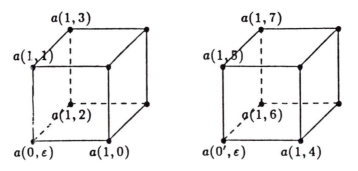

Figure 4.41   A two-rooted $\tilde{P}(1,3)$ hyper-pyramid embedded in a B_4 cube with dilation 3. [44]

Figure 4.42 shows the cube addresses of the nodes of the hyper-pyramid $\hat{P}(2,2)$.

The expansion of the embedding function f_k is less than 2 (except for $k = 0$). The two new roots and $2^d - 2$ of the children nodes of the root to be selected as the real root are in the same B_d cube. One of the other children nodes is adjacent to one of the two new roots. The other child node is adjacent to the other new root. Hence the connections from the real root to its children requires $2^d - 2$ paths to be routed from one node to other nodes in a d-cube. Moreover, for the embedding we do not use the edge between the two roots for the paths between the real root and its children. The routing to the children in the $d - 1$-cube formed by the dimensions of the other edges is done by a balanced spanning tree in a $d - 1$-cube. Routing to the remaining children is made by first routing to their "buddies" in the same dimension as the dimension in which the two roots are connected; then this dimension is routed. Hence the congestion at the root is bounded from above by $\sim 2\frac{2^d}{d+1}$. The routing between the children in the d-cube can be made such that it does not increase this bound. For the two children outside the cube, the routing is made through an unused cube node [44].

Lemma   24   *[44] An upper bound of the congestion for embedding a $\hat{P}(k,d)$ hyper-pyramid in a B_{kd+1} Boolean cube is $\frac{2^{d+1}}{d+1}$.*   □

Lemma   25   *[44] A lower bound of the congestion for embedding a $\hat{P}(k,d)$ hyper-pyramid in a B_{kd+1} Boolean cube is $1 + \lceil \frac{2^d - d}{kd+1} \rceil$.*

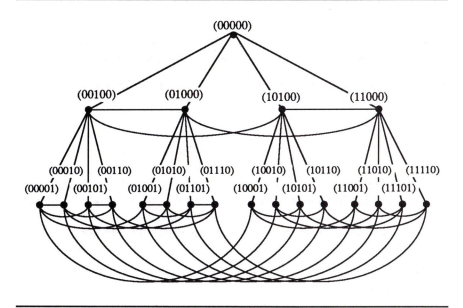

Figure 4.42   The cube addresses of the nodes of an embedded hyper-pyramid $\hat{P}(2,2)$. [44]

Proof   The nodes at level $k-1$ of a hyper-pyramid $\hat{P}(k,d)$ have degree $1+(k-1)d+2^d$. The degree of a B_{kd+1} cube is $kd+1$. Hence a lower bound of the congestion is

$$\left\lceil \frac{1+(k-1)d+2^d}{kd+1} \right\rceil = 1 + \left\lceil \frac{2^d-d}{kd+1} \right\rceil . \quad \square$$

For the *active degree*, note that some nodes at level k of the pyramid use all edges of the node they are mapped to. For example, consider nodes, $a(1,0)$ or $a(1,2^{d-1})$ in Figures 4.40 and 4.41. For these nodes, all the $d+1$ cube edges are used after the first induction step. For each induction step d, new cube edges are used. Hence the active-degree is $kd+1$.

In order to determine the *node load* β, consider the following properties:

(1) For $k=1$, β_i for the root to become a spare root is 0, and for the root to become the real root it is 2^d, and the maximum β_i for a node at level one is $\leq \frac{2^{d+2}}{d+1}+d$.

(2) A spare root after an induction step is also a spare root before the induction step. The β_i for the spare root remains 0.

(3) A real root after an induction step has $\beta_i = 2^d$.

(4) A node at level 1 after an induction step is a real root. The β_i for the node increases by at most $\frac{2^{d+2}}{d+1} + d$.

(5) A node at level i, $i > 1$, after an induction step is a node at level $i-1$ before the induction step. The β_i for the node increases by d.

By these properties, one can show by induction that β_i of a node at level $k - 1$ is maximum, for $k \geq 2$. The node-load is

$$\beta \leq \begin{cases} \max\{2^d, \frac{2^{d+2}}{d+1} + d\}, & k = 1, \\ (1 + \frac{4}{d+1})2^d + (k-1)d\}, & k \geq 2. \end{cases}$$

The above embedding generalizes and improves on earlier embeddings by Stout [110], who first showed that a dilation 2 embedding of a pyramid in a Boolean cube existed, and by Lai [74]. Leighton [78] has recently also reported dilation 2, congestion 2 embeddings of pyramids in Boolean cubes.

4.9.2   Embedding Multiple Pyramids with Unit Expansion

Even though minimal expansion (i.e., expansion < 2) is achieved in the hyper-pyramid embedding above, $2^d - 2$ cube nodes are not used in each induction step. It is possible, however, to embed one $\hat{P}(k, d)$ hyper-pyramid and $2^d - 2$ $\hat{P}(k-1, d)$ hyper-pyramids into a B_{kd+1} cube, such that only one cube node is not used.

Theorem   16   [44] *A hyper-pyramid $\hat{P}(k, d)$ together with $(2^d - 2)$ hyper-pyramids $\hat{P}(k-1, d)$, $k \geq 1$ and $d \geq 2$, can be embedded in a B_{kd+1} Boolean cube with expansion ≈ 1 (only one cube node is not used) and dilation $d + 1$.* □

A proof of the theorem can be found in [44]. For the proof it is convenient to consider one two-rooted hyper-pyramid $\tilde{P}(k, d)$ and $2^d - 2$ hyper-pyramids $\hat{P}(k-1, d)$ (with single roots). The induction hypothesis is that the mapping function f_k for $k \leq n$ satisfies these conditions:

(1) A two-rooted hyper-pyramid $\tilde{P}_0(k, d)$ and $(2^d - 2)$ hyper-pyramids $\hat{P}_j(k-1, d)$, $2 \leq j < 2^d$, $k \geq 1$ and $d \geq 2$, can be embedded in a B_{kd+1} Boolean cube with dilation $d + 1$.

(2) $Hamming(f_k(a_x(0, \varepsilon)), f_k(a_{x\{m\}}(0, \varepsilon))) = 1$ for all $0 \leq x < 2^d$, $0 \leq m < d$, i.e., all the 2^d roots are mapped to a B_d subcube in the B_{kd+1} cube, and the two roots of $\tilde{P}_0$ are mapped to adjacent cube nodes.

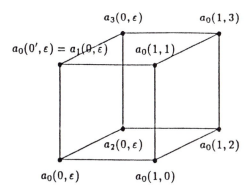

Figure 4.43   The basis: A two-rooted hyper-pyramid $\tilde{P}_0(1,2)$, a hyper-pyramid $\hat{P}_2(0,2)$ and a hyper-pyramid $\hat{P}_3(0,2)$ embedded in a B_3 cube with dilation 3.

In [44] it is also shown that the dilation $d+1$ embedding yields a *congestion* of at most $(d+1)(\frac{2^{d+1}}{d+2}+1)$. However, a dilation $2d$ embedding yields a congestion of $\lceil\frac{2^d}{d}\rceil + \frac{2^{d+1}}{d+2}+1$, which is the same as the congestion for the embedding of a single hyper-pyramid in the same sized Boolean cube. It is fairly easy to show that the *active degree* is $kd+1$. The arguments for proving the maximum *node load* are similar to the arguments for the single hyper-pyramid embedding.

4.10   Permutations

A cyclic shift of objects along an axis in a lattice defines a particular permutation that is a very useful programming primitive. Matrix multiplication by Cannon's algorithm [24, 56] is an example. Butterfly emulation requires communication between nodes of binary addresses i and $i \oplus 2^k$ in the embedding described earlier. But other operations on the same data set may preferably be performed on Gray code encoded data structures. For some computations it may be important to change the allocation from binary encoding to Gray code encoding, or vice versa. Matrix transposition is one of the classical permutations. The data motion required for changing between different schemes for allocating virtual processors defines a class of permutations. These permutations, and matrix transposition, are special cases of dimension permutations (assuming all axes are of a length equal to some power of two).

4.10.1   Cyclic Shifts Along an Axis of Arrays Embedded by Gray Code Encoding

In a Boolean n-cube, i and $(i + j)mod2^n$ are at a distance of at most n. In an array embedded by Gray code encoding of the indices, a cyclic shift by i steps in the direction of an axis implying a routing $G(n)_j \rightarrow G(n)_{(i+j)mod2^n}$, $j = \{0, 1, \ldots, 2^n - 1\}$. The minimal number of routing steps for each j is $\|G(n)_j \oplus G(n)_{(i+j)mod2^n}\|$.

A cyclic shift of i steps can be decomposed into a sequence of shifts of length 2^r by considering the binary encoding of i [56]. By lemma 4, j and $(j + 2^r)mod2^n$ differ in precisely two dimensions for $r > 0$ and in one dimension for $r = 0$. Of the two dimensions that have different bit values, one dimension is the same for all j. The other dimension depends on j. A cyclic shift may require $2n - 1$ nearest neighbor communications. A cube dimension may be routed twice, and no dimension is routed more than twice. Some of the communications are redundant. An element may traverse the same edge in both directions, as for instance the element located in processor 0 subject to a shift of length $i = 6$. For this element the routing is dimension 0 and dimension 1 for a shift of length 2, and dimension 1 and dimension 2 for a shift of length 4.

If the rotation is implemented as a sequence of reflections, i.e., by correcting successive bits that differ in G_j and $G_{(i+j)mod2^n}$, then clearly an element is subject to at most n routing steps. However, two matrix elements may traverse the same edge. For instance, consider a rotation of three steps and elements 0 and 1 in a cube of dimension at least 3. We claim without proof that at most two elements traverse any edge in a given direction.

Performing reflections on the different bits in order from lowest to highest, or the reverse, for all nodes and dimensions that need to be routed results in a traversal of some edges by two elements for some rotations. At this point we conjecture that it is possible to find routing algorithms that result in edge-disjoint paths for all the different source-destination pairs. We have generated such routing schemes for up to five-dimensional cubes.

4.10.2   Conversion Between Binary Encoding and Binary-Reflected Gray Code Encoding

By construction, the highest order bits in the binary reflected Gray code encoding of an integer and its binary encoding coincide. The two encodings of the integer $2^n - 1$, differ in $n - 1$ bits, and the maximum routing distance is $n - 1$. To carry out the conversion in $n - 1$ routing steps, no two elements must compete for the same communications link at any given time.

Lemma   26   *[56] A Gray code encoding can be rearranged to a binary encoding in $n-1$ routing steps.*

Proof   The proof is by induction. The lemma is clearly true for a 1-cube. Assume it is true for a k-cube. Then for a $k+1$-cube the Gray codes $G(k+1)$ of the processors with addresses encoding the integers $\{0, 1, \ldots, 2^k - 1\}^T$ are the same as $G(k)$ with a 0 appended as the highest order bit. For the second half of the processors, the highest order bit is 1, and the Gray codes $G(k)$ reversed. The highest order bit in the Gray code encoding and in the binary encoding coincide. By an exchange operation in dimension $k-1$ for the processors with bit k equal to 1 (dimensions are labeled $0-k$), the Gray codes for the second half of the processors are the same as in the first half. By the induction assumption, a k bit Gray code can be changed to a binary code in $k-1$ routing steps, and the lemma follows.   ☐

Corollary   9   *The directed routing paths are edge-disjoint.*

Proof   A dimension is only routed once, and the routing is an exchange operation.   ☐

Corollary   10   *Conversion of an n-bit binary-reflected Gray code encoding with M elements per node to a binary encoding, or vice versa, can be performed in $n + M - 2$ communication steps on an n-cube.*   ☐

Table 4.16 illustrates the sequence of reflections that converts a 4-bit Gray code to a binary code. Processor addresses are given in binary code, and the integers stored in a processor are given in decimal representation.

The algorithm depicted in Table 4.16 can be described by the following pseudo code.

```
/* An algorithm converting a binary-reflected Gray code to
Binary code starting from the highest order bit */
for j := m - 2 downto 0 do
        if a_{j+1} = 1 then
                exchange contents between processors
                (a_{m-1}a_{m-2} ... a_{j+1}0a_{j-1} ... a_0)
                and (a_{m-1}a_{m-2} ... a_{j+1}1a_{j-1} ... a_0)
        endif
enddo
```

Table 4.16   Conversion of Gray code to binary code.

Gray code	reflection on bit 2	reflection on bit 1	reflection on bit 0
00 0000	00 0000	00 0000	00 0000
01 0001	01 0001	01 0001	01 0001
02 0011	02 0011	02 0011	03 0011
03 0010	03 0010	03 0010	02 0010
04 0110	04 0110	07 0110	06 0110
05 0111	05 0111	06 0111	07 0111
06 0101	06 0101	05 0101	05 0101
07 0100	07 0100	04 0100	04 0100
08 1100	15 1100	12 1100	12 1100
09 1101	14 1101	13 1101	13 1101
10 1111	13 1111	14 1111	15 1111
11 1110	12 1110	15 1110	14 1110
12 1010	11 1010	11 1010	10 1010
13 1011	10 1011	10 1011	11 1011
14 1001	09 1001	09 1001	09 1001
15 1000	08 1000	08 1000	08 1000

```
/* An algorithm converting a binary-reflected Gray code to
Binary code starting from the lowest order bit */
for j := 0 to m − 2 do
        if a_{k−1} ⊕ a_{k−2} ⊕ ... ⊕ a_{j+1} = 1 then
                exchange contents between processors
                (a_{m−1}a_{m−2}...a_{j+1}0a_{j−1}...a_0)
                and (a_{m−1}a_{m−2}...a_{j+1}1a_{j−1}...a_0)
        endif
enddo
```

The following two algorithms convert a Binary code to a binary-reflected Gray code.

```
/* An algorithm converting a binary code to binary-reflected
Gray code starting from the highest order bit */
for j := m − 2 downto 0 do
        if a_{k−1} ⊕ a_{k−2} ⊕ ... ⊕ a_{j+1}) = 1 then
```

exchange contents between processors
$(a_{m-1}a_{m-2}\ldots a_{j+1}0a_{j-1}\ldots a_0)$
and $(a_{m-1}a_{m-2}\ldots a_{j+1}1a_{j-1}\ldots a_0)$
    endif
enddo

/* An algorithm converting a Binary code to binary-reflected
Gray code starting from the lowest order bit */
for $j := 0$ to $m - 2$ do
    if $a_{j+1} = 1$ then
        exchange contents between processors
        $(a_{m-1}a_{m-2}\ldots a_{j+1}0a_{j-1}\ldots a_0)$
        and $(a_{m-1}a_{m-2}\ldots a_{j+1}1a_{j-1}\ldots a_0)$
    endif
enddo

4.10.3   Matrix Transposition

The transpose A^T of a $P \times Q$ matrix A is defined by the relation $a^T(u,v) = a(v,u)$, where $a^T(u,v)$ is the element in row u and column v of A^T, and $a(u,v)$ is the element of A in row u and column v. Formally,

Definition   24   The matrix transposition operation is the permutation $(u_{p-1}u_{p-2}\ldots u_0v_{q-1}v_{q-2}\ldots v_0) \leftarrow (v_{q-1}v_{q-2}\ldots v_0u_{p-1}u_{p-2}\ldots u_0)$.   ∎

A $P \times Q$ matrix with $P > Q$ is extended to a square matrix by introducing *virtual elements* corresponding to $P - Q$ columns. The extension is made similarly if $P < Q$. The extension can be made either by adding columns corresponding to high or low order dimensions of the column address space or by mixing columns of virtual elements with columns of real elements.

Definition   25   A *shuffle* operation, sh^1 on a set of elements $\mathcal{W}$ with addresses w, $w \in \{0,1,\ldots,2^m - 1\}$ encoded in binary representation $(w_{m-1}w_{m-2}\ldots w_0)$ is a permutation defined by a one step *left cyclic shift*, $(w_{m-1}w_{m-2}\ldots w_0) \leftarrow (w_{m-2}w_{m-3}\ldots w_0w_{m-1})$, $w \in \{0,1,\ldots,2^m - 1\}$. An *unshuffle* operation, sh^{-1} is defined by a one step *right cyclic shift*. $sh^k = sh\ sh^{k-1}$ is a k step left cyclic shift.   ∎

Clearly, $sh^1sh^{-1} = I$, where I is the identity operator.

Lemma   27   *Let A be a $2^p \times 2^q$ matrix. $A^T \leftarrow sh^p A$, or $A^T \leftarrow sh^{-q}A$.*

□

One-dimensional partitioning of matrices

We will first consider transposition of matrices stored by one-dimensional partitioning, and then treat two-dimensional partitioning. Consider a one-dimensional cyclic column partitioning such that $p = q > n_0 = n$. Then, for the transposition each processor divides its set of rows into 2^n parts, each of 2^{p+q-2n} elements, and sends a part to every processor. The communication is *all-to-all personalized communication* [41, 66]. The address field before transposition is

$$\big(\underbrace{u_{p-1}u_{p-2}\ldots u_0\; v_{q-1}v_{q-2}\ldots v_n}_{vp}\; \underbrace{v_{n-1}\ldots v_0}_{rp}\big),$$

and after transposition

$$\big(\underbrace{v_{q-1}v_{q-2}\ldots v_0\; u_{p-1}u_{p-2}\ldots u_n}_{vp}\; \underbrace{u_{n-1}\ldots u_0}_{rp}\big),$$

which in the orignal address field is

$$\big(\underbrace{u_{p-1}u_{p-2}\ldots u_n}_{vp}\; \underbrace{u_{n-1}\ldots u_0}_{rp}\; \underbrace{v_{q-1}v_{q-2}\ldots v_0}_{vp}\big).$$

The set of dimensions used for real processors before and after the transposition are disjoint, i.e., $\mathcal{R}^b \cap \mathcal{R}^a = \phi$. If $q < n \leq p$ and the initial assignment is by columns, then only 2^q processors are used before the transposition, but all 2^n processors are used after the transposition. The number of virtual processors per real processor before the transposition is 2^p, and after the transposition 2^{p+q-n}. The row address field is divided into 2^n partitions. The address fields for real processors before and after the transposition are disjoint. The transposition is accomplished by all 2^q processors holding matrix elements sending a unique set of data to each of the 2^n processors. The communication is *some-to-all personalized communication*. In the case of transposing a vector, the communication is *one-to-all* or *all-to-one personalized communication*.

Two-dimensional partitioning of matrices

In a two-dimensional partitioning of a matrix with the same number of processors assigned to rows and columns, and the same assignment scheme (cyclic or consecutive) for rows and columns, the logic dimensions assigned to real processor dimensions before and after the transposition are the same, i.e., $\mathcal{R}^b = \mathcal{R}^a$. The communication is between distinct pairs of processors. This case is the basic two-dimensional matrix transposition. The two-dimensional matrix transposition with $P = 2^p$ and $Q = 2^q$ and all processors used before and after the transposition is a special case of *stable dimension permutations* [46, 45, 64]. A recursive matrix transposition algorithm, as illustrated in Figure 4.44, naturally yields a *single path transpose* (SPT) algorithm for a Boolean cube [56, 53, 82]. The recursion is performed on successive dimension pairs in the binary representation of the row and column indices. An exchange takes place if the two bits have different values. The algorithm can easily be modified such that two paths are used between each pair of nodes, which leads to a *dual paths transpose* (DPT) algorithm [42, 63]. It is still possible to make use of the fact that there are more than two paths between any pair of nodes in a Boolean cube. The *multiple paths transpose* (MPT) algorithm [42, 63] makes use of all minimum length paths between the source/destination pair of processors in a matrix transposition. Whereas all directed paths in the SPT and DPT algorithms are edge-disjoint, the paths in the MPT algorithm are not disjoint. With the appropriate scheduling of messages along the paths, a pipelining effect can be achieved without congestion. An algorithm similar to the MPT algorithm is due to Stout [111, 112].

4.10.4   The Single Path Transpose (SPT) Algorithm

In the basic case, the SPT algorithm assumes that rows and columns both are encoded in binary code. In the case of virtual processors, the allocation scheme (e.g., cyclic or consecutive) is the same for both axes. The modifications necessary when one of the axes is encoded in Gray code is presented after the basic algorithms are given. No modification is necessary when both axes are encoded in (the same) Gray code. For simplicity we also assume that the number of processors in the row and column direction are the same: $n_1 = n_0 = \frac{n}{2}$, n even. In the ith exchange step, data is exchanged between all processors for which $a_{d_i} \oplus a_{f(d_i)} = 1$, where $d_i < q, d_i \in \mathcal{R}$ and $q \leq f(d_i) < p + q, f(d_i) \in \mathcal{R}$. Hence d_i is a dimension

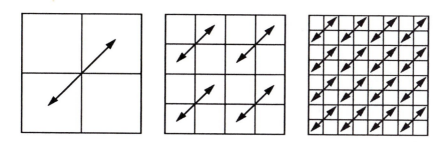

Figure 4.44   Recursive transposition of a matrix. [56]

in the binary encoding of the column indices that is assigned to a real processor dimension, and $f(d_i)$ is the dimension in the binary encoding of the row indices that corresponds to the column dimension d_i. Dimension $f(d_i)$ is also assigned to a real processor dimension.

The routing order for the dimensions is the same for all nodes—for instance, highest to lowest order for both row and column encoding, i.e., $f(d_{\frac{n}{2}-1}), d_{\frac{n}{2}-1}, f(d_{\frac{n}{2}-2}), d_{\frac{n}{2}-2}, \dots, f(d_0), d_0$. For each source-destination pair there is a single path. Let $tr(w) = (v||u)$ with $w = (u||v)$ and let $H_r(w) = Hamming_r(w, tr(w))$. The length of the path for element w is $H_r(w)$.

Lemma 28   *[56] The path for any element w is edge-disjoint from the paths for all other elements.* $\square$

Lemma 28 is immediate since a dimension is routed only once. Elements on the antidiagonal are at distance n. The first packet for each node on the antidiagonal arrives after n routing steps. Additional packets arrive every cycle thereafter. The total number of routing steps is $\lceil \frac{PQ}{BN} \rceil + n - 1$.

4.10.5   The Dual Paths Transpose (DPT) Algorithm

The DPT algorithm is obtained by defining a second path from each node by simply interchanging the order in which row and column dimensions are routed. With the first path as given above, the second path would be $d_{\frac{n}{2}-1}, f(d_{\frac{n}{2}-1}), d_{\frac{n}{2}-2}, f(d_{\frac{n}{2}-2}), \dots, d_0, f(d_0)$. The two directed paths for a particular w are edge-disjoint (as observed in [51], where the same communication pattern occured in the solution of tridiagonal systems on Boolean cubes). Moreover, the two directed paths for any w are edge-disjoint with respect to the paths for all other w [63].

4.10.6   The Multiple Paths Transpose (MPT) Algorithm

The MPT algorithm makes use of all minimum length paths between w and $tr(w)$. The key observation is that the row/column dimension pairs can be routed in an arbitrary order. For instance, instead of starting with the highest order row/column dimensions, it is possible to start with the second highest order pair and proceed in decreasing order and finish with the highest order pair. The number of sequences defined in this manner is equal to the number of pairs. For each such sequence there is the option of either starting with the row or column dimension. For example, if $w = (1001||0100)$ and $n = 8$, then $tr(w) = (v||u) = (0100||1001)$ and $H_r(w) = 6$. The six paths for the MPT algorithm are defined as follows:

$$path\ 0 = 7,3,6,2,4,0. \qquad path\ 3 = 3,7,2,6,0,4.$$
$$path\ 1 = 4,0,7,3,6,2. \qquad path\ 4 = 0,4,3,7,2,6.$$
$$path\ 2 = 6,2,4,0,7,3. \qquad path\ 5 = 2,6,0,4,3,7.$$

Path 0 starts from the source node (10010100) and goes through nodes (00010100), (00011100), (01011100), (01011000), and (01001000) and reaches the destination node (01001001). Path p can be derived by a two steps right rotation of path $p-1$, if $0 < p < H_r(w), p \neq \frac{1}{2}H_r(w)$. Path 0 and $\frac{1}{2}H_r(w)$ are the paths of the DFT algorithm. All paths between a source/destination pair of nodes have the same length and are edge-disjoint. But the paths are not disjoint from paths for other node pairs [63]. Figure 4.45 shows the six edge-disjoint paths from node $w = (000111)$ to node $tr(w)$. The label on an edge is the dimension it traverses.

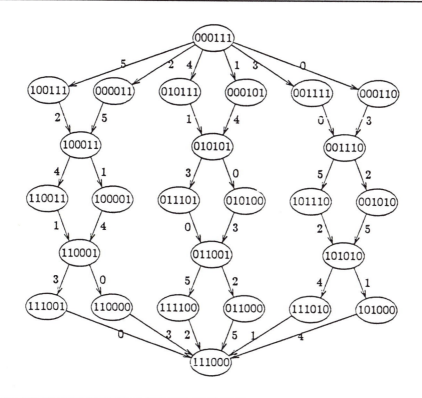

Figure 4.45   Six edge-disjoint paths from node $w = (000111)$ to node $tr(w) = (111000)$. [63]

Theorem 17 *[63] The total matrix transpose time by the MPT algorithm is*

$$
T_{min} = \begin{cases}
(n+1)\tau + (1+\frac{1}{n})\frac{PQ}{2N}t_c & \text{if } n \geq \sqrt{\frac{PQt_c}{N\tau}} \text{ approx.} \\[2mm]
(\frac{n}{2}+3)\tau + (1+\frac{2}{n+4})\frac{PQ}{2N}t_c & \text{if } \sqrt{\frac{PQt_c}{2N\tau}} < n \leq \sqrt{\frac{PQt_c}{N\tau}} \text{ approx.} \\
& \text{and } \frac{n}{2} \text{ is even;} \\[2mm]
(\frac{n}{2}+2)\tau + (1+\frac{2}{n+2})\frac{PQ}{2N}t_c & \text{if } \sqrt{\frac{PQt_c}{2N\tau}} < n \leq \sqrt{\frac{PQt_c}{N\tau}} \text{ approx.} \\
& \text{and } \frac{n}{2} \text{ is odd;} \\[2mm]
(\sqrt{\tau} + \sqrt{\frac{PQt_c}{2N}})^2 & \text{if } n \leq \sqrt{\frac{PQt_c}{2N\tau}},
\end{cases}
$$

and the optimum packet size is

$$
B_{opt} = \begin{cases}
\lceil \frac{PQ}{N(n+4)} \rceil & \text{for even } \frac{n}{2} \text{ and } n > \sqrt{\frac{PQt_c}{2N\tau}}; \\[2mm]
\lceil \frac{PQ}{N(n+2)} \rceil & \text{for odd } \frac{n}{2} \text{ and } n > \sqrt{\frac{PQt_c}{2N\tau}}; \\[2mm]
\sqrt{\frac{PQ\tau}{2Nt_c}} & \text{for } n \leq \sqrt{\frac{PQt_c}{2N\tau}}.
\end{cases}
$$

Proof   The minimum number of start-ups is determined by the largest distance, which is n. Nodes on the main antidiagonal are at distance n. For a lower bound on the required time for data transfer, consider the upper right $\frac{\sqrt{N}}{2} \times \frac{\sqrt{N}}{2}$ submatrix. There are $\frac{N}{4}$ nodes. Each node has to send $\frac{PQ}{N}$ data to some node outside the submatrix. Two dimensions per node connect to nodes outside of the submatrix, i.e., a total of $\frac{2N}{4}$ links. Hence, the data transfer requires a time of at least $\frac{PQ}{2N}t_c$. □

Gray code encoded matrices

We have assumed that the matrix elements are assigned to processors through a binary encoding of row and column indices. However, the transpose algorithms described above also apply to matrices assigned to processors through Gray code encoding [56]. For a binary encoding of row and column indices, matrix element (u,v) is stored in processor $w = (u||v)$ and matrix element (v,u) is stored in processor $tr(w) = (v||u)$. For Gray code encoding of row and column indices, matrix element (u,v) is stored in processor $(G(u)||G(v))$ and matrix element (v,u) is stored in processor $(G(v)||G(u))$. The two-dimensional transpose algorithms described above perform the operation $(u||v) \leftarrow (v||u), \forall u \in \{0,1,\ldots,P-1\}, \forall v \in \{0,1,\ldots,Q-1\}$. No change of the encoding within each field takes place.

Matrix transposition with Gray code to binary code conversion

The transposition and code conversion can be made independently. With the row indices encoded in binary code and the column indices encoded in

Gray code, a binary-to-Gray code conversion can first be done concurrently within each column subcube in $\frac{n}{2} - 1$ steps [53]. The Gray-to-binary code conversion within each row subcube requires an additional $\frac{n}{2} - 1$ steps. Finally, the transposition requires n steps. The total number of routing steps is $2n - 2$. However, n routing steps suffice. With the rows encoded in binary code and the columns in Gray code, matrix block (u, v) is stored in processor $(u \| G(v))$ and matrix block (v, u) is stored in processor $(v \| G(u))$. The transposition with permutation is an exchange of data between processor $(u \| G(v))$ and processor $(G^{-1}(G(v)) \| G(u))$ for all u and v, where $G^{-1}()$ is the inverse Gray code.

An algorithm combining transposition with code conversion consisting of $\frac{1}{2}n$ iterations, each requiring two routing steps, based on the SPT algorithm is described next [63]. In iteration $i \in \{0, 1, \ldots, \frac{1}{2}n - 1\}$, dimensions (bits) $\frac{n}{2} - i - 1$ of the row and column indices are routed. During the first iteration, the upper right block $(0xx..x \| 1xx..x)$ and the lower left block $(1xx..x \| 0xx..x)$ are exchanged in two steps. The code conversion does not affect the first exchange because the highest order bit is the same in the binary and binary-reflected Gray code. For the next iteration, the transposition requires a conversion from Gray code to binary code of the column indices. An exchange should take place on the second highest order bit of the column index. After the first iteration, all these column indices are located in the second half of the row address space. Hence the first step in converting the Gray code to binary code for the column indices requires a horizontal exchange within blocks for the second half of the block rows. Similarly, the conversion from binary code to Gray code of the row indices requires an exchange for the second half of the row indices, which after the first exchange step are in the upper half of the column address space. The binary code encoding of the row indices forces a vertical exchange for the second half of the block columns. In general, the binary code to binary-reflected Gray code conversion starting at the highest order bit is controlled by the parity of the higher order address bits. The Gray code to binary code conversion proceeding from the highest order bit to the lowest order bit is controlled by a single bit (see the preceding section). The transposition in the second iteration requires an antidiagonal exchange within all four blocks. The combined permutation pattern is shown in Figure 4.46.

In general, in the ith iteration the Gray code encoding of the columns causes a horizontal exchange within blocks for all block rows with bit $\frac{n}{2} - i - 1$ in the row index equal to one. The binary code encoding causes a vertical exchange within all block columns such that the parity of the binary encoding of bits $\frac{n}{2} - 1$ through $\frac{n}{2} - i$ is odd. Figure 4.47 shows the four iterations for $n = 8$. c means *clockwise rotation* and cc means *counterclockwise rotation*. The algorithm [63] is presented next.

ALGORITHMS FOR MATRIX TRANSPOSITION

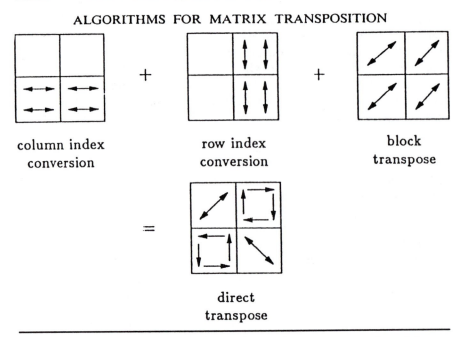

column index   row index   block
conversion   conversion   transpose

direct
transpose

Figure 4.46   Transposition of a matrix stored by binary code encoding of the row indices and Gray code encoding of column indices. [63]

```
/* Note: my-addr is of the form (u||v). */
/* The second argument of "send" and "recv" represent the
cube dimension */
/* and 'buf' contains the data to be transposed initially. */
even-block-row := true;
even-parity-block-column := true;
for j := n/2 - 1 downto 0 do
        case (even-block-row, even-parity-block-column,
        bit j + n/2, bit j) of
                (TT00), (TT11), (FF01), (FF10):
                        recv (tmp, j + n/2); send (tmp, j);
                (TT01), (TT10), (FF00), (FF11), (TF01),
                (TF10), (FT00), (FT11):
                        send (buf, j + n/2); recv (buf, j);
                (TF00), (TF11), (FT01), (FT10):
                        send (buf, j); recv (buf, j + n/2);
        endcase
        even-block-row := (bit j + n/2 = 0);
```

```
              if (bit j = 1) then
                      even-parity-block-column
                      := not even-parity-block-column;
              endif
      enddo
```

To transpose a matrix stored by binary encoding of row and column indices into a transposed matrix with row and columns encoded in Gray code, a combined conversion–transpose algorithm similar to the one above can be applied to accomplish the task in n routing steps. The algorithm above needs only to be modified such that the column operations are controlled by even-block-columns (instead of even-parity-block-columns). Similarly, to transpose a matrix with both row and columns encoded in Gray code into a transposed matrix with rows and columns encoded in binary code, the control of the row operations is changed from even block rows to even-parity block rows.

Remarks

An arbitrary permutation on an n-cube can be realized by all-to-all personalized communication twice, if the size of messages to be permuted is the same for all processors and at least N (per processor) [112, 111]. Since transposing a matrix with two-dimensional partitioning and $n_0 = n_1$ is a permutation, one can also realize it by all-to-all personalized communication twice. However, the communication complexity is higher than that of the best transpose algorithm for the two-dimensional partitioning either for *one-port* communication, or for *n-port* communication.

The correspondence between dimensions for matrix transposition is $f(i) = i$, $g(i) = i + q, \forall i \in \{0, 1, \ldots, q - 1\}$. By changing the correspondence to $f(i) = i$, $g(i) = 2q - 1 - i, \forall i \in \{0, 1, \ldots, q - 1\}$ a *bit-reversal* permutation is realized. A bit-reversal permutation is defined by

$$(a_{n-1}a_{n-2}\ldots a_0) \leftarrow (a_0 a_1 \ldots a_{n-1}).$$

4.10.7   Dimension Permutations

A *dimension permutation* is a permutation defined on the bits of the address field, while an arbitrary permutation is a permutation on the address field. There are $(\log_2 M)!$ possible dimension permutations compared to $M!$ arbitrary permutations for an address space of size M. In the preceding sections we have already seen several examples of dimension permutations: k-shuffle/unshuffle permutations, matrix transposition [56, 63], bit-reversal [67], and conversion between various data allocations such as between consecutive and cyclic storage [56], [63]. Shuffle operations can be used to reconfigure a two-dimensional partitioning to a three-dimensional

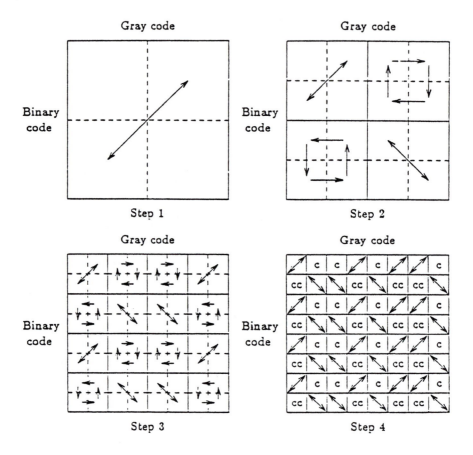

Figure 4.47   Transposition of a matrix stored by mixed encoding of rows and columns in an 8-cube. [63]

partitioning of a matrix for multiplication with maximum concurrency [60], and for data (re)alignment for certain Fast Fourier Transform algorithms [57, 67].

In this section we will give lower bounds for stable dimension permutations [46, 64] and some algorithms for implementing such permutations on Boolean cubes [114, 45, 64]. A *stable* permutation is a permutation such that the same processors are used before and after the permutation. Flanders [30] devised algorithms for stable dimension permutations on mesh-connected array processors. A *generalized shuffle* permutation [64] corresponds to a cyclic shift on a properly ordered index set. A dimension permutation is a generalization of shuffle permutations. It may consist of several shuffle permutations on independent index sets [62]. In *unstable* dimension permutations [59], the set of processors holding data before and after the permutation is not necessarily identical.

Classification of dimension permutations

A dimension permutation implies a change in allocation from $\pi^b(\mathcal{M})$, before the permutation, to $\pi^a(\mathcal{M})$ after it. Let $\mathcal{R}^b$ be the set of logic dimensions mapped to processor dimensions before the permutation and $\mathcal{R}^a$ the set of logic dimensions mapped to processor dimensions after it. $\mathcal{V}^b$ and $\mathcal{V}^a$ are the set of logic dimensions mapped to virtual dimensions before and after the permutation, respectively. The sets of machine dimensions used before and after the permutation are denoted $\Gamma^b = \Gamma_p^b \cup \Gamma_s^b$ and $\Gamma^a = \Gamma_p^a \cup \Gamma_s^a$, where $\Gamma_p^b, \Gamma_p^a \subseteq \mathcal{Q}_p$ and $\Gamma_s^b, \Gamma_s^a \subseteq \mathcal{Q}_s$. Clearly $|\Gamma^b| = |\Gamma^a|$, since the number of elements is conserved. If $\Gamma^b = \Gamma^a$ (i.e., $\Gamma_p^b = \Gamma_p^a$ and $\Gamma_s^b = \Gamma_s^a$), then the dimension permutation is *stable*; otherwise, it is *unstable*. Note that we classify the case where $\Gamma_p^b = \Gamma_p^a$ and $\Gamma_s^b \neq \Gamma_s^a$ as an unstable dimension permutation. The restriction of *stable* dimension permutations to use the same local address space before and after the permutation is made for notational convenience. The algorithms for stable dimension permutations as defined here also work for the case $\Gamma_p^b = \Gamma_p^a$ and $\Gamma_s^b \neq \Gamma_s^a$ with the same complexity as in the stable case, if the time for local data rearrangement is ignored. For stable dimension permutations we let $\Gamma = \Gamma^b = \Gamma^a$, $\Gamma_p = \Gamma_p^b = \Gamma_p^a$ and $\Gamma_s = \Gamma_s^b = \Gamma_s^a$. We use the notation $(v_{m_s-1} v_{m_s-2} \cdots v_0 | r_{m_p-1} r_{m_p-2} \cdots r_0)$ for w when we want to stress the separation of logic dimensions mapped to real and virtual dimensions. Element w is allocated to location a, where $a_i = w_{\pi^{-1}(i)}$ if $i \in \Gamma$, and $a_i = 0$, otherwise. We arbitrarily define the unassigned address fields to be 0.

Definition 26 [64] A *stable dimension permutation* (SDP), δ, on a subset of the machine address space Γ is a one-to-one mapping $\Gamma \rightarrow \Gamma$ with $\delta = \pi^a \circ \pi^{-b}$. (π^{-b} denotes $(\pi^b)^{-1}$.) The *index set* $\mathcal{J}$ of the dimension permutation is the subset of Γ such that $\{i | \delta(i) \neq i\} = \mathcal{J}$. The *order* of the dimension permutation is $\sigma = |\mathcal{J}|$. Alternatively, one can define a *dimension permutation*, δ', on the set of logic dimensions, i.e., $\delta' : \mathcal{M} \to \mathcal{M}$ (for the stable case) with $\delta' = \pi^{-a} \circ \pi^b$. ∎

The *identity permutation* I is defined by $\delta_0(i) = i, \forall i \in \Gamma$ or $\mathcal{J} = \phi$. The order of the identity permutation is 0. A subscript σ on δ, δ_σ, is used to denote the order of the SDP being σ. The permutation function δ applies to the subset of machine dimensions. For convenience, we use $\delta(a)$, $a \in \mathcal{A}$, to denote $(a_{\delta(q-1)} a_{\delta(q-2)} \ldots a_{\delta(0)})$, where δ is extended to a function of $\mathcal{Q} \to \mathcal{Q}$ with $\delta(i) = i$, $i \in \mathcal{Q} - \mathcal{J}$. In an SDP, a logic dimension k assigned to machine dimension $i = \pi^b(k)$ is reassigned to machine dimension $\delta(i) = \pi^a(k)$. Clearly, $\delta = \pi^a \circ \pi^{-b}$.

Figure 4.48 shows a shuffle permutation in which machine dimension i becomes $\delta(i) = (i + 1) \bmod 5$. The processor with global address $(a_4 a_3 a_2 a_1 a_0)$ sends its contents to the processor with global address $(a_{\delta^{-1}(4)} a_{\delta^{-1}(3)} a_{\delta^{-1}(2)} a_{\delta^{-1}(1)} a_{\delta^{-1}(0)}) = (a_3 a_2 a_1 a_0 a_4)$. Following [46, 64], we introduce the following definitions.

Definition 27   A *full-cube permutation* (FCP) is an SDP for which the data set is allocated to all real processors (but not necessarily the entire memory): $\Gamma_p = \mathcal{Q}_p$. An *extended-cube permutation* (ECP) is an SDP for which the data set occupies only a fraction of the cube: $\Gamma_p \subset \mathcal{Q}_p$. ∎

Definition 28   A *generalized shuffle permutation* (GSH) of order σ, gsh_σ, is an SDP such that $\delta_\sigma(\alpha_0) = \alpha_1, \delta_\sigma(\alpha_1) = \alpha_2, \ldots, \delta_\sigma(\alpha_{\sigma-1}) = \alpha_0$, $\alpha_i \neq \alpha_j$, $i \neq j$, $\alpha_i, \alpha_j \in \mathcal{J}$, $0 \leq i, j < \sigma$, and $\delta_\sigma(i) = i$, $\forall i \in \Gamma - \mathcal{J}$. ∎

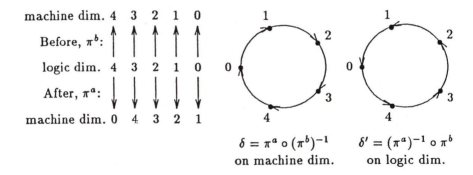

$$\delta = \pi^a \circ (\pi^b)^{-1} \qquad \delta' = (\pi^a)^{-1} \circ \pi^b$$
$$\text{on machine dim.} \qquad \text{on logic dim.}$$

Figure 4.48   The definition of δ and δ' and cycles formed by traversing δ and δ'.

Let the index set $\mathcal{J} = \{\alpha_0, \alpha_1, \ldots, \alpha_{\sigma-1}\}$ be ordered. An SDP consists of a number of independent GSH's (or cycles). The number of independent GSH's for a given SDP is denoted β, and the order of the ith GSH by σ_i, $0 \leq i < \beta$; $\sigma = \sum_{i=0}^{\beta-1} \sigma_i$.

Definition   29   A *real* GSH is a GSH such that $\mathcal{J} \subseteq \Gamma_p$ and $\mathcal{J} \neq \phi$. A *virtual* GSH is a GSH such that $\mathcal{J} \subseteq \Gamma_s$ and $\mathcal{J} \neq \phi$. A GSH that is neither real nor virtual and $\mathcal{J} \neq \phi$ is a *mixed* GSH. ∎

Definition   30   An SDP is *separable* if it can be decomposed into *independent* GSH's (i.e., with their index sets disjoint) that are either real or virtual; it is *nonseparable* otherwise. A *real* SDP is an SDP consisting only of all real GSH's. ∎

Definition   31   The *real order* σ_p of an SDP is $|\{j | j \neq \delta(j), j \in \Gamma_p\}| = |\mathcal{J} \cap \Gamma_p|$, and its *virtual order* σ_s is $|\{j | j \neq \delta(j), j \in \Gamma_s\}| = |\mathcal{J} \cap \Gamma_s|$. ∎

A *real* GSH preserves the local address map. Data is moved between processors in the subcube defined by the set of dimensions Γ_p. All processors in this set have identical address maps. A *real* GSH is a generalization of the *collinear planar exchanges* described by Flanders [30]. A *virtual* GSH (called *vertical exchanges* in [30]) involves only local data movement or a change of local address map in the set of processors defined by the set Γ_p. The same change is made for every processor in this set. A *mixed* GSH (*planar-vertical exchanges* [30]) involves both processor and local storage dimensions in the index set $\mathcal{J}$. For an SDP such that $\mathcal{R}^b \cap \mathcal{R}^a = \phi$, the permutation is an *all-to-all personalized communication* [66, 63]. An SDP such that $\delta^2(i) = \delta(\delta(i)) = i \ \forall i \in \Gamma$ is its own inverse. Examples of such an SDP are a bit-reversal permutation and the transposition of a matrix partitioned into $\sqrt{N} \times \sqrt{N}$ blocks.

Some properties of stable dimension permutations
The time complexity for the different SDP's are denoted $T^*_{type}(ports, \sigma_p, m_p, K)$, where *type* is the type of SDP, such as *gsh* for a generalized shuffle permutation or *sdp* for a stable dimension permutation. The superscript $*$ is either lb for a lower bound or an algorithm identifier for an upper bound. The first argument for T_{sdp} is the number of ports per processor used concurrently, the second argument the real order of the SDP, the third argument the number of processor dimensions being used, and the last argument the data volume per "allocated" processor, i.e., $K = 2^{m_s}$.

A GSH of order σ permutes blocks of $2^{m-\sigma}$ elements, and a GSH of real order σ_p consists of $2^{m_p-\sigma_p}$ independent permutations in disjoint subcubes, each of size 2^{σ_p} [46]. The blocks communicated between processors are of size 2^{m-m_p} elements. If a permutation involves only a subset of the dimensions in the cube, then there is an unused communication capacity. But, as lemma 29 shows, it cannot be used to reduce the time for the permutation. Note that lemma 29 addresses only the communication within the subcube used for the logic address space. It does not address the extended-cube permutation case.

Lemma 29 *[46] An SDP of real order $\sigma_p < m_p$ cannot be improved by communication in the $m_p - \sigma_p$ processor dimensions that are not included in the SDP, if the σ_p dimensions are fully utilized.*

Proof   Let P and P' be two problems such that $\mathcal{V} = \mathcal{V}'$ and subject to the same SDP, δ_{σ_p}, on a σ_p-cube and an m_p-cube, respectively. Let T be the communication complexity of a lower bound algorithm A for P, i.e., $T = \frac{B}{L}$, where B is the total bandwidth required and L the available bandwidth per unit time. Suppose there exists an algorithm A' for problem P' with communication complexity $T' < T$. Consider a new problem P'': an SDP δ_{σ_p} on a σ_p-cube with the bandwidth of each cube link and the data set per node expanded by a factor of $2^{m_p-\sigma_p}$ of the ones in P. Now, map the nodes in P' to nodes in P'' such that the corresponding bits of the σ_p dimensions for problem P' are used to identify processors in P''. Every algorithm for problem P' can be converted to an algorithm for problem P'' with a communication complexity that is at most the same. So, $T = \frac{B}{L} > T' \geq T''$, where T'' is the communication complexity of the corresponding algorithm for P''. However, $T'' \geq \frac{2^{m_p-\sigma_p}B}{2^{m_p-\sigma_p}L} = \frac{B}{L}$, which in turn is equal to T, and we have a contradiction. □

The matrix transposition algorithms presented in the previous section were all based on dimension exchanges. The number of communication steps and data transfers were of the same order as the lower bound. The algorithms for stable dimension permutations presented in this section are also based on dimension exchanges and of optimal order. Let the SDP consist of β *real* or *mixed* GSH's, and let the corresponding index sets be $\mathcal{J}_0, \mathcal{J}_1, \ldots, \mathcal{J}_{\beta-1}$, where $\mathcal{J}_i \cap \mathcal{J}_j = \phi$, $i \neq j$, $\mathcal{J} = \cup_{i=\{0,1,\ldots,\beta-1\}} \mathcal{J}_i$, and $\mathcal{J}_i = \{\alpha_{i0}, \alpha_{i1}, \ldots, \alpha_{i(\sigma_i-1)}\}$ and $\delta_\sigma(\alpha_{ij}) = \alpha_{i((j+1) \bmod \sigma_i)}$, $\forall(i,j) \in \{0,1,\ldots,\beta-1\} \times \{0,1,\ldots,\sigma_i-1\}$. The index sets for different GSH's out of which the SDP is composed are disjoint. Each GSH defines a cycle on its index set. A separable SDP consists of only *real* GSH's, and a nonseparable SDP consists of only *real* and *mixed* GSH's. Let oJ be the number of sets such that σ_i is odd and $\mathcal{J}_i \subset \Gamma_p$. Also, let σ_{p_i} be the *real order* of the GSH defined by $\mathcal{J}_i$. Clearly, $\delta_\sigma(i) = i$, $\forall i \in \Gamma - \mathcal{J}$, and $\sigma_p = \sum_{i=0}^{\beta-1} \sigma_{p_i} \leq \sum_{i=0}^{\beta-1} \sigma_i = \sigma$. Furthermore, let $\sigma_{max} = \max_i \{\sigma_{p_i}\}$ and $\sigma_{min} = \min_i \{\sigma_{p_i}\}$ for all $0 \leq i < \beta$.

Lemma 30 *[46] The lower bound for a full-cube, real GSH of order* σ_p, $\sigma_p > 0$, *on an* n-*cube is*

$$T_{gsh}^{lb}(1, \sigma_p, n, K) = \begin{cases} \max(\frac{\sigma_p K}{2} t_c, \sigma_p \tau), & \sigma_p \text{ is even,} \\ \max(\frac{\sigma_p K}{2} t_c, (\sigma_p - 1)\tau), & \sigma_p \text{ is odd,} \end{cases}$$

for one-port *communication, and*

$$T_{gsh}^{lb}(n, \sigma_p, n, K) = \begin{cases} \max(\frac{K}{2} t_c, \sigma_p \tau), & \sigma_p \text{ is even,} \\ \max(\frac{K}{2} t_c, (\sigma_p - 1)\tau), & \sigma_p \text{ is odd,} \end{cases}$$

for n-port *communication.*

Proof Consider the minimum number of start-ups first. Let a be such that $a_{\alpha_i} = 0$, if $i \bmod 2 = 0$, and $a_{\alpha_i} = 1$ otherwise, $0 \le i < \sigma_p$. It follows that $Hamming_r(a, gsh(a)) = \sigma_p$, if σ_p is even; and $\sigma_p - 1$, if σ_p is odd. To show that $\sigma_p - 1$ is the maximum Hamming distance if σ_p is odd, we show that a Hamming distance of σ_p is impossible. This is easily seen since $a_{\alpha_i} \ne a_{\alpha_{(i+1) \bmod \sigma_p}}$, $\forall 0 \le i < \sigma_p$, is impossible if σ_p is odd.

The minimum data transfer time is bounded from below by the required bandwidth divided by the available bandwidth. By lemma 29, we can consider the bandwidths for each σ_p-cube. For each $j \in \mathcal{J} \cap \Gamma_p = \mathcal{J}$ with $\delta^{-1}(j) = i$, $i \ne j$, only half of the nodes ($a_i \ne a_j$) need to send elements across cube dimension j, and therefore the bandwidth requirement for each permutation in subcubes of dimension σ_p is $\sigma_p 2^{\sigma_p - 1} K$. The available bandwidth per routing cycle of a σ_p-cube is 2^{σ_p} for *one-port* and $\sigma_p 2^{\sigma_p}$ for *n-port* communication. $\square$

Lemma 31 *[46] The lower bound for a full-cube, mixed GSH of real order* σ_p, $0 < \sigma_p < \sigma$, *on an* n-*cube is*

$$T_{gsh}^{lb}(1, \sigma_p, n, K) = \max\left(\frac{\sigma_p K}{2} t_c, \sigma_p \tau\right)$$

for one-port *communication, and*

$$T_{gsh}^{lb}(n, \sigma_p, n, K) = \max\left(\frac{K}{2} t_c, \sigma_p \tau\right)$$

for n-port *communication.* $\square$

Theorem   18   *[46] The lower bound for the communication complexity for a full-cube SDP of real order σ_p on an n-cube is*

$$T_{sdp}^{lb}(1, \sigma_p, n, K) = \max\left(\frac{\sigma_p K}{2} t_c, (\sigma_p - oJ)\tau\right)$$

for one-port *communication, and*

$$T_{sdp}^{lb}(n, \sigma_p, n, K) = \max\left(\frac{K}{2} t_c, (\sigma_p - oJ)\tau\right)$$

for n-port *communication.* (Recall that oJ is the number of sets such that σ_i is odd and $\mathcal{J}_i \subset \Gamma_p$.)

Proof   The data transfer time is bounded from below by the total bandwidth requirement divided by the maximum number of links available per routing cycle. By lemmas 30 and 31, the total bandwidth required for each permutation in subcubes of dimension σ_p (identified by $\mathcal{J}$) is $\sigma_p 2^{\sigma_p - 1} K$. The number of available links per routing step for each subcube is 2^{σ_p} for *one-port* and $\sigma_p 2^{\sigma_p}$ for *n-port* communication.

The minimum number of start-ups is $\max\{Hamming_r(a, \delta(a))\}, \forall a \in \mathcal{A}$, which is obtained by maximizing the *real* distance for each index set $\mathcal{J}_i$. By lemmas 30 and 31, the maximum real distance is $\sigma_p - oJ$.   □

Corollary   11   *[46] The minimum number of start-ups for a full-cube, separable SDP of real order σ_p is at least $\frac{2\sigma_p}{3}$ for any SDP and at most σ_p for some SDP's (as a tight bound).*   □

Corollary   12   *[46] With one-port communication, an SDP can be performed as a sequence of GSH's with disjoint index sets without loss of efficiency, assuming the algorithm chosen for the GSH is optimum.*

Proof   The minimum data transfer time (start-up time) of an SDP is the sum of the minimum data transfer time (start-up time) for each of the GSH's of which the SDP consists.   □

The lower bound in theorem 18 can be improved for some combinations of K, τ, t_c, and σ_i by considering the sum of the lower bounds of each GSH of which the SDP consists. An extended cube permutation of real order σ_p can be improved by communication in the $n - m_p$ processor dimensions not used for the allocation of the data array. A possible algorithm is a composition of a subcube expansion, full cube permutation, and subcube compression algorithms. The permutation is then performed on a data set reduced by a factor of 2^{n-m_p}. The subcube expansion permutation is of type *one-to-all personalized communication* [41].

Corollary   13   *The lower bound for the communication complexity for an extended-cube SDP $(m_p < n)$ of real order σ_p on an n-cube is*

$$T_{sdp}^{lb}(1, \sigma_p, m_p, K) = \max\left((K + \sigma_p - oJ - 1)t_c, \frac{\sigma_p \cdot K}{2^{n-m_p+1}}t_c, (\sigma_p - oJ)\tau\right)$$

for one-port *communication, and*

$$T_{sdp}^{lb}(n, \sigma_p, m_p, K) = \max\left(\left(\frac{K}{n - m_p + \sigma_p} + \sigma_p - oJ - 1\right)t_c,\right.$$
$$\left. \frac{K}{2^{n-m_p+1}}t_c, (\sigma_p - oJ)\tau\right)$$

for n-port *communication.*

Proof   By lemma 29, the lower bound for an SDP of real order σ_p on a data set of $2^{m_s+m_p}$ elements on an n-cube is the same as the lower bound of the same SDP of real order σ_p on a data set of $2^{m_s+\sigma_p}$ elements on an $(n - m_p + \sigma_p)$-cube. We now prove the lower bound for the latter problem. The first argument of the *max* function is derived by considering the minimum time required to send out the K elements for any processor that needs to send data and the propagation delay for the last element sent out. From the proof of theorem 18, the bandwidth required is $\sigma_p 2^{\sigma_p-1}K$. The "effective" bandwidth available is $\sigma_p 2^{n-m_p+\sigma_p}$ for *n-port* communication and $2^{n-m_p+\sigma_p}$ for *one-port* communication . The former can be shown by collapsing the $(n - m_p + \sigma_p)$-cube into a σ_p-cube identified by the σ_p dimensions in the set $\mathcal{J}$ and the bandwidth of each link increased by a factor of 2^{n-m_p}. $\square$

Algorithms for dimension permutations

If an optimal algorithm for a GSH is known for the *one-port* communication case, then an optimal algorithm for an SDP with *one-port* communication is obtained simply by executing the algorithms for the different GSH's making up the SDP, for one GSH after the other, corollary 12. In the *n-port* communication case an optimal algorithm for an SDP is obtained by using an optimal algorithm for each GSH independently, if either there is no start-up time, or all the GSH's have the same real order. The data set is split into β equal parts, one for each GSH. The ith part participates in the GSH's specified by the sequence of index sets: $\mathcal{J}_i, \mathcal{J}_{(i+1) \bmod \beta}, \ldots, \mathcal{J}_{(i-1) \bmod \beta}$. For each partition the data is further subdivided into σ_{p_i} parts for maximum bandwidth utilization. Hence in every step σ_p dimensions are used, which is optimum for full-cube permutations.

Definition 32 A *dimension exchange* function $E(i,j)$ is a GSH with $\mathcal{J} = \{i,j\}$. Data is exchanged between pairs of locations (global addresses) as defined by

$$(a|a_i = 0, a_j = 1) \Longleftrightarrow (a|a_i = 1, a_j = 0). \quad \blacksquare$$

Lemma 32 *A shuffle permutation $sh(a)$ is realized through a sequence of dimension exchanges on cyclically and monotonically decreasing dimensions, and an unshuffle permutation $sh^{-1}(a)$ is realized through a sequence of cyclically and monotonically increasing dimension exchanges. A generalized shuffle permutation $gsh(a)$ of order σ is performed by a sequence of dimension exchanges on a properly ordered index set $\mathcal{J}$.*

$$sh(a) = E((i + 2) \bmod \sigma, (i + 1) \bmod \sigma) \circ \cdots \circ E((i - 1) \bmod \sigma,$$
$$(i - 2) \bmod \sigma) \circ E(i, (i - 1) \bmod \sigma) = \coprod_{j=i}^{i+2} E(j, (j - 1) \bmod \sigma)$$

$$sh^{-1}(a) = E((i - 2) \bmod \sigma, (i - 1) \bmod \sigma) \circ \cdots \circ E((i + 1) \bmod \sigma,$$
$$(i + 2) \bmod \sigma) \circ E(i, (i + 1) \bmod \sigma) = \prod_{j=i}^{i-2} E(j, (j + 1) \bmod \sigma)$$

$$gsh(a) = E(\alpha_{(i+2)\bmod\sigma}, \alpha_{(i+1)\bmod\sigma}) \circ \cdots \circ E(\alpha_{(i-1)\bmod\sigma}, \alpha_{(i-2)\bmod\sigma})$$
$$\circ E(\alpha_i, \alpha_{(i-1)\bmod\sigma}) = \coprod_{j=i}^{i+2} E(\alpha_j, \alpha_{(j-1)\bmod\sigma}) \quad \Box$$

Lemma 32 is easily proved by induction. The dimension exchange operations are performed in a right to left order. The starting dimension for the exchange sequence is arbitrary, but the direction (increasing or decreasing dimension) is important. A shuffle or generalized shuffle permutation can also be realized by a sequence of exchange operations between a fixed dimension and cyclically and monotonically increasing dimensions.

Lemma 33 *A generalized shuffle permutation $gsh(a)$ can be realized by a sequence of dimension exchanges between an arbitrary, fixed, dimension and a sequence of dimensions of a properly ordered index set $\mathcal{J}$.*

$$gsh(a) = E(\alpha_i, \alpha_{(i-1)\bmod\sigma}) \circ \cdots \circ E(\alpha_i, \alpha_{(i+2)\bmod\sigma}) \circ E(\alpha_i, \alpha_{(i+1)\bmod\sigma})$$
$$= \prod_{j=i+1}^{i-1} E(\alpha_i, \alpha_j) \quad \Box$$

Lemma 33 can be proved by induction. The number of exchanges that are required is $|\mathcal{J}| + 1$. The first dimension is routed twice. The "$\prod$" "$\coprod$" signs both denote the composition of a sequence of functions. The ordered index set for $\prod_{j=i}^{i'}$ is $i \bmod \sigma, (i+1) \bmod \sigma, \ldots, i' \bmod \sigma$. For $\coprod_{j=i}^{i'}$ the ordered index set is $i \bmod \sigma, (i - 1) \bmod \sigma, \ldots, i' \bmod \sigma$. A dimension exchange $E(i,j)$, $i \neq j$, requires $Hamming_r(a', a'')$ routing cycles, where $a'_k = a''_k, k \neq i, k \neq j$, and $a'_i \neq a''_i$ and $a'_j \neq a''_j$. The number of routing cycles is 2 if $i, j \in \Gamma_p$; 0 if $i, j \in \Gamma_s$; and 1, otherwise.

In the case of a *mixed* GSH, the fixed dimension should be chosen such that $\alpha_i \in \mathcal{J} \cap \Gamma_s$. Each exchange operation then becomes *one* nearest-neighbor communication step. For a *real* GSH $\mathcal{J} \cap \Gamma_s = \phi$, the index set $\mathcal{J}$ can be extended by a virtual dimension $v \in \Gamma_s$, which is used as the fixed dimension.

The memory of each processor is partitioned into two equal parts with respect to dimension v, i.e., one with $a_v = 0$ (the first half) and one with $a_v = 1$ (the second half). If $v \in \Gamma_s$, then the data is evenly distributed between the two halves, but if $v \in Q_s - \Gamma_s$, then all the data is in the first half. By moving the data of the processors with $a_{\alpha_{i-1}} = 1$ into the second half of their storage, $a_v = a_{\alpha_{i-1}}$ for all processors. Under this condition the last exchange step $E(v, \alpha_i)$, which exchanges $a_{\alpha_{(i-1)\bmod\sigma}}$ and a_v, becomes unnecessary. Hence one exchange step is saved, but the entire data set in a processor is communicated in every dimension exchange, instead of half of the data set.

The **base algorithm** [114, 46] is a direct application of lemma 33 with the fixed dimension being a virtual dimension. During each step, all communications occur in the same dimension of the cube. Processors in subcube 0 exchange the second half of the data with the first half of the data of the processors in subcube 1. The communication is bidirectional. The sequence of exchange steps for a shuffle permutation can be illustrated as follows:

$$(v_{k-1}v_{k-2}\ldots v_1 v_0 | r_{n-1}r_{n-2}\ldots r_1 \underline{r_0})$$
$$\rightarrow (\underline{r_0}v_{k-2}\ldots v_1 v_0 | r_{n-1}r_{n-2}\ldots r_2 \underline{r_1}v_{k-1})$$
$$\rightarrow (\underline{r_1}v_{k-2}\ldots v_1 v_0 | r_{n-1}r_{n-2}\ldots \underline{r_2}r_0 v_{k-1})$$
$$\rightarrow \cdots \rightarrow (r_{n-2}v_{k-2}\ldots v_1 v_0 | \underline{r_{n-1}}r_{n-3}\ldots r_1 r_0 v_{k-1})$$
$$\rightarrow (\underline{r_{n-1}}v_{k-2}\ldots v_1 v_0 | r_{n-2}r_{n-3}\ldots r_1 r_0 \underline{v_{k-1}})$$
$$\rightarrow (v_{k-1}v_{k-2}\ldots v_1 v_0 | r_{n-2}r_{n-3}\ldots r_1 r_0 r_{n-1}).$$

The procedure for generating different exchange sequences that can operate concurrently with *n-port* communication is based on the ordered index set $\mathcal{J} = \alpha_0, \alpha_1, \ldots, \alpha_{\sigma-1}$. Let L be the left rotation operator: $L(\mathcal{J}) = \alpha_1, \ldots, \alpha_{\sigma-1}, \alpha_0$. Then σ exchange sequences are defined by

$\text{Seq}_i = L^i(\mathcal{J}), \alpha_i, \ 0 \leq i < \sigma$. Note that α_i is also the first dimension of Seq_i. The sequences for $\sigma = 3$ are,

$$\text{Seq}_0 = \alpha_0, \alpha_1, \alpha_2, \alpha_0.$$
$$\text{Seq}_1 = \alpha_1, \alpha_2, \alpha_0, \alpha_1.$$
$$\text{Seq}_2 = \alpha_2, \alpha_0, \alpha_1, \alpha_2.$$

During any routing cycle, different sequences use edges in different dimensions. Figure 4.49 shows the four exchange steps in a 3-cube that realizes the shuffle permutation. Figure 4.50 shows the data allocation as a function of exchange step in a 4-cube.

Note that after n exchange steps, every processor has received half of the data it should receive. The other half of the data is received in the last exchange step. If instead of choosing $v \in \Gamma_s$, as assumed above, $v \in Q_s - \Gamma_s$, then n exchange steps suffice. But the amount of data communicated during every step is 2^{vp} instead of 2^{vp-1}. Since each dimension is routed only once, the data is sent along shortest paths. With *one-port* communication, the data transfer time nKt_c is exactly twice the lower bound. In the *n-port*

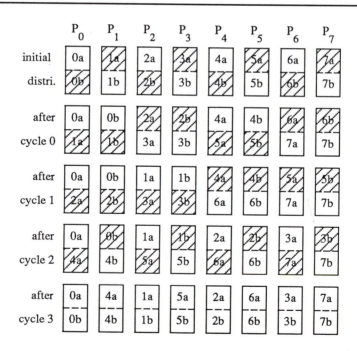

Figure 4.49   The four exchange steps in a 3-cube. The shaded areas are the parts of data subject to exchange during the next step.

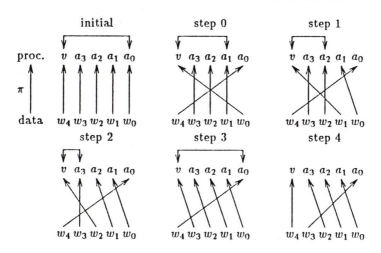

Figure 4.50 The changes of data allocations along time.

version the data is partitioned into n equal sized subsets for a permutation of order n. For $\log_2 n = \eta$ is an integer the exchange sequence i, $0 \le i < n$, can be represented as

$$(v_{k-1} \ldots v_{k-\eta}\underline{v_{k-\eta-1}}v_{k-\eta-2} \ldots v_0 | r_{n-1}r_{n-2} \ldots r_{i+2}r_{i+1}\underline{r_i}r_{i-1} \ldots r_0)$$
$$\rightarrow (v_{k-1} \ldots v_{k-\eta}\underline{r_i}v_{k-\eta-2} \ldots v_0 | r_{n-1}r_{n-2} \ldots r_{i+2}\underline{r_{i+1}}v_{k-\eta-1}r_{i-1} \ldots r_0)$$
$$\rightarrow (v_{k-1} \ldots v_{k-\eta}\underline{r_{i+1}}v_{k-\eta-2} \ldots v_0 | r_{n-1}r_{n-2} \ldots \underline{r_{i+2}}r_i v_{k-\eta-1}r_{i-1} \ldots r_0) \rightarrow \cdots$$
$$\rightarrow (v_{k-1} \ldots v_{k-\eta}\underline{r_{i-2}}v_{k-\eta-2} \ldots v_0 | r_{n-2}r_{n-3} \ldots r_i v_{k-\eta-1}\underline{r_{i-1}}r_{i-3} \ldots r_0 r_{n-1})$$
$$\rightarrow (v_{k-1} \ldots v_{k-\eta}\underline{r_{i-1}}v_{k-\eta-2} \ldots v_0 | r_{n-2}r_{n-3} \ldots r_i \underline{v_{k-\eta-1}}r_{i-2} \ldots r_0 r_{n-1})$$
$$\rightarrow (v_{k-1} \ldots v_{k-\eta}v_{k-\eta-1}v_{k-\eta-2} \ldots v_0 | r_{n-2}r_{n-3} \ldots r_i r_{i-1}r_{i-2} \ldots r_0 r_{n-1}),$$

where $(v_{k-1}v_{k-2} \ldots v_{k-\eta}) = i$. The assumption of $\log_2 n$ being an integer is only required for notational convenience. For *n-port* communication, the exchange sequences $\text{Seq}_i = L^i(\mathcal{J})$, $0 \le i < n$, Node a is active (during the intermediate steps) for Seq_i if $a_i = a_{(i-1)\bmod n}$. During routing cycle j, $1 \le j \le n-2$, node a exchanges data along dimension i if $a_{(i-j)\bmod n} = a_{(i-j-1)\bmod n}$.

Full-cube, separable dimension permutation algorithms
Lemma 34 *[46] With* one-port *communication, a full-cube, separable SDP of real order σ_p on an n-cube can be performed in a time of at most*

$$T^*_{sdp}(1, \sigma_p, n, K) \le \sum_{i=0}^{\beta-1} T^*_{gsh}(1, \sigma_i, n, K). \quad \square$$

The SDP is simply obtained by performing β GSH's in sequence. The ith GSH consists of $2^{n-\sigma_i}$ independent permutations in subcubes of dimension σ_i, concurrently. If the algorithms chosen for the GSH's are optimum, then the resulting algorithm for the SDP remains optimum.

If $\sigma_i = 2$, $0 \le i < \beta = \frac{\sigma_p}{2}$, then the SDP is equivalent to a matrix transposition, and the base algorithm that communicates only once in any dimension degenerates to the Single Path Transpose Algorithm, if the dimensions are relabeled in a proper way.

With n-*port* communication, σ_i ports can be used concurrently for each GSH, and all GSH's can be performed concurrently. For $\sigma_i = 2$ the base algorithm degenerates to the Dual Paths Transpose Algorithm.

Lemma   35   *[46] With* n-port *communication, a full-cube, separable SDP of real order σ_p on an n-cube can be performed in a time of at most*

$$T_{sdp}^*(\sigma_{max}, \sigma_p, n, K) \le \sum_{i=0}^{\beta-1} T_{gsh}^*(\sigma_i, \sigma_i, n, K)$$

by exploiting concurrency within each GSH, and in time

$$T_{sdp}^*(\sigma_p, \sigma_p, n, K) \le \beta \times \max_{i=0}^{\beta-1}\{T_{gsh}^*(\sigma_i, \sigma_i, \sigma_i, \frac{K}{\beta})\}$$

by exploiting the concurrency fully.   □

These results are independent of the algorithm chosen for the GSH.

Permutations by all-to-all personalized communication

An arbitrary permutation can be performed as two successive *all-to-all personalized communications* in subcubes of dimension σ_p [112]. If the number of local storage dimensions $q - n \ge \sigma_p$, then for each dimension $j \in \mathcal{J}$ there exists a unique dimension $k_j \in \Gamma_s$, such that the SDP can be performed in two steps, that is, $E(k_j, j), \forall j \in \mathcal{J}$ followed by $E(\delta(j), k_j), \forall j \in \mathcal{J}$. Assume for simplicity that the σ_p lowest order storage dimensions are used in the permutations and that the SDP involves the σ_p lowest order processor dimensions. Then the two *all-to-all personalized communications* can be illustrated by

$$(v_{s-1}v_{s-2}\ldots v_{\sigma_p}v_{\sigma_p-1}\ldots v_0 | r_{n-1}r_{n-2}\ldots r_{\sigma_p}\underline{r_{\sigma_p-1}r_{\sigma_p-2}\ldots r_0})$$

$$\rightarrow (v_{s-1}v_{s-2}\ldots v_{\sigma_p}\underline{r_{\sigma_p-1}\ldots r_0} | r_{n-1}r_{n-2}\ldots r_{\sigma_p}\underline{v_{\sigma_p-1}v_{\sigma_p-2}\ldots v_0}),$$

and

$$(v_{s-1}v_{s-2}\ldots v_{\sigma_p}\underline{r_{\sigma_p-1}\ldots r_0}|r_{n-1}r_{n-2}\ldots r_{\sigma_p}v_{\sigma_p-1}v_{\sigma_p-2}\ldots v_0)$$

$$\rightarrow (v_{s-1}v_{s-2}\ldots v_{\sigma_p}v_{\sigma_p-1}\ldots v_0|r_{n-1}r_{n-2}\ldots$$

$$r_{\sigma_p}\underline{r_{\delta-1}(\sigma_p-1)r_{\delta-2}(\sigma_p-2)\ldots r_{\delta-1}(0)}).$$

Permutations by matrix transposition

An SDP can be implemented by exchanging subsets of dimensions recursively in $\lceil \log_2 \sigma_p \rceil$ steps for a GSH of order σ_p, and $\lceil \log_2 \sigma_{max} \rceil$ steps for an SDP. Each step is equivalent to a matrix transposition with two-dimensional partitioning [63]. The complexity of permutation based on matrix transposition is in general an order of $\log \sigma_p$ higher than the lower bound. But for the special case where $\sigma_i = 2$ for all i's, the SDP is indeed a matrix transposition, and an optimal transpose algorithm should be used.

Permutations by shuffle permutations

One way in which a dimension permutation algorithm can be created to make efficient use of the available bandwith with *n-port* communication is to find several independent shuffle operations, each using the base algorithm. But the version of the base algorithm that communicates half the data set in each exchange routes the starting dimension twice. By performing the SDP by this version of the base algorithm for each GSH, β start-ups in excess of the lower bound are required, and the communication bandwidth of the Boolean cube is not used optimally. By applying the algorithm concurrently to all GSH's of the SDP, the communication bandwidth is fully utilized, but the number of start-ups is $\beta(\sigma_{max} + 1)$.

Next we present three different ways in which the trade-off between utilization of communication bandwidth and start-up time can be approached. In the first algorithm the number of concurrent exchange sequences is maximized subject to the constraint that in any step the number of exchanges in any dimension is at most two. The algorithm requires $\sigma_p + \beta$ start-ups and performs $\sigma_p + \beta + \sigma_{min} - 1$ permutations concurrently. In the second algorithm the value of σ_{max} is reduced by partitioning a GSH into several GSH's that are performed concurrently, followed by one GSH that couples the parts. In the third algorithm the GSH's are grouped into sets such that the number of start-ups is proportional to the number of sets and the maximum number of start-ups required for any set. The grouping of GSH's is performed recursively. The first SDP algorithm with concurrent application of the base algorithm use the following exchange sequences:

$$\begin{aligned} \text{Seq}_0 &= \hat{\mathcal{J}}_0, \alpha_{00}, \hat{\mathcal{J}}_1, \alpha_{10}, \ldots, \hat{\mathcal{J}}_{\beta-1}, \alpha_{(\beta-1)0}. \\ \text{Seq}_k &= d_0, d_1, \ldots, d_{\sigma_p+\beta-1}, \\ \text{Seq}_{k+1} &= d_1, d_2, \ldots, d_{\sigma_p+\beta-1}, \delta(d_0), \qquad \forall 0 \le k < \sigma_p + \beta + \sigma_{min} - 2. \end{aligned}$$

Table 4.17   Combining different GSH's by Algorithm A2 with *n-port* communication .

Cycle	0	1	2	3	4	5	6	7	8	9	10	11
Seq_0	0	1	2	3	0	4	5	6	4	7	8	7
Seq_1	1	2	3	0	4	5	6	4	7	8	7	1
Seq_2	2	3	0	4	5	6	4	7	8	7	1	2
Seq_3	3	0	4	5	6	4	7	8	7	1	2	3
Seq_4	0	4	5	6	4	7	8	7	1	2	3	0
Seq_5	4	5	6	4	7	8	7	1	2	3	0	1
Seq_6	5	6	4	7	8	7	1	2	3	0	1	5
Seq_7	6	4	7	8	7	1	2	3	0	1	5	6
Seq_8	4	7	8	7	1	2	3	0	1	5	6	4
Seq_9	7	8	7	1	2	3	0	1	5	6	4	5
Seq_{10}	8	7	1	2	3	0	1	5	6	4	5	8
Seq_{11}	7	1	2	3	0	1	5	6	4	5	8	7
Seq_{12}	1	2	3	0	1	5	6	4	5	8	7	8

Note that $\delta(d_0)$ is not necessarily d_1, (Table 4.17). The number of paths that can be generated by the left shift and appending the first exchange dimension subject to the constraint of at most two communications in any dimension during any routing cycle is $\sigma_p + \beta + \sigma_{min} - 1$.

Sequence 0 is the dimension exchange sequence for a real SDP and *one-port* communication. Sequence $k + 1$ is defined in terms of sequence k by a left cyclic shift of Seq_k, followed by the application of the δ function to the last entry. In most cases, the last entry is the same as the first entry, but it is different when the first GSH in the sequence is completed. Table 4.17 shows an example of thirteen sequences of processor dimensions, which are used to perform the real SDP: $\hat{\mathcal{J}}_0 = 0, 1, 2, 3$, $\hat{\mathcal{J}}_1 = 4, 5, 6$, and $\hat{\mathcal{J}}_2 = 7, 8$. Row i applies to GSH i, and column j to cycle j. The table entry is the processor dimension subject to communication.

The number of start-ups is minimal for the base algorithm. The data transfer time is also of optimum order and higher than minimum by at most a factor of two, mostly because two communications are needed for some dimensions in every cycle. This fact is used to generate a number of exchange sequences that exceeds the number of dimensions that need to be routed, thereby improving somewhat on the use of the available communication bandwidth. If the SDP in fact is a GSH, then this algorithm is indeed the *n-port* version of the base algorithm.

Another idea is to reduce the number of start-ups by partitioning each GSH of order greater than $\gamma \leq \sigma_{max}$, to be chosen optimally, into multiple GSH's of order γ. An interpartition GSH is required to realize the original GSH.

As an example, consider the shuffle operation

$$(a_8a_7a_6a_5a_4a_3a_2a_1a_0) \rightarrow (a_7a_6a_5a_4a_3a_2a_1a_0a_8),$$

which can be realized as

$$(a_8a_7a_6a_5a_4a_3\underline{a_2}a_1a_0) \rightarrow (a_8a_7a_6\underline{a_5a_4a_3}a_1a_0a_2) \rightarrow (\underline{a_8a_7a_6}a_4a_3a_5a_1a_0a_2)$$

$$\rightarrow (a_7a_6\underline{a_8}a_4a_3\underline{a_5}a_1a_0\underline{a_2}) \rightarrow (a_7a_6a_5a_4a_3a_2a_1a_0a_8).$$

The index set $\mathcal{J} = \{0,1,2,3,4,5,6,7,8\}$ in the example is partitioned into three sets $\mathcal{J}_0 = \{0,1,2\}$, $\mathcal{J}_1 = \{3,4,5\}$, and $\mathcal{J}_2 = \{6,7,8\}$ with a combining set $\mathcal{J}_3 = \{3,6,0\}$.

Lemma 36   *A GSH with index set $\mathcal{J}$ of order σ can be performed as μ independent GSH's on index sets $\mathcal{J}_j$, $0 \leq j < \mu$, where $\mathcal{J}_j = \{\alpha_{f(j)}, \alpha_{f(j)+1}, \alpha_{f(j)+2}, \ldots, \alpha_{f(j+1)-1}\}$ and $0 = f(0) < f(1) < \cdots < f(\mu) = \sigma$, followed by a GSH on the index set $\mathcal{J}_\mu = \{\alpha_{f(1)}, \ldots, \alpha_{f(\mu-1)}, \alpha_{f(0)}\}$.*

Proof   Define GSH_j, $0 \leq j \leq \mu$, to be a GSH with index set $\mathcal{J}_j$. Also, let σ_j be the order of GSH_j. Then by definition, GSH_j, $0 \leq j < \mu$, realizes $\alpha_{f(j)+i} \rightarrow \alpha_{f(j)+(i+1)\bmod\sigma_j}$, $0 \leq i < \sigma_j$, which is equal to

$$\begin{cases} \alpha_i \rightarrow \alpha_{(i+1)\bmod\sigma}, & \text{if } i \neq f(j+1)-1, 0 \leq j < \mu, \\ \alpha_i = \alpha_{f(j+1)-1} \rightarrow \alpha_{f(j)}, & \text{if } i = f(j+1)-1, 0 \leq j < \mu. \end{cases}$$

But GSH_μ realizes $\alpha_{f(j)} \rightarrow \alpha_{f((j+1)\bmod\mu)}$, $0 \leq j < \mu$, and the proof is complete. □

If each GSH is performed by the base algorithm, then a direct application would yield $\sum_{i=0}^{\mu-1}(\sigma_i+1)+\mu+1 = \sigma+2\mu+1$. It is unnecessary to perform the last restoring dimension exchange for *any* of the index sets, however, as illustrated next:

$$(v_2v_1\underline{v_0}|r_8r_7r_6r_5r_4r_3r_2r_1\underline{r_0}) \rightarrow (v_2\underline{v_1}r_0|r_8r_7r_6r_5r_4\underline{r_3}r_2r_1v_0)$$

$$\rightarrow (\underline{v_2}r_3r_0|r_8r_7\underline{r_6}r_5r_4v_1r_2r_1v_0) \rightarrow (r_6r_3\underline{r_0}|r_8r_7v_2r_5r_4v_1r_2\underline{r_1}v_0)$$

$$\rightarrow (r_6\underline{r_3}r_1|r_8r_7v_2r_5\underline{r_4}v_1r_2r_0v_0) \rightarrow (\underline{r_6}r_4r_1|r_8\underline{r_7}v_2r_5r_3v_1r_2r_0v_0)$$

$$\rightarrow (r_7r_4\underline{r_1}|r_8r_6v_2r_5r_3v_1\underline{r_2}r_0v_0) \rightarrow (r_7\underline{r_4}r_2|r_8r_6v_2\underline{r_5}r_3v_1r_1r_0v_0)$$

$$\rightarrow (\underline{r_7}r_5r_2|\underline{r_8}r_6v_2r_4r_3v_1r_1r_0v_0) \rightarrow (r_8r_5\underline{r_2}|r_7r_6v_2r_4r_3\underline{v_1}r_1r_0v_0)$$

$$\rightarrow (r_8\underline{r_5}v_1|r_7r_6\underline{v_2}r_4r_3r_2r_1r_0v_0) \rightarrow (\underline{r_8}v_2v_1|r_7r_6r_5r_4r_3r_2r_1r_0\underline{v_0})$$

$$\rightarrow (v_0v_2v_1|r_7r_6r_5r_4r_3r_2r_1r_0r_8).$$

Note that a local shuffle is required to restore the original local storage map. To implement the strategy, GSH_i of the SDP is partitioned into shuffles on subsets according to lemma 36, such that $|\mathcal{J}_{i_j}| = \gamma$, $0 \leq i < \beta$, $0 \leq j < \mu_i$ where $\mu_i = \lceil \frac{\sigma_i}{\gamma} \rceil$ is the number of subsets for the ith GSH. If σ_i is not a multiple of γ, then "dummy" dimensions ϕ are added to $\mathcal{J}_i$ such that $|\mathcal{J}_i|$ is a multiple of γ. The total number of GSH's after the partitioning is $\mu = \sum_{i=0}^{\beta-1} \mu_i$, each of order γ. For example, for an SDP with $\beta = 2$, $\mathcal{J}_0 = \{0,1,2,3,4,5,6,7\}$, $\mathcal{J}_1 = \{8,9,10\}$ and γ chosen to be 3, then $\mathcal{J}_0$ become $\{0,1,2,3,4,5,6,7,\phi\}$, $\mathcal{J}_1$ is still $\{8,9,10\}$, $\mu_0 = 3$, $\mu_1 = 1$, and $\mu = \mu_0 + \mu_1 = 4$. $\mathcal{J}_{0_0} = \{0,1,2\}$, $\mathcal{J}_{0_1} = \{3,4,5\}$, $\mathcal{J}_{0_2} = \{6,7,\phi\}$, and $\mathcal{J}_{1_0} = \{8,9,10\}$. A way of constructing $\gamma\mu$ sequences that can be performed concurrently is given in [46].

Since every data element has to be permuted according to all GSH's making up the SDP, performing all GSH's concurrently implies $\beta(\sigma_{max}+1)$ start-ups, because the GSH of maximum order has to be performed β times. By grouping the GSH's it may be possible to reduce the number of start-ups and the total communication time. Ideally, all groups require the same time. For instance, if a real SDP consists of three real GSH's of order 7, 3, and 3, then the data set is partitioned into two parts of $\frac{K}{2}$ elements each (assuming that the base algorithm is used for each real GSH). One part is permuted through the GSH of real order 7. The other part is partitioned further into $\frac{K}{4}$ elements each, one for each real GSH of order 3. All three GSH permutations are performed concurrently, and for each GSH the data set is further subdivided for maximum concurrency within the GSH. By using the *n-port* version of the base algorithm, each part of $\frac{K}{2}$ elements requires the same number of start-ups. When the permutation for all sets is complete, the data is repartitioned for a new set of permutations by applying the permutation defined by the next set of GSH's, modulo the number of sets, to the data. Hence the data permuted for the real GSH of order 7 is then permuted according to the two GSH's of order 3 each, and conversely the data already permuted according to these two GSH's will be permuted according to the GSH of order 7. For nonseparable SDP's it may not be necessary to extend the set of dimensions for the base algorithm. Figure 4.51 illustrates the data movement for this example, where D0 to D3 are the four equal partitions of the data set.

Specifically, let $\mathrm{SDP}_0, \mathrm{SDP}_1, \ldots, \mathrm{SDP}_{\mu-1}$ be some μ-partitioning of the given SDP, i.e., the β GSH's are distributed in the μ-partitioned SDP's and

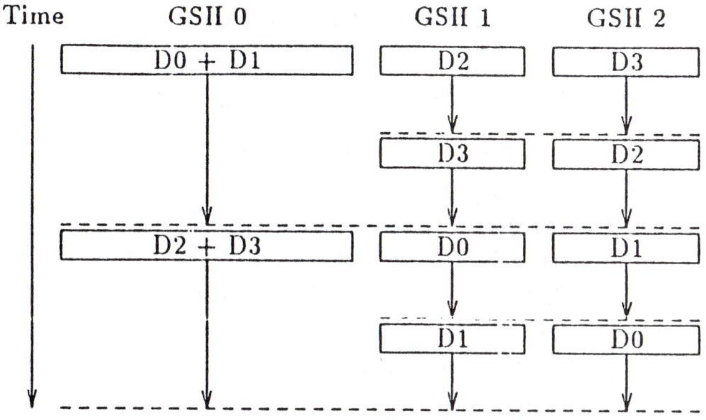

Figure 4.51   Hierarchical partitioning of the data and an SDP consisting of three GSH's of order 7, 3, and 3.

$\mu \le \beta$. Then the given SDP can be solved recursively in a time

$$T(\text{SDP}, K) = \mu * \max\{T(\text{SDP}_i, \frac{K}{\mu}), \forall 0 \le i < \mu\},$$

where $T(\text{SDP}, K)$ is the time for an SDP with data volume K. This is derived simply by partitioning the data volume into μ equal parts. Let different parts participate in different sequences of SDP_i's and all sequences of SDP_i's are performed concurrently. A partitioning strategy is given in [46].

Full cube, nonseparable dimension permutation algorithms
With *one-port* communication, the dimension permutation is simply obtained by performing β GSH's in sequence. Let β' be the number of real GSH's with orders σ_0, σ_1, ..., $\sigma_{\beta'-1}$, respectively. The communication complexity is

$$T_{sdp}^{\text{A2}}(1, \sigma_p, n, K) = (\sigma_p + \beta')\frac{K}{2}t_c + (\sigma_p + \beta')\lceil\frac{K}{2B}\rceil\tau,$$

by using the base algorithm.

With *n-port* communication, one can employ the same algorithm as if there were no virtual dimension. The communication complexity is determined by the number of real dimensions (since the time for local data movement is ignored). The number of virtual dimensions introduced for the exchange sequences is $\sigma_p - m'_s$, where m'_s is the number of virtual dimensions that have a succeeding real dimension in the index set. The number of exchange sequences that can be run concurrently without any extra storage requirement (extra virtual dimensions) is m'_s.

4.10.8   Extended-cube Permutation Algorithms

For an extended-cube permutation, we adopt a three-phase scheme: subcube expansion, full-cube permutation, and subcube compression. In the first phase, each processor with data partitions it into 2^{n-m_p} pieces, and all processors concurrently perform a *one-to-all personalized communication* to each of the 2^{m_p} distinct subcubes of dimension $n - m_p$. In the second phase, an algorithm for a full-cube permutation is used concurrently in the 2^{n-m_p} subcubes, with the data volume reduced by a factor of 2^{n-m_p}. The third phase is the reverse of the first phase, i.e., data are gathered (compressed) into the original active subcube. The complexities of the first phase and the third phase are the same, and for the best known algorithm [41, 66] the complexity of each is

$$ K\left(1 - \frac{1}{2^{n-m_p}}\right)t_c + (n - m_p)\tau $$

for *one-port* communication, and

$$ \frac{K}{n - m_p}\left(1 - \frac{1}{2^{n-m_p}}\right)t_c + (n - m_p)\tau $$

for *n-port* communication. With *n-port* communication, if the algorithm used in the second phase is optimal, then the total data transferred is $\approx \frac{K}{2^{n-m_p+1}} + \frac{2K}{n-m_p}$ compared to $\frac{K}{2}$ for an optimal algorithm using links of the active subcube only. The speedup of the data transfer time is about a factor of $\frac{n-m_p}{4}$, but the start-ups compare as $2(n - m_p) + \sigma_p$ to σ_p. For *one-port* communication, the data transferred is $\approx \frac{\sigma_p K}{2^{n-m_p+1}} + 2K$ compared to $\frac{\sigma_p K}{2}$. The speedup of the data transfer time is about a factor of $\frac{\sigma_p}{4}$.

4.11   Emulation with Wafer Scale Integration

In the preceeding sections we presented techniques and bounds for optimum communication in Boolean cube networks. For the choice of network it is important to consider the layout and timing characteristics of the technology. The time for signal propagation along a wire does not scale well with decreased feature sizes; and the wiring area for most networks dominates the area for logic. Accounting for wire delays and routing area can change the relative order of the networks with respect to performance. In this section we compare the performance of two-dimensional meshes, Boolean cubes, and Cube Connected Cycles (CCC) networks [95] on three emulations [97]: a two-dimensional mesh, a butterfly network, and a spanning tree. Following Dally [23], we assume the same number of processors for all networks, the same memory size per processor, and the same total area, i.e., the hypercube and the CCC have narrower communication paths than a mesh. We consider three timing models: the capacitive, the resistive, and a constant delay. In the third model the clock period is determined by some other piece of the design.

4.11.1   Host Architectures

The processors are layed out on a grid. Each processor has sides s and a local memory of size M^2. All communication channels lie outside this area and enter and leave the processor at a corner. All networks have $P^2 = 2^{2p}$ processors numbered 0 through $P^2 - 1$. For the CCC, $P^2 = (P'^2 \log P'^2)$, with $\log P' = p'$, and for simplicity we assume that p' is an integer and $2p'$ a perfect square. The processors communicate using channels of width w_g, w_h, and w_c for the mesh, the hypercube, and the CCC. We assume that there are separate channels for communication in different directions.

Normal layout of the hypercube and the mesh
The processors are laid out in a square region of side S, in P rows and P columns separated by communication channels. For both networks each processor directly communicates only with processors in its row or column. For meshes, it is only necessary to have two tracks per row, one per channel in each direction. Thus, $S = P(s + 2w_g)$. For hypercubes, each row of processors forms a smaller hypercube of p dimensions. The communication channels for the ith dimension can be accomodated in $2 \cdot 2^i$ tracks, and thus the total number of tracks required is at most $\sum_{i=0}^{i=p-1} 2 \cdot 2^i = 2(P - 1)$ tracks. Thus, $S \approx P(s + 2Pw_h)$. This layout is known to have optimal area.

Layout for the CCC
We embed a cycle of length $2p'$ in a square grid of side $\sqrt{2p'}$, with dilation 1 or 2 ($\sqrt{2p'}$ odd). A point on the cycle is arbitrarily designated as 0, and the subsequent points are numbered 1 through $2p' - 1$. Let this numbering be denoted by $path(x, y)$ for grid point (x, y). The layout for the CCC can be obtained from the layout for the $2p'$ dimensional hypercube, each hypercube processor being replaced by the $2p'$ processors in each cycle of the CCC, using the $path$ function. Thus, the processor at position $(x|p'v|p - p', y|p'w|p - p')$ of the layout is numbered $x|y|\text{path}(v, w)$. The number of horizontal tracks required per row of cycles is at most $2P'$.

Figure 4.52   Normal layout for a hypercube. [97]

Each row of cycles requires at most $2\sqrt{2p'}$ horizontal tracks to layout the intracycle channels. Given that there are P' rows of cycles, the total height of the layout is $S = P'(\sqrt{2p'}s + 2\sqrt{2p'}w_c + 2P'w_c) \doteq Ps + 2Pw_c + P^2 w_c/p' \approx Ps + P^2 w_c/p$. Because the cycle can be embedded with dilation 2, the maximum distance between adjacent processors in a cycle can be at most $2S/P$. The maximum distance between arbitrary processors is $S/2$, when the cycles differ in the most significant bit.

Gray code derived layouts
Gray code derived layouts have better performance for meshlike communication, without significantly sacrificing performance for other communication patterns. The total area and maximum wire length are the same as

for the previous layout. For hypercubes, the processor occupying position (x, y) is numbered $G(x)|G(y)$. Processors communicate with others only in the same row or column. Since $G(x)$ and $G(x + 1)$ ($x < P$) differ only in one bit, the processor at (x, y) is directly connected to the processor at $(x + 1, y)$. Thus this layout contains an embedded mesh for which all the channels have the same length (see Figure 4.53). For the CCC a node in the hypercube layout is replaced by a cycle.

In the layout of section 4.11.1 all the channels corresponding to a given dimension in the hypercube have the same length, $\frac{S}{P} 2^{i \bmod p}$ for dimension i. However, in the Gray code layout, channels along the ith dimension have lengths $\frac{S}{P}(2 \cdot 2^{k \bmod p} - 1)$, $k = \{0, 1, \ldots, i\}$. The intercycle channel lengths in the CCC vary similarly. $\frac{S}{P}(2 \cdot 2^{k \bmod p'} - 1)$, $k = \{0, 1, \ldots, i\}$. For FFT-like communication patterns the performance of Gray code derived layouts

Figure 4.53   Gray code derived layout for a hypercube. [97]

may be at most a factor of two lower than the normal layout. Meshlike communication may be a factor of $P/2$ higher.

Channel width and cycle time comparison

We have $S = Ps + 2Pw_g \approx Ps + 2P^2w_h \approx Ps + P^2w_c/p$. Thus $w_h = w_g/P$ and $w_c = 2pw_g/P$. With the normal layout the maximum signal propagation distance is $S/2$ for the hypercube and the CCC and S/P for the mesh. With the Gray code derived layouts for the hypercube and the CCC, the maximum distance is S. For the constant, capacitive, and resistive delay models, t_1, t_2, and t_3 are the times taken for signal propagation over a distance S/P. Hence signal propagation over distance l requires time t_1, $c_2 \log l$, and $t_3 l/(\frac{S}{P})$, with $t_2 = c_2 \log \frac{S}{P}$. Table 4.18 summarizes the delay along the longest channel for the different layouts.

4.11.2   Emulation of Two-Dimensional Meshes

We consider the naive means of emulating meshes in different host architectures. For Boolean cube host networks, only a single path is used between a pair of processors. For a CCC host, submeshes of size $\sqrt{2p'} \times \sqrt{2p'}$ are mapped onto each cycle. For a mesh with more processors than nodes in the host graph the consecutive assignment is used.

Suppose each node of the guest mesh communicates L bits with its neighbors. Then for the mesh host, each of its four channels handles $2^{vp}L$ bits. Similarly, each utilized channel of the hypercube transfers $2^{vp}L$ bits.

Table 4.18   Channel delay and width. [97]

	Channel width	Max. length	Delay along longest channel		
			Const.	*Cap.*	*Res.*
Mesh	w	S/P	t_1	t_2	t_3
Hypercube	w/P	$\frac{1}{2}S$	t_1	$t_2 + c_2 p$	$\frac{1}{2}Pt_3$
Hypercube$_{Gray}$	w/P	S	t_1	$t_2 + c_2 p$	Pt_3
CCC	$2wp/P$	$\frac{1}{2}S$	t_1	$t_2 + c_2 p$	$\frac{1}{2}Pt_3$
CCC$_{Gray}$	$2wp/P$	S	t_1	$t_2 + c_2 p$	Pt_3

The total time required for the communication depends on the widths of the data paths and the rate at which signals can travel along the paths (see Table 4.19). On the CCC, there is an extra level in the hierarchy. The submeshes held by the $2p'$ processors in each cycle together form a larger submesh of size $\sqrt{2p'}2^{vp} \times \sqrt{2p'}2^{vp}$. Neighboring submeshes of the larger size lie in cycles directly connected by an intercycle channel. Thus the neighbors of the submesh held by each processor may lie within the cycle or in a neighboring cycle. The data movement has three phases. In the first, the processors communicate with their neighbors within the cycle. Processors whose neighbors lie outside the cycle also send the required data to the processor that can communicate with the appropriate cycle. In the second phase, the data collected in the first phase for intercycle communication is actually sent over to the other cycle. Finally, in the last phase the data received by each cycle is forwarded to the appropriate processor.

The first-phase communication can be accomplished by a simple scheme in which the data associated with all boundaries of the smaller meshes is circulated around the cycle. This communication requires $4 \cdot 2^{vp} L \cdot 2p'/2w_c \approx 2P2^{vp}L/w$ transmissions utilizing the bidirectional channels, $w_c = 2wp/P$, and assuming $p \approx p'$. Each transmission is over a distance of at most $2S/P$. The final phase can easily be accomplished in the same time as the first phase. Thus the combined time for these phases for the three

Table 4.19   Time for emulating one meshlike communication. [97]

	Delay Model		
	Constant	*Capacitive*	*Resistive*
Mesh	$2^{vp}Lt_1/w$	$2^{vp}Lt_2/w$	$2^{vp}Lt_3/w$
Hypercube	$P2^{vp}Lt_1/w$	$(t_2 + c_2p)$ $\cdot P2^{vp}L/w$	$P^2 2^{vp}Lt_3/2w$
Hypercube$_{Gray}$	$P2^{vp}Lt_1/w$	$P2^{vp}Lt_2/w$	$P2^{vp}Lt_3/w$
CCC	$4P2^{vp}Lt_1/w$	$(4t_2 + c_2\sqrt{\frac{1}{2}p})$ $\cdot P2^{vp}L/w$	$(P/\sqrt{8p} + 8)$ $\cdot P2^{vp}Lt_3/w$
CCC$_{Gray}$	$4P2^{vp}Lt_1/w$	$4P2^{vp}Lt_2/w$	$10P2^{vp}Lt_3/w$

timing models can be estimated to be $4P2^{vp}Lt_1/w$, $4P2^{vp}L(t_2+c_2)/w$ and $8P2^{vp}Lt_3/w$. In the second phase all four intercycle channels can transmit concurrently. The number of transmissions required is $2^{vp}L\sqrt{2p'}/w_c = P2^{vp}L\sqrt{2p'}/2wp$. The Gray code layout guarantees that the length of each channel is at most $2S/P' = 2\sqrt{2p'}S/P$. Thus the time required for this phase is $P2^{vp}Lt_1/(\sqrt{2p}w)$, $(t_2 + c_2\log 2\sqrt{2p'})P2^{vp}L/(\sqrt{2p}w)$, and $2P2^{vp}Lt_3/w$ respectively for the three timing models. The time for the first and last phases dominates, except for the resistive model. For the normal layout, the longest channel has a length $S/2$. The results of this analysis are summarized in Table 4.19.

4.11.3   Butterfly-Based Algorithms

The algorithm has n steps, 1 through n. At the beginning of step j, nodes in stage j receive data from nodes in stage $j-1$, which they use to compute during that step. Thus at the jth step only the nodes in stage j are active. The $N = 2^{2vp}P^2$ nodes are mapped onto the P^2 processors so that the initial $2vp$ steps are executed independently by each processor (*consecutive* mapping). The time required for the first $2vp$ steps is $2vp2^{2vp}$. The remaining steps are similar to a butterfly algorithm with P^2 inputs.

Butterfly emulations
For the hypercube, in step $2vp + i$, $1 \le i \le 2p$ of the algorithm, the communication is over a distance of $2^{i\bmod p}\frac{S}{P}$ for the normal layout with delays t_1, $t_2 + c_2(i-1) \bmod p$, and $t_3 2^{i-1\bmod p}$. The total time, computed by summing over i, is $2pt_1 2^{2vp}L/w_h$, and $2pt_2 2^{2vp}L/w_h + c_2p(p-1)2^{2vp}L/w_h$, and $2Pt_3 2^{2vp}L/w_h$. The time required for the Gray code derived layout is computed in a similar manner.

For the mesh, the communication in step $2vp + i$ is between processors that are at a distance of $2^{i\bmod p}\frac{S}{P}$. The number of transmissions required is $2^{i\bmod p}2^{2vp}L/w$. The total number of transmissions over the course of the algorithm is

$$\sum_{i=1}^{i=2p} 2^{i-1\bmod p}2^{2vp}L/w \le 2P2^{2vp}L/w.$$

For the CCC, the first $2vp$ local steps are followed by $\log 2p'$ steps requiring communication within cycles. This communication can be done in $2p'$ iterations [95], each iteration requiring $2^{2vp}L$ elements to be communicated between nearest neighbors in the cycle. The maximum communication distance in this phase is $2\frac{S}{P}$, and the number of transmissions in this phase is $2p'2^{2vp}L/w_c$. The computation takes $\log 2p'$ iterations, with each processor computing the results for $2^{2vp}L$ nodes in each iteration. The final phase consists of $2p'$ steps and requires intercycle communication in $4p'$

iterations [95]. In each iteration, the maximum number of data transmitted is $2^{2vp}L$, and the maximum distance of transmission is $S/2$ with the normal layout, and S with the Gray code layout. Each iteration requires processors to compute results for $2^{2vp}L$ nodes. These results are summarized in Table 4.20.

The naive butterfly emulation is unable to utilize the channels in the hypercube efficiently. Using some of the techniques described in the section on butterfly network emulation, the performance of the hypercube layout can be improved by a factor of at most $2p$, under proper conditions, thus making its performance competitive with that of the mesh and the CCC for the constant delay model.

4.11.4   Summary

Meshes are superior in emulating meshes if the data volume is large ($2^{vp}L > w/P$) for all three timing models. The hypercube layout performs comparably to the mesh layout if the data volume is low and if the constant delay model is applicable, and also under the capacitive and resistive model if the Gray code layout is used. The performance of the CCC layout is inferior to the hypercube layout by a constant factor.

The emulation of butterfly networks on a mesh layout yields a higher performance than on the CCC and the hypercube for all three timing models if the data volume is high ($2^{2vp}L > w$). The performance of the CCC is only off by a small constant factor (1.5) in the constant delay case. The Gray code layout is almost as effective as the normal layout in the capacitive case, and inferior by a factor of 2 in the resistive case. For a small data volume, the wide channels in the mesh cannot be utilized fully. Under the constant (capacitive) delay model, the narrow channels in the CCC can be kept busy, and there is no (low) penalty for having long channels. In these cases the CCC performs better than the mesh. The CCC is in general superior to the hypercube for butterfly emulation.

Table 4.21 summarizes the performance on broadcasting L bits using spanning trees. The conclusions are similar to those for butterfly emulation. Note that for the resistive delay model and hypercubes it is preferable not to use a spanning tree of minimum height but instead to form a spanning tree of the wires forming a mesh embedded in the cube. However, the mesh layout is faster because of its wider data paths. If latency is important, then the mesh is superior by a constant factor only for the resistive model.

We conclude that with a high data volume, the mesh layout yields the best performance for all the emulations under all timing models. For small data volumes, the mesh is comparable to the hypercube and CCC if the Gray code layout is used for emulating meshes. It is inferior to both the hypercube and the CCC for emulating butterfly networks and spanning

Table 4.20   Time for emulating a butterfly network. [97]

	Communication Time			Computation
	Constant Delay	*Capacitive Delay*	*Resistive Delay*	*Time*
Mesh	$2Pt_1\lceil 2^{2vp}L/w\rceil$	$2Pt_2\lceil 2^{2vp}L/w\rceil$	$2Pt_3\lceil 2^{2vp}L/w\rceil$	$2^{2vp}L(2m+2p)$
Hypercube	$2pt_1\lceil P2^{2vp}L/w\rceil$	$p(2t_2+c_2(p-1))\cdot\lceil P2^{2vp}L/w\rceil$	$2Pt_3\cdot\lceil P2^{2vp}L/w\rceil$	$2^{2vp}L(2m+2p)$
Hypercube$_{Gray}$	$2pt_1\lceil P2^{2vp}L/w\rceil$	$p(2t_2+c_2(p+1))\cdot\lceil P2^{2vp}L/w\rceil$	$4Pt_3\cdot\lceil P2^{2vp}L/w\rceil$	$2^{2vp}L(2m+2p)$
CCC	$6pt_1\lceil P2^{2vp}L/2wp\rceil$	$2p(3t_2+c_2(2p-1))\cdot\lceil P2^{2vp}L/2wp\rceil$	$2p(P+2)t_3\cdot\lceil P2^{2vp}L/2wp\rceil$	$2^{2vp}L(2m+4p)$
CCC$_{Gray}$	$6pt_1\lceil P2^{2vp}L/2wp\rceil$	$2p(3t_2+c_2(2p+1))\cdot\lceil P2^{2vp}L/2wp\rceil$	$4p(P+1)t_3\cdot\lceil P2^{2vp}L/2wp\rceil$	$2^{2vp}L(2m+4p)$

Table 4.21   Time for broadcast using naive spanning trees of minimum height. [97]

	Constant Delay	Capacitive Delay	Resistive Delay
Mesh	$(\lceil L/w\rceil - 1 + 2P)t_1$	$(\lceil L/w\rceil - 1 + 2P)t_2$	$(\lceil L/w\rceil - 1 + 2P)t_3$
Hypercube	$(\lceil PL/w\rceil - 1 + 2p)t_1$	$(\lceil PL/w\rceil - 1)(t_2 + c_2(p-1)) + 2pt_2 + c_2p(p-1)$	$(\frac{1}{2}(\lceil PL/w\rceil - 1) + 2)Pt_3$
Hypercube (Gray)	$(\lceil PL/w\rceil - 1 + 2p)t_1$	$(\lceil PL/w\rceil - 1)(t_2 + c_2p) + 2pt_2 + c_2p(p+1)$	$(\lceil PL/w\rceil - 1 + 4)Pt_3$
CCC	$(\lceil PL/wp\rceil - 1 + 4p)t_1$	$(\lceil PL/wp\rceil - 1)(t_2 + c_2(p-1)) + 4pt_2 + c_2p(p+1)$	$(\frac{1}{2}(\lceil PL/wp\rceil - 1) + 2)Pt_3$
CCC (Gray)	$(\lceil PL/wp\rceil - 1 + 4p)t_1$	$(\lceil PL/wp\rceil - 1)(t_2 + c_2p) + 4pt_2 + c_2p(p+3)$	$(\lceil PL/wp\rceil - 1 + 4)Pt_3$

trees if the constant or capacitive delay models apply. Even for small data volumes, meshes are superior if the resistive model applies.

For the emulations considered, the benefit of the Gray code layouts in emulating meshes is higher than the drawback in emulating butterfly networks or trees.

4.12   Shared Memory

In the previous sections a great deal of knowledge about data structures and the algorithms used for the computations at hand was assumed. Under such circumstances, optimization of address maps with respect to locality is possible. With complex operations on data structures, or if the data structures are dynamically created, such an optimization may be difficult or impossible. A different strategy can then be adopted: provide a shared memory programming model. The difficulty with providing a shared memory model is that current technology is communication limited. The main problems to be resolved are conflicts in accessing variables in shared address space, either because concurrent access is attempted to different variables in the same storage unit or multiple concurrent accesses to the same variable, or because of communication conflicts during routing of messages.

The most general shared memory models in the literature, the concurrent-read concurrent-write parallel random-access machines (CRCW PRAM's), allow an arbitrary number of processors to read or write a common memory location in one time step. Complex communications operations—broadcast and multicast, for example—can be implemented in one step. Abstracting complex communications patterns into unit steps greatly simplifies the task of designing algorithms and writing programs. For this reason, CRCW PRAM models are favored over weaker abstract machine models for which most, if not all, of the programming effort is spent synchronizing the movement of data.

Multiple concurrent accesses to a shared variable can be realized without congestion if messages are combined en route to the storage module of the shared variable. Combining in some form has been proposed, or included, in some recent architectures such as the Columbia CHoPP [113], the NYU Ultracomputer [105], the RP3 by IBM [92], and the Connection Machine System CM-2. In the RP3, associative memories are used for message combining. Neither the RP3 nor the Connection Machine model CM-2 guarantees complete combining. But the routing algorithm described in this section guarantees complete combining using simple hardware.

The probability of accessing different variables in the same storage unit is minimized by distributing the variables evenly over all storage units and either randomizing the address map or replicating shared variables among several storage units. Upfal [122] showed that a randomized address map requires an access time of at most $O(\log^2 N)$. Melhorn and Vishkin [84] explored the idea of replicated variables using a deterministic address map and randomized address maps on architectures with completely interconnected processors. For such architectures they provided an $O(\log N)$ bound for the randomized address map. The routing algorithm by Ranade [96], together with a randomized address map, guarantees a worst-case behavior of $O(\log N)$ with high probability for networks of diameter $O(\log N)$.

The congestion in the network has been studied in terms of permutation routing. Borodin and Hopcroft [12] have shown that for *oblivious* routing algorithms the worst-case routing time is of order $\Omega(\sqrt{N})$. Bitonic sort [5] provides an $O(\log^2 N)$ deterministic solution to the routing problem. Valiant [124, 123] showed that a deterministic address map and randomized routing guaranteed completion of the message delivery in time $O(\log N)$ with high probability on Boolean cubes of N nodes, and with communications buffers of size $O(\log N)$. Aleliunas [2] and Upfal [121] both extended the results of Valiant to bounded degree networks. Pippenger [93] showed that queues of size $O(1)$ were sufficient for permutation routing in time $O(\log N)$ on networks of diameter $O(\log N)$, with high probability. The results are summarized in Table 4.22 [98], which also includes the emulation algorithm described below.

Architectures with a physically distributed memory providing shared memory programming features have been proposed and built. Examples are the Connection Machine by Thinking Machines Corporation [39, 40], the BBN Butterfly [6] and Monarch, the IBM RP3 [92], and the NYU

Table 4.22   Randomized routing and PRAM emulation.

	Upfal 82 Aleliunas 82	Pippenger 84	Karlin and Upfal 86	Ranade 88
Time	$O(\log N)$	$O(\log N)$	$O(\log N)$	$O(\log N)$
Queuesize	$O(\log N)$	$O(1)$	$O(\log N)$	$O(1)$
Queue type	Priority	FIFO	Priority	FIFO
Combining	-	-	Sorting	Implicit
# Random bits	$N \log N$	$N \log N$	$N \log N$	$O(\log^2 N)$

Ultracomputer [105, 33]. The Connection Machine model CM-2 has hardware support for concurrent-read as well as concurrent-write operations. The Connection Machine and the NYU Ultracomputer/RP3 efficiently support the scan operation [11]. The Ultracomputer and the RP3 also support the fetch-and-add operation, but the switching hardware is expensive, and experiments reveal poor performance because of "hot spots" [76, 92].

Ranade [96] has devised a scheme for emulating shared memory on bounded degree networks. The emulation also provides very powerful instructions, such as multiprefix and set operations. The storage access conflict problem is resolved by randomizing the address space and thereby reducing the risk that related variables are allocated to the same storage unit, and by combining in the communication system reducing the number of storage accesses to a shared variable to one. The randomized address map also guarantees with high probability that the number of messages waiting at any given time for a communications channel is bounded by a constant. The communication distance is not necessarily minimized. For shared memory on a bounded degree network, the instruction time is of optimal order, with very high probability.

Next we first describe the routing algorithm due to Ranade [96], and then we present the Fluent abstract machine, a shared memory machine based on this routing algorithm. Its instruction set includes multiprefix and set operations. The hardware requirements for message routing are minimal. The routing function at each node is that of merging presorted message streams, combining messages destined for the same location. The *multiprefix* primitive requires some additional hardware in that it is necessary to remember what messages were combined, what the combining operation was, and the values involved in the combining operation. With its high-level instruction set, the Fluent abstract machine is suited as a target for compiling very high-level languages.

4.12.1 Shared Memory Emulation

Routing

We describe the emulation on a butterfly network and then show how the emulation can also be carried out on other networks. The routing algorithm by Ranade in combination with a randomized address map guarantees an $O(\log N)$ time emulation with high probability. The butterfly network has 2^n nodes in each of $n + 1$ columns, for a total of $N = (n + 1)2^n$ nodes. Each node is labeled with a string (c, r) ($0 \leq c \leq n$, $0 \leq r < 2^n$) formed by concatenating the binary representations of the column number c and the index r of the node within the column. Each node (c, r) ($c < n$) is connected by forward links to the nodes $(c + 1, r)$ and $(c + 1, r \oplus 2^c)$, where $\oplus$ denotes bitwise exclusive-or. Each node in the butterfly network contains a processor, a memory module, and communication logic. The routing algorithm proceeds in six phases that correspond to six passes through the butterfly network. Without a wraparound feature, every other phase is forward and every other backward, i.e., phases one, three, and five are backward and phases two, four, and six are forward. With one physical network, the communication logic in each node contains queues and switches for all six phases. Each switch has two inputs and two outputs. Every input into a switch enters a first-in first-out queue.

A request by a processor (c, r) for a variable located in storage unit (c', r') involves six phases through the butterfly network for forward and backward routing. Figure 4.54 shows the six phases. In phase 1, the message issued at node (c, r) is directed to node $(0, r)$. In phase 2, the message follows the unique (forward) path in the butterfly from node $(0, r)$ to node (n, r'). This path is determined at each switch by looking at the appropriate bit of the destination field. In phase 3, the message reaches the node (c', r'), where it acquires the required data. The next three phases simply retrace the path traced thus far, back to the source processor (c, r). The access is now complete.

The combining of messages is guaranteed by scheduling the messages arriving at each switch such that they are sorted by destination. In the butterfly network there is a unique path from source to destination. The paths of all messages to a given location must intersect at some node along the paths. The paths form a tree, as shown in Figure 4.55. The combining can be accomplished by recording what messages have been forwarded to which destinations, and the combining operation performed by a table

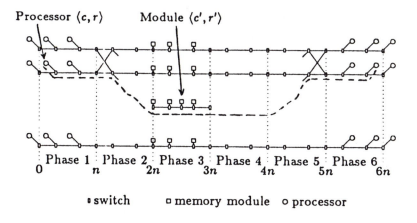

Figure 4.54   Logical network. [99]

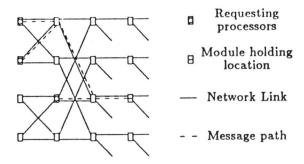

Figure 4.55   Message paths to a common location form a tree. [99]

look-up. This approach is used in the RP3 architecture. In the routing algorithm by Ranade [96], each switch serves as a stream merger. By scheduling message delivery only when an element is present in each of the streams being merged, combining can be guaranteed.

Each node receives messages sorted according to destination addresses and places them into FIFO queues, one for each input. At each step the node compares the destination addresses of the messages at the heads of the two queues. The message with the smaller destination address is transmitted forward. If both messages are destined for the same location, they are combined, and only one request is sent out. If only one queue has a message waiting and the other queue is empty, no message is sent out.

In our snapshot at time T, node A in Figure 4.56 selects the message destined for location 35. Then it waits until the message to location 48 arrives, at which point it discovers that the messages at the heads of both the queues are to location 48 and can be combined.

The reply message returns backward along each edge of the tree and reaches every requesting processor. For backrouting, it suffices to store two *direction bits* at each node. The bits encode along which branch the request arrived, or if it was combined. Since messages are kept sorted throughout the six phases, *replies at each node arrive in the same order as the requests were sent out.* Therefore the direction bits can be stored in a 2-bit wide FIFO queue. This simple idea is more efficient than the associative memories proposed earlier [34].

The simple idea of keeping message streams sorted has one deficiency: If only one of the incoming buffers has a message, then the node may wait for a long time, possibly only for an EOS message. This waiting state may cause messages upstream to be delayed if the buffer holding a message

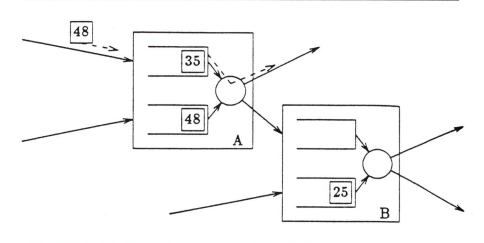

Figure 4.56   Combining messages by merging streams. [99]

becomes full. Consider Figure 4.56 again. At time T, processor B cannot transmit the message it holds for location 25. However, when A selects the message to location 35 for transmission, it can send a *ghost* message labeled 35 to B. When B receives the ghost message, it knows that future messages along that edge will be destined for locations greater than 35. At the next time step, therefore, B can forward the message waiting in the lower queue. Ghost messages notify nodes of the minimum location to which subsequent messages can be destined. Ghosts are not used for any other purpose. Ghost messages do not accumulate in queues. If a ghost message arrives at a switch in which the last message is another ghost message, it replaces it. If a real message arrives at a switch in which the last message is a ghost message, the new message replaces the ghost. There is at most one ghost message in a queue.

Immediately following a request, each processor also issues an EOS message. It notifies a switch that no more requests will follow. The switch can safely forward the requests on the other edge and eventually forward the EOS messages themselves. EOS messages form a wavefront, which guarantees that every instruction will terminate. The EOS message behaves as if its destination was ∞.

Address map

The optimum worst-case performance of the routing algorithm with combining is guaranteed by the choice of address map. The routing is deterministic, but the address map is randomized. The M shared variables are distributed randomly with M/N variables per processor by a hash function $\mathcal{H}$. The randomized address map guarantees that bottlenecks are unlikely. A random $O(\log N)$ degree polynomial can be proved to have this property [71, 96], but simple first degree polynomials perform well in practice [99, 98]. The hash function maps a $\log M$ bit address to a $\log N$ bit node address. A second function $\mathcal{M}$ computes the address ($\log(M/N)$ bits) within the memory of node $\mathcal{H}(x)$. The physical address of shared variable x is given by the concatenation $(\mathcal{H}(x), \mathcal{M}(x))$.

Performance
The technique for proving a worst-case performance of order $O(\log N)$ is based on constructing message paths causing a maximum delay and determining the likelihood that such paths might occur. The theorem below can be proved by such an analysis.

Theorem 19   *[96] Assuming a perfect random address map, the probability that any memory reference takes more than $15 \log N$ steps is less than N^{-20} with buffers of size order $O(1)$.*   □

Every routing algorithm has a worst-case routing time of at least $4 \log N$ steps. The provable performance is only about a factor of four from this worst-case lower bound. Our simulations indicate that the routing time is approximately $12 \log N$, i.e., three times the worst-case lower bound. A message can be delayed for one of the following three reasons. One reason is that the switch selects another message for transmission because it is destined for a processor with a lower address. Another reason a message may not be forwarded is that the queue in the receiving switch is full. The third reason for a delay of a message en route is that the switch does not have any message in the other queue. The proof proceeds by considering the last message that is delivered, tracing it backward to the point where it was last delayed, and then tracing the message that delayed it back to the point where it was last delayed, etc. For a large delay, many messages must have paths that intersect in a way causing delay. By a fan-in argument it is clear that the number of messages that might contend for a switch increases from the input toward the output of the butterfly network, but at the same time, because of message combining a fan-out argument limits the number of messages that might intersect at a switch as the last column is approached. For the proof see [96, 98].

The proof requires the address map to be computed by a universal hash function. Computing the address map by such a function may be quite expensive, and it is desirable to find hash functions that are easy to evaluate yet yield optimal or close to optimal performance. Figure 4.60 gives timing results from simulations of the routing algorithm. In the simulations, different hash functions, queue sizes, and memory access patterns were tried. Linear congruential hash functions such as $\mathcal{H} = ax + b \bmod M$, where the size of the address space M is a prime number and a and b are constants, worked well in all simulated cases. Matrix access, binary and fibionacci tree access, shuffle permutations, and random permutations were tried. The routing time for random permutations was always somewhat longer than for the other cases. The average time is approximately $12 \log N$, even with queues of size 2. Increasing queue size did not appreciably affect performance. The simulations also showed that concurrent access is faster than permutation routing. The extreme case is when all processors read the same variable. The number of steps reduces from 154 (see Figure 4.60) to 85 because there is no buffering delay.

Other networks

The key feature of the routing algorithm is that there is a directed, acyclic path from source to destination for every source-destination pair. The algorithm can be used for any network once such a graph is defined for it. This property was observed by Leighton et al. [78], who also adapted the algorithm to such networks. Ranade [98] showed that if each message has an identifier that allows each node along the route to decide on which subset of its outgoing edges the message should be sent, then the communication with high probability can be completed in time $O(c + d + \log N)$ with bounded queue sizes. d is the diameter of the acyclic graph and c the *congestion*, i.e., the maximum number of messages routed through the same edge for the entire communication. Messages are combined in the nodes as in the implementation on the butterfly network. The number of random bits required are $O(\log{(c + N)}\log{(c + d + \log N)})$ [98].

4.12.2   The Fluent Abstract Machine

Based on the shared memory emulation above, a computer with a high-level instruction set can be defined. The concurrent-read and concurrent-write instructions are deterministic since the messages are sorted. READ(A) returns the value of A to the requesting processor. WRITE($A, v, \otimes$) is the concurrent-write instruction. The order in which messages are combined is always the same and is determined by the order in which the processors are labeled. The combining and deterministic property of the routing algorithm makes it possible to define a *multiprefix* operation $MP(A, v, \otimes)$. This instruction is more powerful than the nondeterministic fetch-and-$\otimes$ operation [34]. Other operations that can be implemented as unit time operations include the common set operations, insertion, deletion, union, intersection, difference, prefix, reduce, copy, enumerate, and function applications. A set is identified by a location in shared memory. At each step, every processor can execute a set instruction. If in a given instruction several processors insert elements into the same set, then at the end of the instruction the set will be updated to include the new elements. Different processors may operate on the same set concurrently, or on different sets. For instance, two key operations in event driven simulation are WAIT and SIGNAL. Wait causes a process to wait for the specified event, while SIGNAL causes all waiting processes to resume. The two events are readily modeled by set operations. WAIT is realized by an insertion into the set of processes waiting for a specified event, and SIGNAL is realized by a set application. Several processes may simultaneously execute wait operations on the same or different events. The result of concurrent set operations is as if the individual operations were executed atomically in some arbitrary, unspecified serial order.

In addition to the high-level, shared memory instructions, the Fluent machine also incorporates the standard instructions of any architecture. The processor synchronization is made through the EOS messages. They implement a distributed global clock. There is one EOS message injected into the communication system with each message. One EOS message per instruction passes through each switch. EOS messages inserted with different instructions cannot pass each other in the routing algorithm above. Each switch determines the global time by keeping a count of the number of EOS messages that have passed through. At any given instant different switches may have different counts, but if two instructions in the same instruction cycle access a common location, then the one that arrives first will wait for the slower one to reach an intermediate switch for combination. The EOS message guarantees that instructions for different instruction cycles are separated, and the sorting feature of the switches guarantees that only one request for access will be passed into the memory module that holds the variable. The effect is the same as if all the processors were operating synchronously.

For example, our implementation guarantees that for the code of Figure 4.57, processors 1 and 2 will respectively read 10 and 20, provided no other processor writes a and b in the meantime. This ordered behavior is guaranteed in spite of the fact that both processors might issue all three instructions without waiting for any to complete. The distributed global clock implementation also allows each processor to stop the clock if necessary, for instance, if it detects an error. Withholding the end-of-stream message stops the clock.

The multiprefix instruction

In the multiprefix operation $MP(A, v, \otimes)$, A is a shared variable, v is a value, and $\otimes$ is a binary associative operator. Any number of processors can execute a multiprefix operation at any time, with the constraint that if processor P_i and processor P_j execute $MP(A, v_i, \otimes_i)$ and $MP(A, v_i, \otimes_j)$, then $\otimes_i = \otimes_j$. The semantics of the multiprefix operator are as follows:

TIME	PROCESSOR 1	PROCESSOR 2
1	A=20	B=10
2	Read B	Read A
3	A=30	B=40

Figure 4.57   Synchronization guarantee. [99]

At time T let $P_A = \{p_1 \ldots p_k\}$ be the set of processors referring to variable A, such that $p_1 < p_2 < .. < p_k$. Suppose that $p_i \in P_A$ executes instruction $MP(A, v_i, \otimes)$. Let a_0 be the value of A at the start of time T. Then, at the end of time cycle T, processor p_i will receive the value $a_0 \otimes v_1 \otimes \cdots \otimes v_{i-1}$ and the value of variable A will be $a_0 \otimes v_1 \otimes \cdots \otimes v_k$.

Thus when a set of processors performs a multiprefix operation on a common variable, the result is the same as if a single prefix operation were performed with the processors ordered by their index. For example, suppose that processors numbered 25, 32, and 65 execute the instructions $MP(A, 4, +), MP(A, 7, +)$ and $MP(A, 11, +)$ respectively at time T, and suppose that variable A initially contains the value 5. Then, at the end of the Tth cycle, processor 25 will receive 5, processor 32 will receive 9, processor 65 will receive 16, and the variable A will equal 27.

The fetch-and-op operation [34] also calculates a set of prefixes, but the order of inputs is undetermined before execution. Multiprefix is a determinate implementation of the fetch-and-op, and is more powerful. The *segmented scan* operation [11] is a special case of the multiprefix in that the different sets on which a multiprefix operation is performed need to form contiguous, disjoint sets in the address space. READ and WRITE are special cases of the multiprefix instruction, as is observed in [34].

4.12.3   A Radix Sort by Multiprefix

The multiprefix operation can be used very effectively to express and implement radix sort. The multiprefix provides a stable radix sort that can be iterated for longer keys. N keys, each $k \log N$ bits long, can be sorted in $O(k)$ multiprefix instructions [99].

For the program below, the keys to be sorted are stored in an array KEY. First we give a stable algorithm that sorts N keys of length $\log N$, with one key per processor. The number of distinct key values is N. We first count the number of occurrences of $KEY[i]$ that lie in processors indexed less than i, then add to that the cumulative sum of the counts for keys less than $KEY[i]$.

```
SHORTSORT()
     COUNT[i] := 0
     CUMULATIVE[i] := 0
     TEMP := 0
     MP(COUNT[KEY[i]], 1, +)
     CUMULATIVE[i] := MP(TEMP, COUNT[i], +)
     return MP(CUMULATIVE[KEY[i]], 1, +)
```

Because the multiprefix operation is ordered by processor indices, the sort above is stable and can be iterated for larger keys by dividing them into blocks. The primitive operation $LSBLOCK(w, j)$ below returns the least significant jth block of $\log N$ bits of location w, that is, bits $(j-1)\log N + 1$ through $j \log N$.

```
SORT:
     KEYPTR[i] := i        ; initialize pointer to self
     FOR j=1 to k DO
KEY[i] := LSBLOCK(KEYPTR[i], j)
RANK[i] := SHORTSORT()
KEYPTR[RANK[i]] := KEYPTR[i]
     ENDDO
```

The fetch-and-add can be used as well as the multiprefix to sort N keys in a constant number of steps, if there are less than $\log N$ bits per key. But the fetch-and-add operation cannot be used iteratively to sort longer keys because it is nondeterministic and does not yield a stable sort [11].

Implementation issues

The message format for the Fluent is (dest, type, data). The destination dest is the physical address of varible x. The type field denotes the kind of access requested, e.g., READ, WRITE, MP, EOS, or GHOST.

4.12.4   Functionality of the Switches

We first describe the implementation for fetch-and-add [34]. Let s be an arbitrary switch in phase 1 (or 2). Suppose that the messages at the heads of the queues are $m_1 = (l, \text{fetch-add}, v_1)$ and $m_2 = (l, \text{fetch-add}, v_2)$, respectively. As shown in [34], the switch must forward a message $m = (l, \text{fetch-add}, v_1 + v_2)$ in place of m_1 and m_2. If the reply to m is a value v, then the corresponding switch in phase 6 (or 5) returns v as a reply to m_1, and $v + v_1$ as a reply to m_2. Thus the switch must remember the value v_1 received on its top queue for each pair of fetch-and-add messages that it combines.

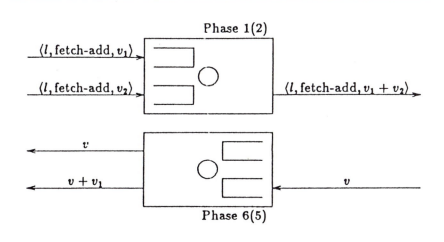

Figure 4.58   Fetch-and-add. [99]

Notice that this is equivalent to a serial execution of the message received on the top input (m_1) before the message received on the bottom input (m_2). Thus, if we ensure that messages received on the top input always originate in a processor with a smaller number than those received at the bottom input, we effectively have an implementation for the multiprefix operation, with addition replaced by the prefix operator. By numbering processor (c, r) $(n + 1)r + c$ this property is guaranteed. Phase 1 is a backward phase in that messages are routed toward column 0. By connecting the top input to the local processor and the bottom input to the switch in the next column, it is guaranteed that messages on the bottom input always originate in processors with a larger address. Phase 2 is a forward phase. The numbering scheme orders the processors rowwise. Every processor in row $r_2 > r_1$ has an address greater than any processor in row r_1. Hence the bottom input should be connected to $(c - 1, r \oplus 2^{c-1})$ if $r \oplus 2^{c-1} > r$; otherwise it should be connected to $(c - 1, r)$.

As noted earlier, the only extra requirement over a read instruction is that, in addition to the two direction bits, each switch must remember a value (*partial sum*) for every combination that occurs at that switch. Figure 4.59 shows a pair of switches with the required queues.

4.12.5   Message Decoding and Scheduling

Each message can be transmitted bit-serially in a pipelined manner, in a way analogous to the Connection Machine router [39] and the wormhole

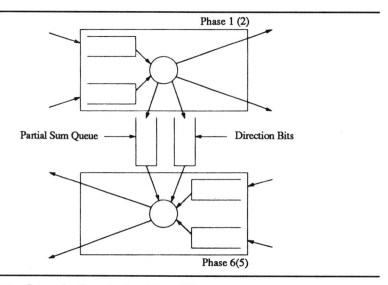

Figure 4.59   Internals of a pair of switches. [99]

router of Dally and Seitz [23]. Message transmission can be pipelined because

(1) address comparison can be done bit-serially, provided the addresses
are received most significant bit first;

(2) message combination can be done bit-serially; for operators like +,
the data must be transmitted least significant bit first. Also see on-
line arithmetic [118];

(3) when a message leaves a switch, the corresponding **GHOST** message
(whose **dest** is identical to the real message) can be generated bit-
serially.

Each message is transmitted with the **dest** field first (most significant
bit leading), followed by the **type** field, and finally the **data** field (least significant bit leading). A switch begins operating when (1) each input queue
contains at least one message, and (2) the input queues of the receiving
switches are not full. The operation of a switch in phase 2 is

(1) **Transmit dest:** The minimum of the destinations of the two messages in the input queues is transmitted along both outputs. The
minimum is discovered only after the transmission, so till then both
destinations must be retained in the input queues.

(2) **Transmit type:** While transmitting the destination, the switch detects which output link the request must be routed on. This requires checking one fixed bit in the **dest** field. The type of the message with the minimum destination is transmitted on the output, while on the other, type GHOST is transmitted.

(3) **Transmit data:** This is relevant for messages like MP-⊗ or WRITE. In either case, the message type indicates how messages must be combined when necessary. Again, the data fields can be combined and transmitted as they arrive.

The operation of switches in other phases can be specified similarly. The ability to pipeline messages speeds up message delivery considerably when there are no queueing delays. The message delivery time reduces from (network latency) × (message length) to (network latency) + (message length). The expected latency of each switch is about 4 (message enters an input queue, passes through the ALU, is sent to the output queue, and then transmitted), giving a total latency of $4 \times 6n$ for the logical network. Assuming 100-bit long messages and 4-bit wide data paths, the time for a 13-dimensional butterfly is $(4 \times 6 \times 13) + 100/4 = 337$ steps.

4.12.6   Area Requirements for a Switch

Each switch consists of message queues, an ALU (for address comparison, message combination, etc.), counters to maintain the message FIFO queues, memory for storing partial sums, and direction bits for reply routing. We assume that messages are 100 bits wide and that partial sums are 64 bits wide.

Switches in phases 2 and 5 have two input queues, while others only have one input queue. The total number of message queues per node is eight. Simulations (section 4.12.7) indicate that for the 100,000 node machine each message queue need hold only three messages. The total memory requirement for message queues thus equals $8 \times 3 \times 100 = 2400$ bits, or roughly 1.2 $M\lambda^2$ at $500\lambda^2$ per bit using layout estimates from [88].

Simulations also strongly indicate that no switch will ever transmit more than 40 messages along its outputs. For reply routing, we need 2-bit wide direction queues and 64-bit wide partial sums. Long partial sum queues are maintained only in phase 2 so that the total memory requirement adds up to $40 \times 64 + 6 \times 2 \times 40 = 3040$ bits, or 1.52 $M\lambda^2$.

Each queue requires three counters, except for the message queues, which require four. Assuming 8-bit wide counters, the total memory is 424 bits. With 3000 λ^2 per counter bit, total area requirement is 1.28 $M\lambda^2$.

Assuming 8-bit wide data paths, each ALU requires around 1.2 $M\lambda^2$, for a total of 7.2 $M\lambda^2$ per node. The total area requirement is thus approximately 11.2 Mλ^2. Including miscellaneous overhead, 15 M$\lambda^2$is a conservative estimate for six switches per node.

4.12.7   Global Architecture

It is estimated that a Fluent machine based on a 13-dimensional butterfly network would have a peak performance in excess of 1 trillion floating-point operations, or instructions, per second. A 13–dimensional butterfly network has 2^{13} nodes in each of 14 ranks for a total of 114,688 nodes. Such a network can be divided into 256 boards, each housing a 6–dimensional butterfly. Assuming two nodes per chip, each board has 224 chips and 448 nodes. In addition to the two processors, each chip also has routing switches for the two nodes, and for a peak performance of 1 trillion operations per second one floating-point unit, and memory. Table 4.23 summarizes the breakup of the chip area.

With the current packaging technology the number of pins at the board boundary is 500–1000, and the number of pins on a chip is at most 200–300. The data paths between nodes vary in width, depending on whether the path is on-board or across boards. With 128 4-bit wide data paths, connecting different boards requires 512 pins. A 6-dimensional butterfly has 64 nodes in the last rank, each with two forward links. On-board paths may be 8 bits wide without violating the restrictions on pin limitations per chip. Phases 2 and 5 require six off-chip channels, each with two nodes per chip. The other phases require two off-chip channels. The total number of off-chip channels with two nodes on a chip is twenty, which with 8-bit wide paths require 160 pins.

Table 4.23   Chip specification. [99]

switches	30 Mλ^2
2 32-bit RISC Processors	40 Mλ^2
Floating point unit	100 Mλ^2
128 Kbytes memory per processor	200 Mλ^2
Total area requirement per chip	370 Mλ^2

The difference in channel widths on-chip, on-board, and between boards was not accounted for in the analysis of the router performance. The performance of the routing algorithm changes somewhat with narrow channels. The off-board channels also have to be multiplexed over the six phases of the logical network, while on-board channels are replicated.

Simulation results

The behavior of the communication system has been simulated for up to 57,344 processors. The influence of the hash function and various access patterns are already mentioned. The variation in the channel widths and the multiplexing at the board boundary is shown in Figure 4.60. The *narrowness* is the ratio of the width of the off-board channels to the on-board channels. Switches in lower phases are given higher priority in accessing channels. Each channel first allows phase 0 messages to pass, followed by phase 1 messages, and so on. From the plot we can conclude that the performance degrades by a factor of 1.7 over the ideal case (no narrow channels and no multiplexing). The time goes up from 154 steps, as in Figure 4.60, to about 260 steps (extrapolated for 114,688 processors from Figure 4.60).

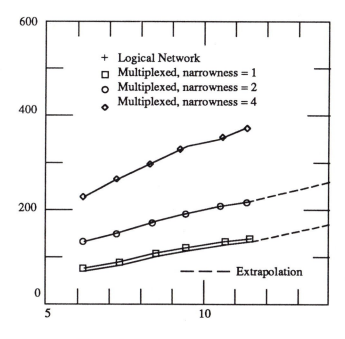

Figure 4.60   Average routing time (50 randomly chosen access patterns). [99]

Suppose that messages are 100 bits long (64 data bits, 32 address bits, and 4 type bits). If every channel was 8 bits wide, sending a message across one link would require $100/8 \approx 13$ steps. From the simulations we estimate that, with narrow channels and multiplexing, an arbitrary permutation can be routed in $260 \times 13 = 3380$ steps. With a 50 nanosecond clock rate, the time is about 169 μsec. If all processors access a single variable, then 337 cycles are required, or about 17 μsec.

The multiprefix based radix sort for 3.5 billion 16-bit numbers (32K numbers in each node) can be performed in 8.5 seconds. Only one iteration of the procedure SHORTSORT is needed. For each number being sorted, three shared memory instructions are executed (the others are local). The instructions can be packed into 50-bit messages. The total number of steps required is $3 \times 32K \times 3380/2$, or about 169 million for a total time of 8.5 seconds. If the numbers are 32 bits long, the time is about 17 seconds.

Bibliography

[1] Aho, A. V., J. E. Hopcroft, and J. D. Ullman. *The Design and Analysis of Computer Algorithms.* Reading, Mass.: Addison-Wesley, 1974.

[2] Aleliunas, R. Randomized parallel computation. *ACM Symposium on Principles of Distributed Computing*, 60–72, 1982.

[3] Aleliunas, R., and A. L. Rosenberg. On embedding rectangular grids in square grids. *IEEE Trans. Computers* 31(9):907–913, September 1982.

[4] Baillie, C., S. L. Johnsson, L. Ortiz, and G. S. Pawley. QED on the Connection Machine. *The Third Conference on Hypercube Concurrent Computers and Applications*, 1288–1295. ACM Press, January 1988.

[5] Batcher, K. E. Sorting networks and their applications. *Spring Joint Computer Conf.*, 307–314. IEEE, 1968.

[6] BBN Laboratories, Inc. *Butterfly Parallel Processor Overview*, 1985.

[7] Bhatt, S. N., F. R. K. Chung, F. T. Leighton, and A. L. Rosenberg. Optimal simulations of tree machines. *Proc. 27th IEEE Symposium*

Foundations Computer Science, 274–282. IEEE Computer Society, 1986.

[8] Bhatt, S. N., and I. I. F. Ipsen. How to embed trees in hypercubes. Tech. Rep. YALEU/CSD/RR-443, Dept. of Computer Science, Yale Univ., New Haven, Conn., December 1985.

[9] Bhatt, S. N., and C. E. Leiserson. Minimizing the longest edge in a VLSI layout. Tech. Rep. MIT VLSI Memo 82-86, MIT, 1982.

[10] Bhatt, S. N., and C. E. Leiserson. *How to Assemble Tree Machines*, vol. 2, 95–114. JAI Press Inc., 1984.

[11] Blelloch, G. E. Scans as primitive parallel operations. *The 1987 International Conf. on Parallel Processing*, 355–362. IEEE Computer Society Press, 1987.

[12] Borodin, A., and J. E. Hopcroft. Routing, merging and sorting on parallel models of computation. *Proc. of STOC 82*, 338–344, 1982.

[13] Brandenburg, J. E., and D. S. Scott. Embeddings of communication trees and grids into hypercubes. Tech. Rep. no. 280182-001, Intel Scientific Computers, 1985.

[14] Brandt, A. Multilevel adaptive solutions to boundary-value problems. *Mathematics of Computation*, Vol. 31, 333–390, 1977.

[15] Brent, R. P., and H. T. Kung. On the area of binary tree layouts. *Information Processing Letters* 11(1):44–46, 1980.

[16] Callaway, D. J. E., and A. Rahman. *Phys. Rev. Lett.* 49:613, 1982.

[17] Cannon, L. E. *A Cellular Computer to Implement the Kalman Filter Algorithm.* Ph. D. thesis, Montana State Univ., 1969.

[18] Cantoni, V., and S. Levialdi, eds. *Pyramidal Systems for Computer Vision.* Springer-Verlag Berlin Heidelberg, 1986.

[19] Chan, M. Y. Dilation-2 embeddings of grids into hypercubes. Tech. Rep. UTDCS 1-88, Computer Science Dept., Univ. of Texas at Dallas, 1988.

[20] Chan, T. Personal communication, 1986.

[21] Chan, T. F., and Y. Saad. Multigrid algorithms on the hypercube multiprocessor. *IEEE Trans. Computers* 35(11):969–977, November 1986.

[22] Chang, J. H., O. H. Ibarra, T. C. Pong, and S. M. Sohn. Two-dimensional convolution on a pyramid computer. *1987 International Conf. on Parallel Processing*, 780–782. IEEE Computer Society, 1987.

[23] Dally, W. J. *A VLSI Architecture for Concurrent Data Structures*. Ph. D. thesis, Calif. Institute of Tech., 1986.

[24] Dekel, E., D. Nassimi, and S. Sahni. Parallel matrix and graph algorithms. *SIAM J. Computing* 10:657–673, 1981.

[25] Denneau, M. M., P. H. Hochschild, and G. Shichman. The switching network of the TF-1 parallel supercomputer. *Supercomputing Magazine* 2(4):7–10, 1988.

[26] Deshpande, S. R., and R. M. Jenevin. Scaleability of a binary tree on a hypercube. *1986 International Conf. on Parallel Processing*, 661–668. IEEE Computer Society, 1986.

[27] Dongarra, J. J. and D. C. Sorensen. A fully parallel algorithm for the symmetric eigenvalue problem. *SIAM J. Scientific and Statistical Computing* 8(2):s139–s153, 1987.

[28] Duane, S. *Nuclear Physic.* B257:652, 1985.

[29] Fischer, M. J. *Efficiency of Equivalence Algorithms*, 153–167. Plenum Press, 1972.

[30] Flanders, P. M. A unified approach to a class of data movements on an array processor. *IEEE Trans. Computers*, 31(9):809–819, September 1982.

[31] Fox, G. C., and D. Jefferson. Concurrent processor load balancing as a statistical physics problem. Tech. Rep. CCCP-172, Calif. Institute of Tech., May 1985.

[32] Gentleman, W. M. Some complexity results for matrix computations on parallel processors. *J. ACM* 25(1):112–115, January 1978.

[33] Gottlieb, A., R. Grishman, C. P. Kruskal, K. P. McAuliffe, L. Rudolph, and M. Snir. The NYU Ultracomputer—designing an MIMD shared memory parallel computer. *IEEE Trans. Computers* 32(2):175–189, 1983.

[34] Gottlieb, A., B. D. Lubachevsky, and L. Rudolph. Coordinating large numbers of processors. *The 1981 International Conf. on Parallel Processing*. IEEE Computer Society, 1981.

[35] Greenberg, D. S. Minimum expansion embeddings of meshes in hypercubes. Tech. Rep. YALEU/DCS/RR-535, Dept. of Computer Science, Yale Univ., New Haven, Conn., August 1987.

[36] Havel, I., and J. Móravek. B-valuations of graphs. *Czech. Math. J.* 22:338–351, 1972.

[37] Hennessey, J. L., N. Jouppi, F. Baskett, and J. Gill. MIPS: A VLSI processor architecture. *VLSI Systems and Computations*, 337–346. Computer Sciences Press, 1981.

[38] Hennessey, J. L., N. Jouppi, S. Przybylski, and C. Rowen. Design of a high performance VLSI processor. *Proc. Third Caltech Conf. on VLSI*, 33–54. Computer Sciences Press, 1983.

[39] Hillis, W. D. *The Connection Machine.* Cambridge, Mass.: MIT Press, 1985.

[40] Hillis, W. D., and G. L. Steel. Data parallel algorithms. *Comm. CACM.* 29:1170–1183, December 1986.

[41] Ho, C.-T., and S. L. Johnsson. Distributed routing algorithms for broadcasting and personalized communication in hypercubes. *1986 International Conf. on Parallel Processing*, 640–648. IEEE Computer Society, 1986.

[42] Ho, C.-T., and S. L. Johnsson. Matrix transposition on Boolean n-cube configured ensemble architectures. Tech. Rep. YALEU/DCS/RR-494, Dept. of Computer Science, Yale Univ., New Haven, Conn., September 1986.

[43] Ho, C.-T., and S. L. Johnsson. On the embedding of arbitrary meshes in Boolean cubes with expansion two dilation two. *1987 International Conf. on Parallel Processing*, 188–191. IEEE Computer Society, 1987.

[44] Ho, C.-T., and S. L. Johnsson. Dilation d embedding of a hyper-pyramid into a hypercube. Supercomputing 89, 294–303, ACM Press.

[45] Ho, C.-T., and S. L. Johnsson. Optimal algorithms for stable dimension permutations on Boolean cubes. *The Third Conference on Hypercube Concurrent Computers and Applications*, pp. 725–736, ACM, 1988.

[46] Ho, C.-T., and S. L. Johnsson. Stable dimension permutations on Boolean cubes. Tech. Rep. YALEU/DCS/RR-617, Dept. of Computer Science, Yale Univ., October 1988.

[47] Ho, C.-T., and S. L. Johnsson. Embedding meshes in Boolean cubes by graph decomposition. *Journal of Parallel and Distributed Computing* 8(4):325–339, 1990.

[48] Ho, C.-T., and S. L. Johnsson. Spanning balanced trees in Boolean cubes. *SIAM J. on Sci. Stat. Comp* 10(4), July 1989.

[49] Hong, J. W., and H. T. Kung. I/O complexity: The red-blue pebble game. *Proc. of the 13th ACM Symposium on the Theory of Computation*, 326–333. ACM, 1981.

[50] Johnsson, S. L. Combining parallel and sequential sorting on a Boolean n-cube. *1984 International Conf. on Parallel Processing*, 444–448. IEEE Computer Society, 1984.

[51] Johnsson, S. L. Odd-even cyclic reduction on ensemble architectures and the solution tridiagonal systems of equations. Tech. Rep. YALE/DCS/RR-339, Dept. of Computer Science, Yale Univ., October 1984.

[52] Johnsson, S. L. Band matrix systems solvers on ensemble architectures. *Algorithms, Architecture, and the Future of Scientific Computation*, 195–216. Austin, Tex.:University of Texas Press, 1985.

[53] Johnsson, S. L. Data permutations and basic linear algebra computations on ensemble architectures. Tech. Rep. YALEU/DCS/RR-367, Dept. of Computer Science, Yale Univ., New Haven, Conn., February 1985.

[54] Johnsson, S. L. Fast banded systems solvers for ensemble architectures. Tech. Rep. YALEU/DCS/RR-379, Dept. of Computer Science, Yale Univ., March 1985.

[55] Johnsson, S. L. Solving narrow banded systems on ensemble architectures. *ACM TOMS* 11(3):271–288, November 1985.

[56] Johnsson, S. L. Communication efficient basic linear algebra computations on hypercube architectures. *J. Parallel Distributed Comput.* 4(2):133–172, April 1987.

[57] Johnsson, S. L. The FFT and fast Poisson solvers on parallel architectures. *Fast Fourier Transforms for Vector and Parallel Computers.* The Mathematical Sciences Institute, Cornell Univ., 1987.

[58] Johnsson, S. L. Solving tridiagonal systems on ensemble architectures. *SIAM J. Sci. Statist. Comput.* 8(3):354–392, May 1987.

[59] Johnsson, S. L. Convergence of substructuring in Poisson's equation. Tech. Rep., Dept. of Computer Science, Yale Univ., New Haven, Conn. (In preparation).

[60] Johnsson, S. L., and C.-T. Ho. Algorithms for multiplying matrices of arbitrary shapes using shared memory primitives on a Boolean cube. *Parallel Processing and Medium Scale Parallel Processors*, pp. 108–156. SIAM, 1989.

[61] Johnsson, S. L., and C.-T. Ho. Multiple tridiagonal systems, the alternating direction method, and Boolean cube configured multiprocessors. SIAM J. on *Sci. Statist. Computing*, vol. II, no. 1, January 1990.

[62] Johnsson, S. L., and C.-T. Ho. Expressing Boolean cube matrix algorithms in shared memory primitives. *The Third Conference on Hypercube Concurrent Computers and Applications*, 1599–1609. ACM, 1988.

[63] Johnsson, S. L., and C.-T. Ho. Matrix transposition on Boolean n-cube configured ensemble architectures. *SIAM J. Matrix Anal. Appl.* 9(3):419–454, July 1988.

[64] Johnsson, S. L., and C.-T. Ho. Shuffle permutations on Boolean cubes. Tech. Rep. YALEU/DCS/RR-653, Dept. of Computer Science, Yale Univ., October 1988.

[65] Johnsson, S. L., and C.-T. Ho. Emulating butterfly networks on gray code encoded data in Boolean cubes. Tech. Rep. YALEU/DCS/RR-764, Dept. of Computer Science, Yale Univ., February 1990.

[66] Johnsson, S. L., and C.-T. Ho. Spanning graphs for optimum broadcasting and personalized communication in hypercubes. *IEEE Trans. Computers*, Vol. 38, no. 9, pp. 1249–1268. September 1989.

[67] Johnsson, S. L., C.-T. Ho, M. Jacquemin, and A. Ruttenberg. Computing fast Fourier transforms on Boolean cubes and related networks. *Advanced Algorithms and Architectures for Signal Processing II*, vol. 826, 223–231. Society of Photo-Optical Instrumentation Engineers, 1987.

[68] S. L. Johnsson, and P. Li. Solution set for AMA/CS 146. Tech. Rep. 5085:DF:83, Calif. Institute of Tech., May 1983.

[69] S. L. Johnsson, and K. K. Mathur. Data structures and algorithms for the finite element method on a data parallel supercomputer. *International Journal on Numerical Methods in Engineering* 29(4):881–908, 1990.

[70] S. L. Johnsson, Y. Saad, and M. H. Schultz. Alternating direction methods on multiprocessors. *SIAM J. Sci. Statist. Comput.* 8(5):686–700, 1987.

[71] Karlin, A., and E. Upfal. Parallel hashing—an efficient implementation of shared memory. *Proc. Symposium on Theory of Computing*, 1986.

[72] Katevenis, M. G. H. *Reduced Instruction Set Computer Architectures for VLSI.* Cambridge, Mass.: MIT Press, 1985.

[73] Knauer, S. C., J. H. O'Neill, and A. Huang. *Self-routing Switching Network*, 424–448. Reading, Mass.: Addison-Wesley, 1985.

[74] Lai, T.-H., and W. White. Embedding pyramids in hypercubes. Tech. Rep., Dept. of Computer and Information Science, Ohio State Univ., November 1987.

[75] Lee, D., Y. Saad, and M. H. Schultz. An efficient method for solving the three-dimensional wide angle wave equation. Tech. Rep. YALEU/DCS/RR-463, Dept. of Computer Science, Yale Univ., October 1986.

[76] Lee, G., C. P. Kruskal, and D. J. Kuck. The effectiveness of combining in shared memory parallel computers in the presence of 'hot spots'. *Proc. International Conf. on Parallel Processing*, 35–41, 1986.

[77] Leighton, F. T. *Complexity Issues in VLSI: Optimal Layouts for the Shuffle-Exchange Graph and Other Networks.* Cambridge, Mass.: MIT Press, 1983.

[78] Leighton, F. T., F. Makedon, and I. H. Sudborough. Personal communication, 1988.

[79] Leiserson, C. E. Area-efficient graph layouts (for VLSI). *Proc. 21st IEEE Symp. Foundations Comput. Sci.*, 270–281. IEEE Computer Society, 1980.

[80] Leiserson, C. E. *Area-Efficient VLSI Computation.* Cambridge, Mass.: MIT Press, 1982.

[81] Lutz, C., S. Rabin, C. L. Seitz, and D. Speck. Design of the mosaic element. *Proc. Conf. on Advanced Research in VLSI*, 1–10. Artech House, 1984.

[82] McBryan, O. A., and E. F. Van de Velde. Hypercube algorithms and implementations. *SIAM J. Sci. Statist. Comput.* 8(2):s227–s287, March 1987.

[83] Mead, C. A., and L. Conway. *Introduction to VLSI Systems.* Reading, Mass: Addison-Wesley, 1980.

[84] Melhorn, K., and U. Vishkin. Granularity of parallel memories. Ultracomputer Note 59, New York Univ., October 1983.

[85] Metcalf, M., and J. Reid. *Fortran 8X Explained.* Oxford Scientific Publications, 1987.

[86] Metropolis, N., A. W. Rosenbluth, M. N. Rosenbluth, A. H. Teller, and E. Teller. *J. Chem Phys* 21:1087, 1953.

[87] Miller, R., and Q. F. Stout. Data movement techniques for the pyramid computer. *SIAM J. Comput.* 16(1):38–60, February 1987.

[88] Newkirk, J., and R. Mathews. *The VLSI Designers Library.* Reading, Mass: Addison-Wesley, 1983.

[89] Olsson, P. and S. L. Johnsson. Solving three-dimensional Navier-Stokes equations by explicit methods on a data parallel computer. Tech. Rep., Thinking Machines Corp. (In preparation 1989).

[90] Parter, S. The use of linear graphs in Gaussian elimination. *SIAM Review* 3(2):119–130, 1961.

[91] Paterson, M. S., W. L. Ruzzo, and L. Snyder. Bounds on minimax edge length for complete binary trees. *Proc. 13th Annual Symposium on the Theory of Computing*, 293–299. ACM, 1981.

[92] Pfister, G. F., W. C. Brantley, D. A. George, S. L. Harvey, W. J. Kleinfelder, K. P. McAuliffe, E. A. Melton, V. A. Norton, and J. Weiss. The IBM research parallel processor prototype (RP 3); introduction and architecture. *1985 International Conf. on Parallel Processing*, 764–771. IEEE Computer Society, 1985.

[93] Pippenger, N. Parallel communication with limited buffers. *Proc. FOCS 84*, 127–136, 1984.

[94] Polonyi, J., and H. W. Wyld. *Phys. Rev. Lett.* 51:2257, 1983.

[95] Preparata, F. P., and J. E. Vuillemin. The Cube Connected Cycles: A versatile network for parallel computation. *Proc. 20th Annual IEEE Symposium on Foundations of Computer Science*, 140–147, 1979.

[96] Ranade, A. How to emulate shared memory. *Proc. 28th Annual Symposium on the Foundations of Computer Science*, 185–194. IEEE Computer Society, October 1987.

[97] Ranade, A., and S. L. Johnsson. The communication efficiency of meshes, Boolean cubes, and cube connected cycles for wafer scale integration. *1987 International Conf. on Parallel Processing*, 479–482. IEEE Computer Society, 1987.

[98] Ranade, A. G. *Fluent Parallel Computation*. Ph. D. thesis, Yale Univ., 1988.

[99] Ranade, A. G., S. N. Bhatt, and S. L. Johnsson. The Fluent abstract machine. *Advanced Research in VLSI, Proc. Fifth MIT VLSI Conf.* 71–93. MIT Press, 1988.

[100] Reingold, E. M., J. Nievergelt, and N. Deo. *Combinatorial Algorithms*. Englewood Cliffs, N.J.: Prentice-Hall, 1977.

[101] Richtmyer, R., and K. W. Morton. *Difference Methods for Initial-Value Problems*. Wiley-Interscience, 1967.

[102] Rosenberg, A. L. Preserving proximity in arrays. *SIAM J. Computing* 4:443–460, 1975.

[103] Saad, Y., and M. H. Schultz. Data communication in hypercubes. Tech. Rep. YALEU/DCS/RR-428, Dept. of Computer Science, Yale University, Conn., October 1985.

[104] Saad, Y., and M. H. Schultz. Topological properties of hypercubes. Tech. Rep. YALEU/DCS/RR-389, Dept. of Computer Science, Yale University, Conn., June 1985.

[105] Schwartz, J. T. Ultracomputers. *ACM Trans. on Programming Languages and Systems* 2:484–521, 1980.

[106] Seitz, C. L. Concurrent VLSI architectures. *IEEE Trans. Comp.* 33(12):1247–1265, 1984.

[107] Seitz, C. L. Experiments with VLSI ensemble machines. *J. VLSI Comput. Syst.* 1(3), 1984.

[108] Siegal, H. J. Interconnection networks for large scale parallel processing, *Lexington Books*, 1985.

[109] Sorensen, D. C., J.-P. Brunet, and S. L. Johnsson. A data parallel implementation of the divide-and-conquer algorithm for computing eigenvalues of tridiagonal systems. Tech. Rep., Thinking Machines Corp. (In preparation 1989).

[110] Stout, Q. F. Sorting, merging, selecting, and filtering on tree and pyramid machines. *1983 International Conf. on Parallel Processing*, 214–221. IEEE Computer Society, 1983.

[111] Stout, Q. F., and B. Wager. Intensive hypercube communication: prearranged communication in link-bound machines. Tech. Rep. CRL-TR-9-87, Computing Research Lab., Univ. of Michigan, Ann Arbor, Mich., 1987.

[112] Stout, Q. F., and B. Wager. Passing messages in link-bound hypercubes. M. T. Heath, ed., *Hypercube Multiprocessors 1987*. Society for Industrial and Applied Mathematics, Philadelphia, Penn., 1987.

[113] Sullivan, H., T. Bashkow, and D. Klappholtz. A large scale homogeneous, fully distributed parallel machine. *Proc. Fourth Annual Symposium on Computer Architecture*, 105–124, 1977.

[114] Swarztrauber, P. N. Multiprocessor FFTs. *Parallel Computing* 5:197–210, 1987.

[115] Thompson, C. D. Area-time complexity for VLSI. *Proc. 11th ACM Symposium on the Theory of Computing*, 81–88. ACM, 1979.

[116] Thompson, C. D. *A complexity theory for VLSI*. Ph. D. thesis, Dept. of Computer Science, Carnegie Mellon University, 1980.

[117] Thinking Machines Corp. *LISP release notes, 1987.

[118] Trivedi, K. S., and M. D. Ercegovac. On-line algorithms for division and multiplication. *IEEE Trans. Computers* 26:681–687, July 1977.

[119] Uhr, L. Parallel, hierarchical software/hardware pyramid architectures. V. Cantoni and S. Levialdi, eds., *Pyramidal Systems for Computer Vision*. Springer-Verlag Berlin Heidelberg, 1986.

[120] Ullman, J. D. *Computational Aspects of VLSI*. Computer Sciences Press, 1984.

[121] Upfal, E. Efficient schemes for parallel computation. *ACM Symposium on Principles of Distributed Computing*, 55–59. ACM, 1982.

[122] Upfal, E., and A. Wigderson. How to share memory in a distributed system. *ACM Symposium on Foundations of Computer Science*, 171–180. ACM, 1984.

[123] Valiant, L. A scheme for fast parallel communication. *SIAM J. on Computing* 11:350–361, 1982.

[124] Valiant, L., and G. J. Brebner. Universal schemes for parallel communication. *Proc. 13th ACM Symposium on the Theory of Computation*, 263–277. ACM, 1981.

[125] Wilson, K. G. *Phys. Rev.* D10:2445, 1974.

[126] Zienkiewicz, O. C. *The Finite Element Method.* New York: McGraw-Hill, 1967.

CHAPTER 5

YASER ABU-MOSTAFA
DAVID SCHWEIZER
California Institute of Technology

Neural Networks

Introduction

These sections address the popular subject of neural networks, a delicate subject because of its history and diversity. *The sections are not, by any means, a comprehensive survey of the subject.* They are intended only to be an introduction to the field. Several important areas and results are not addressed here.

There are two very popular models of neural networks: the feedback model and the feedforward model. I will start with the feedback model because historically that is what triggered the current wave of interest. If we went back twenty-five years we would find interest in the same models. That wave, however, died out, and so did at least one earlier wave. Before we get to the specifics of the feedback model, however, we need some general discussion of neural networks.

5.1 Networks and Neurons

The architecture of neural networks can be described as an undirected graph. We call the nodes neurons and the edges synapses, and we have a neural network. What characterizes a neural architecture in general, whether it is feedback or feedforward, is that the number of neurons is huge, and each neuron does a very simple task. If we are considering neural networks as a subset of all distributed computations or parallel architectures, that is the main characterization. The number of units is very large, and the task of each unit is very simple, much simpler than the tasks performable by a microprocessor, for example. In many models the nodes perform threshold logic only. Furthermore, the number of synapses per neuron is large. Usually a neuron is connected to a large subset of all other neurons, for example, a fraction αN of all neurons, rather than $\sqrt{N}$ or $\log N$, as in other architectures such as the hypercube.

What does a neuron do? In both the feedback model and the feedforward model, the neuron performs a simple threshold function. It has N inputs, called $x_1 \ldots x_N$; and a single output, called y. For the moment, we consider all of these variables to be binary, and, for convenience, take the binary values to be -1 or $+1$ instead of 0 or 1. The neuron calculates a function from $\{-1, +1\}^N$ (i.e., N-tuples of -1's and $+1$'s) to $\{-1, +1\}$. The relation of the output to the input depends on a set of real numbers called the weights. There is one weight for each input variable, and what the neuron does is sum up the variables times the weights and subtract a threshold, t. If the result happens to exceed 0, it sets y to be $+1$; if it is less than 0, it sets y to be -1, and if it happens to hit zero, well, there are many technical variations of what to do.

Given this definition, the set of functions we can implement using a single neuron is well understood: It's the set of threshold functions, also called linearly separable functions. If we take the hypercube $\{-1, +1\}^N$, we can implement any dichotomy, that is, any hyperplane that separates the points on the hypercube.

There are other models for what the neuron can do, but this is the most popular one and the one on which both models, the feedback and the feedforward, are based, so this is the one I am going to consider.

5.2   Feedback Networks

The feedback network [1], which was introduced and reintroduced by several people, is a fully interconnected network of neurons, that is, every neuron is connected to every other neuron by a synapse. Let's call the states of the neurons u_i. Thus we have a vector of states, $(u_1 \ldots u_N)$. The rule is that the future state of neuron i is simply the value y, the output of the threshold function computed by that neuron.

For this particular model, we can express the entire network by the matrix W of weights simply because the thresholds have been chosen to be zero. That's part of the specification of the model, that all thresholds are zero. Furthermore, $w_{ij} = w_{ji}$: The weights are symmetric. The final restriction is that it has a zero diagonal, which means that the neuron is not connected to itself.

Now, how does the network operate? We have the neurons, and we can initialize them to anything we want, so they have initial states. A neuron decides to update its state, so it looks at its surroundings, multiplies its inputs by the right weights, evaluates the sum, and calculates the sign. (Recall that the threshold is zero.) If it happens that this sign is different from the current state, it flips to the new sign. These updates are done by all the neurons asynchronously, that is, no two neurons ever try to update at the same moment.

Conceivably, neurons could continue to do updates forever. It could be the case that we have chosen the weights and the initial states poorly. It might be the case that every time a neuron updates, it changes its sign, and that that change would require some other neuron to change its state, and so on. The network would go around and around, and never settle down to an unchanging state, and so on. Such a process would be useless. However, there is a result [1] that says that a feedback network is guaranteed to go to a stable state after a finite number of updates. There is obviously an assumption of fairness, that is, that every neuron will eventually get to update as many times as it wants.

Do the updates have to be random? For the proof that the network actually gets to a stable state, they must be asynchronous, but not necessarily random. If we were to allow two neurons to evaluate and update simultaneously, the network could end up in a limit cycle. However, we can scan the neurons sequentially, and the network will still reach a stable state.

How do we show that a feedback network will always go to a stable state? We will define a scalar-valued energy function, and show that the energy of the network always decreases. Let the states be represented by a vector of N bits, denoted $\mathbf{u}$. We propose the following function, which is called E for energy:

$$E = -\frac{1}{2}\,\mathbf{u}^T\,W\,\mathbf{u}.$$

A helpful way to organize our thoughts about the energy function is to write the N^2 terms of the summation as a matrix. This matrix corresponds directly to the weight matrix, with each term being $u_i w_{ij} u_j$ instead of just w_{ij}. In order to evaluate E, we add up all the terms in this matrix and multiply that sum by $-1/2$. It is convenient to keep this matrix, illustrated in Figure 5.1, in mind as we go through the proof.

Let's see what happens when neuron i updates. When neuron i updates, the only element of $\mathbf{u}$ that can change is u_i, and thus all the terms in the matrix will remain the same, except possibly the ith row and the ith column because these are the only places u_i appears.

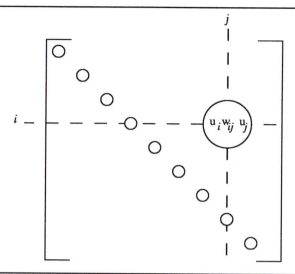

Figure 5.1   Matrix tabulation of the energy function.

Now suppose that neuron i changes state, that is, that neuron i calculates its threshold function and decides to flip the sign of u_i. We know that the new u_i is the sign of $\sum_j w_{ij} u_j$, and, since the neuron changed state, that the old u_i must have been different from that sign. Thus when we multiply the new u_i by the sum, we always get a positive number. When we multiply the old u_i by the sum we always get a negative number. If we put these facts together, and consider summing the entries in the ith row (or the ith column), we will find that ΔE, the change in energy, is always negative.

Why is this useful at all? This is useful because E has a lower bound, and the magnitude of ΔE for actual updates also has a lower bound. Both of these are clear because the matrix is finite, so there is only a finite number of combinations of the signed terms. Therefore, if each time the network updates, E decreases, the network is bound to reach a state where E cannot decrease further, that is, no further updates would change the state of a neuron, that is, a stable state of the entire network. This proves that with any update scheme using the rules we have, we must eventually get to a state where every neuron is happy. Every neuron looks at the states of the other neurons, multiplies them by their weights, sums the results, and finds that the sign of that sum agrees with the state it is already in.

When the feedback model is used for computation, the time evolution from the initial state to the stable state is usually what constitutes the computing. The output of the computation is deciphered from the stable state. One designs the network to perform a certain computation by choosing the weights so that the network dynamics fulfill the specification of some algorithm. Later, we will discuss computation on feedback networks further, but first we need to investigate our control over the stable states.

5.3   Choosing the Stable States

The mere convergence to a stable state is not very useful unless we have some control over the choice of stable states. To perform a computation, we need to be able to demand that particular initial states converge to particular stable states. We need an algorithm for constructing a network with a prescribed set of stable states. If we have a network that has N neurons, the only free parameters left are the weights. We want to take a set of vectors, $\mathbf{u}^1$ through $\mathbf{u}^K$, which we want to be stable states, and produce a set of weights, W, that makes them stable. We will use an algorithm known as the sum-of-outer-products algorithm [1]. This algorithm sets

$$W = \sum_{k=1}^{K} (\mathbf{u}^k \, \mathbf{u}^{k^T} - I),$$

where I is the identity matrix. Each term of this sum is an outer product, that is, a rank 1 matrix with the entry in the ith row and j^{th} column equal to $u_i u_j$. We subtract I from each term to make the diagonal be zero; this is just a technicality. This produces a weight matrix W, but we are supposed to make the $\mathbf{u}^k$'s stable, so we will argue shortly that this choice of W does indeed, with high probability, make them stable.

An interesting feature of this algorithm for the weights is that although it looks like an elaborate algebraic formula, the implementation on a neural network is very simple. In order to modify w_{ij}, we do not have to know the entire state vector, just the two bits of neurons i and j. Consider what neuron i and neuron j see when we load $\mathbf{u}^k$ as one of the states. Each of the two neurons gets a single bit of $\mathbf{u}^k$, and there is weight between them, w_{ij}. If the bits agree in sign, w_{ij} is incremented, and if they disagree in sign, it is decremented. If we do this for every weight, then when we have gone through all the $\mathbf{u}^k$'s, we have the sum of outer products. In a different scheme it could be that two bits do not suffice to calculate what w_{ij} should become, but here we just enter the first vector and adjust each w_{ij} using two bits and we get the first outer product. We input the second vector and accumulate the second outer product, and so on. This is a nice property because it is a local rule. This rule is called the Hebbian rule.

Now we are going to convince you that the $\mathbf{u}^k$'s are stable. Let us assume that the way the $\mathbf{u}^k$'s were chosen in the first place was by flipping a fair coin independently to determine each bit of each vector. This is the same as saying that the bits of the vectors are independent *Bernoulli trials*. Assume W was constructed by the sum-of-outer-products formula. Let's take one of the vectors, say $\mathbf{u}^1$, and check that it is stable. First we multiply W by $\mathbf{u}^1$. This product gives a vector of the signals seen by each of the neurons at the beginning. The first component of the vector is the signal available to neuron 1, the second is the signal available to neuron 2, and so on, up to the signal available to neuron N. If the signal seen by a given neuron is positive, that neuron wants its state to be $+1$; if it is negative, it wants its state to be -1. If these signals happen to agree in sign with the initial state vector that we want to be stable, we are in good shape. We want to go to $\mathbf{u}^1$, and we have a vector of real numbers that happens to agree with the sign of $\mathbf{u}^1$ in every entry. We are all set because when the neurons look at this signal, they will decide to stay where they are, which means that the state is stable.

So we multiply W by $\mathbf{u}^1$. Recall that W is

$$\sum_{k=1}^{K} (\mathbf{u}^k \, \mathbf{u}^{k^T} - I).$$

The first term involves the quantity $\mathbf{u}^1 \mathbf{u}^{1^T}$, which, as it is the inner product of a vector of -1's and $+1$'s with itself, is N. Thus the first term equals

$$N \times \mathbf{u}^1 - I \times \mathbf{u}^1,$$

which reduces to $(N-1) \times \mathbf{u}^1$. That's good news. We want the signals to agree in sign with $\mathbf{u}^1$, and we start with a signal that is $\mathbf{u}^1$ multiplied by a big positive number. Thus, unless the second term provides enough 'noise' to take one of the components all the way to the opposite sign, $\mathbf{u}^1$ will indeed be stable. See Figure 5.2.

When we evaluate the noise term, we find that we add up a number of independent Bernoulli trials to make up each component of the noise vector. For example, some of these Bernoulli trials come up when we compute $\mathbf{u}^1 \mathbf{u}^{2^T}$. Since, by assumption, the bits of $\mathbf{u}^2$ were selected at random independently from those of $\mathbf{u}^1$, we will end up with a variety of ± 1's that are unrelated. (We call these terms noise because the bits are unrelated.) On the average, we will have as many $+1$'s as -1's, and we expect a zero sum. However, we probably won't be that lucky and have the noise cancel out exactly. There will be some variance around zero, and the variance will be controlled by how many Bernoulli trials we add. As a matter of fact, the variance will be proportional to the number of Bernoulli trials, a result that is very simple to derive.

Thus, there are many Bernoulli trials that are not really conspiring against the stability of $\mathbf{u}^1$. They are giving arbitrary values, and although they are not conspiring, if there are too many of them the variance will be

$$Wu^1 = (u^1{u^1}^T - I)u^1 + \underbrace{\sum_{k=2}^{K}(u^k{u^k}^T - I)u^1}$$

$$= \underbrace{(N-1)u^1}_{\text{signal}} + \quad \text{noise}$$

$$\text{noise} = \sum \text{Bernoulli trials}$$
$$\text{variance} \propto \text{Number of trials}$$

Figure 5.2   Noise in the sum-of-outer-products.

so large that there is a possibility that one of them will swing some component all the way to the opposite sign. The number of Bernoulli trials grows with the number of 'parasitic' vectors $u^2, \ldots, u^K$. If this number is small, then the variance of the noise will not be sufficient to affect stability. How small should it be? That's the question of capacity [2]: how many states can be added without disrupting the behavior of the first stable state? The answer is that K can be at most the order of $\frac{N}{\log N}$ vectors [3].

5.4   Feedforward Networks

Feedforward networks are currently receiving more attention than feedback networks. The feedforward network is an architecture where the neurons are grouped in layers. There are directed connections between layers, with no loops anywhere and no connections among the neurons in the same layer. The network starts with inputs; these neurons simply hold the values of the inputs, x_i through x_N. These values are passed to the next layer through weights. The neurons in the second layer perform some computation, then pass that result to the next layer through weights, and so on. There can be many layers. The neurons in the first layer are called input units. They don't really do anything, so we could delete them and show the signals going into the second layer. The neurons in the final layer are called output units; they hold the result of the computation. The neurons in between are called hidden units because they're hidden. The relation between these networks and combination circuits is obvious. The main issue in this case is that the weights will be allowed to be real numbers.

If a feedforward network doesn't have hidden layers, there will be functions it cannot implement. The argument is actually trivial—it's the old

Perceptron [7] argument. Essentially, if the network doesn't have any hidden layers, it computes a single threshold function; if the output happens to be several bits, it computes several threshold functions. So suppose we want to implement the parity function of x_1 through x_N. Let's look at the case of two variables, x_1 and x_2, in Figure 5.3. The values $\{-1,-1\}$ and $\{+1,+1\}$ give even parity, and $\{-1,+1\}$ and $\{+1,-1\}$ give odd parity.

We would like to set the weights and the threshold. When we choose the w's, we will compute the inner product with the vector (x_1, x_2), and that will correspond to a plane. The points on one side of the plane will be mapped to $+1$, and those on the other side will be mapped to -1. We look at the hypercube, which in this case is a very simple one, and we want to be able to find a hyperplane that separates the *black* nodes, which we want to map to $+1$, from the *white* nodes, which we want to map to -1. In the case of the parity function, this is clearly impossible.

Adding a hidden layer turns out to be a very good idea. With a hidden layer, we can implement any Boolean function. The reason is very simple: We can implement any function using a canonical or-of-ands form, and we can implement those gates using threshold functions. (See Figure 5.4.) Thus, if we replace each logic gate with a neuron that computes the same function—and that is a very simple derivation—we end up with a canonical form that can implement any Boolean function of the inputs.

Although one layer suffices, sometimes it is convenient to use more than one. This is a topic in the current research, as is the number of units per layer. At this point, the rules are completely heuristic. We have intuitive arguments for what network structure to use, how many layers and how many units per layer, but no theoretical foundation whatsoever.

Now let's look more closely at the units. Again, we need a threshold

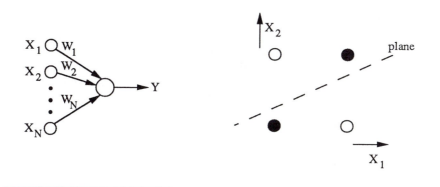

Figure 5.3   The parity is not a threshold function.

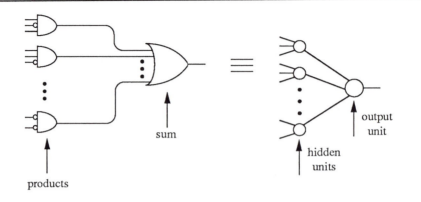

Figure 5.4   One hidden layer suffices.

function, but this time we have a feature that is sometimes very useful. In the feedback model we used a hard threshold: If the weighted sum of the inputs was positive, the output was $+1$; if negative, -1. In this model we are going to make a soft decision. If the weighted sum is very positive, the output will be very close to $+1$; if it's very negative, very close to -1, but if it is only slightly positive or slightly negative, the output will be of smaller magnitude. This nonlinear input-output relation, $y = f(\sum w_i x_i)$, has a sigmoid shape, as shown in Figure 5.5.

Only in the limit do these neurons compute Boolean functions. The final output will be passed through a hard threshold, but we would like to keep the reliability of intermediate decisions as the computation proceeds. There is no reason to make hard decisions in the intermediate stages. There is also another reason for having a soft decision, which will come from the learning rule. We will get there shortly. We also set the threshold to zero. A threshold can be added by having one of the inputs be $+1$ all the time, and using the weight on that input to give the effect of a threshold. Thus there is no loss of generality brought about by omitting the threshold.

Let us look at the learning rule. If all we were to do with multilayer networks was to take a function and implement it on a network, then we would have nothing more to offer than simple circuit design. We would even be restricting ourselves to stupid circuit elements when we might be able to build better ones. The whole idea of using multilayer networks is to be able in some sense to make the circuit build itself. We will start with just the architecture of the network: How many units and how many layers. And then we won't be given the function we want to implement, but only examples from that function. Given an input and output pair, we will use a learning rule to adjust the weights so that the circuit is more likely to give

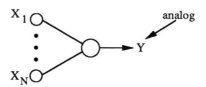

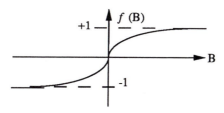

Figure 5.5   Nonlinearity of the neurons.

that output when it sees that input. We'll do that, example after example, until finally, we hope, we have a circuit that actually computes the original function.

Suppose we are training just one unit to learn a function. Assume the weights start at any random initial set of values. Given an input, $(x_1 \ldots x_N)$, suppose we want the output to be $\hat{y}$. We feed the inputs as they are to the circuit. We have no idea what the output will be, but it's a starting point. Now we get something that could, by chance, very closely agree with the desired output, or could disagree terribly. Or, as the output is an analog value, it could be in between. Now we would like to perturb the w's so that after the perturbation, if we re-enter the inputs $(x_1 \ldots x_N)$, the output will be closer to $\hat{y}$.

Unfortunately, these perturbations may disturb a value the circuit has already learned. We use the same noise argument as in the feedback model. It's a small perturbation and is optimally directed toward accommodating a particular input/output pair. If the other pairs are unrelated, we have a good chance that this behaves as noise for them. Empirically, focusing on one pair at a time and doing several runs actually works.

5.5   Back Error Propagation

Now, how do we perturb the weights? Let's assume we have an error
function, not specified at the moment but well defined for any input/output
pair. The error is a function of the weights $w_1, \ldots, w_N$ in the network:

$$E = E(w_1, \ldots, w_N).$$

Some sets of weights will give one value for the error function, and other
sets will give smaller or larger values. We will perturb the weights in a
manner that depends on the error, and would like to make ΔE, the change
in the error, of large magnitude. It's a gradient, so we'll take it negative or
positive. The size of the perturbations permitted is bounded by

$$\sum_i (\Delta w_i)^2 = const.$$

(This limits the distance moved in weight space.)

The problem is how to decide how much to modify each weight to maximize ΔE, subject to the constraint. We begin by assuming that we can
approximate the change in the energy by:

$$\Delta E = \sum \frac{\partial E}{\partial w_i} \Delta w_i.$$

Since we want to maximize ΔE subject to a constraint, we will use the
method of Lagrange multipliers. We set the term

$$\frac{\partial}{\partial \Delta w_i} \left[\sum \frac{\partial E}{\partial w_i} \Delta w_i + \lambda \left(\sum (\Delta w_i)^2 - const. \right) \right]$$

equal to 0 for all i from 1 to N, solve, and end up having to perturb each
weight by an amount proportional to the rate of variation of E with respect
to that weight, but in the opposite direction. That is, we get

$$\Delta w_i \propto -\frac{\partial E}{\partial w_i}.$$

This is a very satisfactory result. If the error is extremely sensitive to one
of the weights, then we will perturb that weight a lot because it will affect
E a lot. If it doesn't matter, then we shouldn't bother because we have a
constraint on the total perturbation. We want to make our modifications
where they will have the most effect. We will go for the bigger ones, and
although we might think at the beginning that we would want to pick the
largest one and perturb it all the way, we now know that isn't optimal. Note
that this is independent of the choice of the error function. We know now

if we have any error function and would like to change the weights, with this restriction on the perturbation, such that we get closer to the correct output, we should pick the perturbation proportional to the negative of the derivative.

This gives rise to what's called the Delta rule [4]. First, we need a specific error function. A good choice is the mean square error,

$$E = \frac{1}{2}\left(\hat{y} - y\right)^2$$

where $\hat{y}$ is the target output, and y is the actual output. We could start with another function and would end up with a different rule. Mostly, however, we use this one. For convenience, we will write θ for the weighted sum of the inputs to the neuron, that is,

$$\theta = \sum_i w_i x_i.$$

We still have a sigmoid for the output; recall Figure 5.5. We already have

$$\Delta w_i = -\alpha \frac{\partial E}{\partial w_i}.$$

Now,

$$-\frac{\partial E}{\partial w_i} = -\frac{\partial E}{\partial \theta}\frac{\partial \theta}{\partial w_i}.$$

(The term $-\frac{\partial E}{\partial \theta}$ is called δ, and that's why this whole procedure is called the Delta rule.) Further,

$$-\frac{\partial E}{\partial \theta} = -\frac{\partial E}{\partial y}\frac{\partial y}{\partial \theta} = -\left(\hat{y} - y\right)f'(\theta),$$

and obviously

$$\frac{\partial \theta}{\partial w_i} = x_i.$$

All of this together yields our rule for changing the weights:

$$\Delta w_i = \alpha\left(\hat{y} - y\right)f'(\theta)x_i = \alpha\delta x_i.$$

Now let us explain this intuitively; we don't have to do the math. First, if we are way out on one of the tails of the sigmoid, we don't care what happens to w because a change in w will barely affect the output. The neuron is in saturation. So if $f'(\theta)$ is close to zero, we just forget about it. But if we are near the origin, that's very important, because perturbing w could drive the output to the positive side or the negative side. In that case, the change is very important. The rest of the result is simply agreement in sign. If the target output minus the actual output agrees with x_i, we would like to increase w_i because that increase will push θ in the direction we already want. If they disagree, the negative sign means the weight will be decreased, and that will drive it in the other direction. Thus, at least in the first-order checks, we see that this is a plausible formula.

Now we can see another reason for maintaining analog values instead of using a hard threshold, and that is to be able to use intermediate information. When we want to adjust things slightly, we would like to have some flexibility in the perturbation. If the output is required to be $+1$ or -1, we either leave it entirely unchanged or flip it completely. There is no measure for how severe or mild the perturbation is. But in this case, we have a spectrum of values to use, which is usually useful to have around in these algorithms.

The generalization of the Delta rule for the multilayer networks is the famous Back Error Propagation Algorithm [4], and it is because of that algorithm that a lot of attention is paid to multilayer networks.

Suppose we have a fragment of a multilayer network that looks like Figure 5.6. We will again be using gradient descent. We start with the equation

$$-\frac{\partial E}{\partial w_{ij}} = -\frac{\partial E}{\partial \theta_j}\frac{\partial \theta_j}{\partial w_{ij}}.$$

Note that since E is indeed affected by a perturbation of w_{ij} (even though in a rather elaborate way), it is legitimate to define the partial derivative of E with respect to w_{ij}. Gradient descent requires that we make Δw_{ij} proportional to that derivative. Note also that the only way w_{ij} comes into the equation is by affecting the input signal to the neuron, so it is legitimate to apply the chain rule as we have.

There are now two cases: Either j is an output unit, or it is a hidden

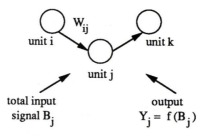

Figure 5.6   Intermediate unit.

unit. If j is an output unit, then we use the Delta rule. Once we have δ_j, we know how to compute Δw. So there is nothing new here, if j is an output unit. The situation becomes very interesting when we are dealing not with an output unit, but with a hidden unit.

In that case, the formula is much more elaborate. As before, we start with

$$\delta_j = -\frac{\partial E}{\partial \theta_j}.$$

We've added some subscripts, but the definition is exactly what it was. We can apply the chain rule and we get

$$-\frac{\partial E}{\partial \theta_j} = -\frac{\partial E}{\partial u_j}\frac{\partial u_j}{\partial \theta_j}.$$

The second term of the product is easy: it's just $f'(\theta_j)$. For the first part, we need to remember that we have more layers after this neuron. We can write

$$\frac{\partial E}{\partial u_j} = \sum_k \frac{\partial E}{\partial \theta_k}\frac{\partial \theta_k}{\partial u_j},$$

where k indexes the neurons in the next layer. The second term of the product is easy: The rate of change of the output of a neuron with respect to one of its inputs is just the weight on that input line, w_{jk}. Now comes the trick. We don't know very much about $\frac{\partial E}{\partial \theta_k}$, but it's enough to be able to write

$$\frac{\partial E}{\partial \theta_k} = -\delta_k.$$

We put it all together and get:

$$\delta_j = f'(\theta_j) \sum_k w_{jk}\delta_k.$$

We already know how to compute the errors at the output layer. Once we've done that, we can move back one layer and compute the δ's there. Notice that the summation we need to compute looks just like the one the neuron uses to compute θ. That's why the algorithm is called back error propagation. We propagate answers in one direction and errors in the other (see Figure 5.7).

So, how well does this actually perform? It performs extremely well on toy examples. It does so extremely well that it not only implements the examples it has seen, but also extrapolates correctly in many cases for examples it hasn't seen [4].

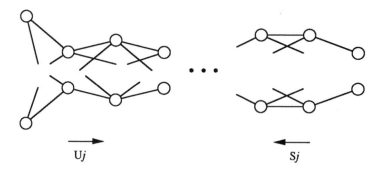

Figure 5.7   Back error propagation.

5.6   Collective Computation

The idea of collective computation [1] is to work with a very large system made up of a lot of simple components. The components may be unreliable: They could execute their instructions inexactly or even die completely. But the hope is that the overall system will reliably perform some sophisticated task despite the individual failures.

Do feedback neural networks perform that kind of computation? In some sense, yes. See Figure 5.8. During the programming phase, when we load in a state we want to be stable, the change in the weight on each

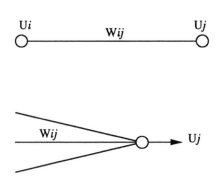

Figure 5.8   Simple rules leading to sophisticated function.

synapse depends only on the states of the neurons at each end of that synapse. This is physically implementable by burning, or not burning, a wire, a simple operation. But collectively, the neurons make a specific set of vectors stable in the network. That may not be too sophisticated, but it is definitely more sophisticated than the individual operations.

During the computation phase we would like to initialize the network to some state and have it converge to one of the stable states. Although there are no theoretical results, we find that experimental [1] networks—where the weights are not very exact, and some of the neurons fail, and some of the neurons are very lazy about updating, and so on—actually operate correctly. The overall operation is reliable, although the individual components are not.

Now let's look at two specific examples of computation on neural networks. We would like to justify the examples with an analogy. Suppose we were to go back in time 40 or 50 years and try to sell someone on the idea of computers. We would like to claim that computers can do wonderful things, but as we don't have a whole system, we can only give an example. So we build a full adder from nand gates. That's simple enough, but it's rather surprising the first time one sees it. It looks very good because the full adder performs an operation that isn't very simple, and it demonstrates the idea of computers. Unfortunately, we couldn't really say that once we have a full adder, we have all kinds of computers. The question for neural networks is similar: We have this model, which does some very specific tasks quite nicely, but can we extrapolate further?

We are not claiming that these two examples are good applications or bad applications. We just want to state what the applications are to give you some idea of the use of these models for computation. We know that the individual components are simple and inexact because we build them that way. The rest of the story comes at the system level.

5.7   Nearest Neighbor Search

Our first example is the nearest neighbor search problem. In this example, the neural network solution will be apparent to you as soon as we state the problem. We have a fixed set of M binary vectors of length N. The problem is this: Given an N-bit binary vector, find the vector in the set that is closest to it in the Hamming sense. This problem comes from the theory of error-correcting codes. The binary vectors are the legitimate code words that could be transmitted over a channel. When a message is sent over a noisy channel, what is received is only an approximate version of what was sent. Finding the code word closest in the Hamming sense to the received message is a maximum likelihood decoding.

So here we have a useful problem: optimal decoding for linear codes.

Linear codes are a subclass of all codes and have the property that the code can be specified without listing all the code words. There is a simple matrix from which all code words can be generated. The whole problem can be expressed by something of size polynomial in N. Performing a maximum likelihood decoding in a linear code is $\mathcal{NP}$-complete in N. We emphasize N because, in the case of linear codes, the list of code words $\mathbf{x}_1$ through $\mathbf{x}_M$, is already of exponential length. Even if the computation runs in time polynomial in M, that's still exponential in N. The problem may still be easy in terms of M, if M is not too large.

The feedback network solution is what one would expect. Take the code words, that is, the vectors $\mathbf{x}_1$ through $\mathbf{x}_M$, and use the sum of outer products algorithm to make them be the stable states of the network. To do maximum likelihood decoding, simply initialize the network to the received vector, let it run, and after a while it will converge to the closest code word. We have not proven to you that it will converge to the closest, but it has been demonstrated experimentally.

The problem is that the capacity of a network of N neurons is of order $N/\log N$. That is, this scheme will work properly as long as the number of code words is, at most, essentially of order $N/\log N$. Let's be generous and call it order N. Now, for a linear code, or any other reasonable code, the number of code words is exponential in N. If we were to apply the sum of outer products to exponentially many code words, we would end up with a very noisy matrix that would exhibit all kinds of irrelevant dynamics.

But the number of code words is big. What can we do? We'll just make N big. We would like the length to be the same as the number, so we will pad every code word with a bunch of extra bits, say +1's. Then apply the sum of outer products to set the weights. Each time we receive a vector, we'll initialize the network to the vector we received, padded with +1's, and let it converge. That will definitely do it, but now we have a problem: The network size is proportional to M, which is exponential in N. The exponential pops somewhere every time we try a new solution. We will argue that this is not a coincidence. *When we try to use neural networks to solve a problem that is inherently exponential, the size of the network has to be exponential.*

5.8   Traveling Salesman Problem

Now let's look at the Traveling Salesman problem [5]. When we looked at the nearest neighbor search problem, we were able to embed the problem directly into a neural network. A typical Traveling Salesman problem is stated in a way that doesn't lend itself to any direct implementation, so we have to do something to make it fit. The problem is this: Given a list of M cities (labeled 1 through M) and a matrix $[d_{xy}]$ of intercity distances, find a minimum length tour of the cities. That is, find a permutation $x_1 \dots x_M$ of $1 \dots M$ that minimizes

$$\sum_{m=1}^{M} d_{x_m x_{m+1}},$$

where we take $M + 1$ to be 1. There may be more than one solution.

How can we solve this on a feedback neural network? In general, if we have a problem that doesn't fit directly into a feedback network, we use the energy as a catalyst to make it fit. The steps are the following.

First, we have to encode the problem into a feedback neural network. Feedback neural networks have weights and initial states, and problems have inputs. We have to make them correspond. Also, problems have solutions and neural networks have stable states, so we must be able to encode the solution to the problem as a stable state. We just have to choose some convention. Sometimes the choice of that convention is crucial.

In this case, we have an optimization problem, so the second step is to find a scalar function such that minimizing that function will yield a solution to the problem. Find just one scalar such that the absolute minimum of that scalar corresponds to a minimum tour.

Finally, consider the scalar function to be the energy function of a network. Since energy has a specific expression in terms of the states and weights of the network, if the scalar function happens to be in the right form, we will end up with a correspondence between the weights and thresholds of the network and the constants that arise in the problem. This is abstract. We will apply it directly to the Traveling Salesman problem and you will see the application.

There is one small change from our earlier discussion of feedback networks. In the proof that a network will converge to a stable state, we didn't have any internal thresholds—they were all set to 0. It gives us some more room to encode different scalar functions if we allow linear threshold terms as well. The new expression for the energy is

$$E = -\frac{1}{2} \left(\mathbf{u}^T W \mathbf{u} - 2 \mathbf{u}^T \mathbf{t} \right),$$

where $\mathbf{t}$ is the threshold vector. The energy function is still monotonically decreasing with every update, and exactly the same analysis applies.

Now we'll look at the feedback network encoding of the Traveling Salesman problem. If there are M cities, we will build a network with M^2 neurons. For the moment, think of the neurons as being arranged in a matrix and labeled by a double index. We can relabel everything in a linear manner later.

The rows of the matrix correspond to the cities and are indexed by x, $1 \le x \le M$. The columns correspond to the rank, that is, the order in the tour of each city, and are indexed by i, $1 \le i \le M$. (See Figure 5.9.) We

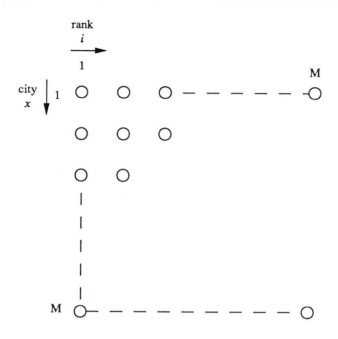

Figure 5.9   Traveling Salesman matrix.

require that the solution be a permutation matrix, that is, that there is exactly one +1 in each row and in each column. If there is a +1 in, for example, the fifth position of the first row, then city 1 is to be visited fifth on the tour. If we want to read off the tour, we find the +1 in the first column, which indicates the first city to visit, then the +1 in the second column, and so on. If the matrix is a permutation matrix, then this will be a legitimate tour, because every city will be visited once and only once.

Now we need to find a function to optimize such that optimizing that function will correspond to solving the Traveling Salesman problem, and then equate that function with an energy function. We might as well call the function an energy function right away. In the expression of the energy function, we are going to use $\{0, 1\}$-valued variables, v_{xi}, instead of the $\{-1, +1\}$-valued u_{xi}. It's easier to express, and at the end we can substitute the expression

$$\frac{1}{2}(u_{xi} + 1)$$

for each occurence of v_{xi}. The energy function will incorporate four conditions, each of which describes some part of the requirements of the problem.

The first condition is that a solution have at most one 1 per row. (We actually need exactly one 1 in each row, but we'll deal with that requirement in a moment.) Consider the expression

$$E_1 = \sum_x \sum_i \sum_{j \neq i} v_{xi} v_{xj}.$$

We sum over all cities over all pairs of distinct columns. If there is at most a single 1 in each row, then every term will be 0. Furthermore, since the v_{xi} are non-negative numbers, $E_1 \geq 0$. So here is a function which is greater than or equal to 0, and furthermore, is exactly 0 if and only if there is at most one 1 per row. That's a function I'd like to minimize.

The second condition is just the same—one 1 per column. We use the same idea, and get

$$E_2 = \sum_i \sum_x \sum_{y \neq x} v_{xi} v_{yi}.$$

Nothing to it.

We need exactly one 1 in each row and each column. If we have at most one 1 in each row, and at most one 1 in each column, and M 1's in the whole matrix, then we're done—we have a permutation matrix. The expression for the third condition is:

$$E_3 = \left(\sum_x \sum_i v_{xi} - M \right)^2.$$

If there are M 1's, this expression is 0. If there are more or fewer, it is positive. So now we have three conditions that we would like to minimize.

We could write an entirely different set of conditions to accomplish the same purpose, and that set might make the network perform better or worse. But the claim we will make later is that once you choose the size of the network, there is no good way of doing it. The network is too small.

The fourth condition is the real condition. The conditions we have written so far take care of only the syntax—they make sure that the solution we read off is a tour. Now we would like to minimize the length of the tour, so we look at this expression:

$$E_4 = \frac{1}{2} \sum_x \sum_{y \neq x} \sum_i d_{xy} v_{xi} \left(v_{y,i+1} + v_{y,i-1} \right),$$

where (in the subscripts) we take $M+1$ to be 1, and $1-1$ to be M. We start with the input matrix of distances d_{xy}. Whenever v_{xi} is 1, the distance will matter, because $v_{xi} = 1$ means that we pass through city x at time i. Now suppose that $v_{y,i+1} = 1$. This means that city y is visited at time $i+1$. The tour goes from city x is to city y, and therefore the distance between x and y matters. The same reasoning applies to the preceding city; the 1/2 in front fixes counting everything twice. This is exactly the way it was done originally, and we are doing it the same way in order to follow experimental results. If we minimize this expression, we will minimize the length of the tour—if the matrix of v_{xi}'s is a permutation matrix. If the matrix has 1's all over the place, then this expression corresponds to complete nonsense; but if we have a tour, and we minimize E_4, then we have the minimum tour.

Now we take a big step. We would like to minimize E_1, E_2, E_3, and E_4. If we absolutely minimize each of them, then we will get an exact, correct solution: the minimum tour. But unfortunately we have only one energy (because we have only one network) so we have to combine the four components and hope for the best. We use a linear combination:

$$E = AE_1 + BE_2 + CE_3 + DE_4.$$

Recall that E_1, E_2, and E_3 make the solution a tour, and E_4 makes it of minimal length. Suppose we set A, B, and C to one million, and set D to 1. We are sure to get a tour. It could be disastrously long, but it will be a tour. If we set A, B, and C to 1, and D to one million, then we would get a wonderful non-tour. We can spend some time choosing the constants optimally, and all kinds of questions about stability and getting stuck in local minima arise, but at least we have an energy function.

If we look at all the expressions, we find that we can write the energy in the canonical form

$$E = -\frac{1}{2}\left[\sum_x \sum_y \sum_i \sum_j \alpha_{xyij} u_{xi} u_{yj} - \sum_x \sum_i \beta_{xi} u_{xi} + \delta \right].$$

Originally, we wanted to design a network. Since we know that the energy function corresponds exactly to the network, we can simply read off the coefficients as the weights and thresholds. Whatever is multiplied by $u_{xi} u_{yj}$ becomes $w_{xi,yj}$, the weight between them. Whatever is multiplied by u_{xi} is t_{xi}, the internal threshold. We can forget the constants, since if we minimize an expression, we minimize that expression plus a constant. That's the solution to the Traveling Salesman problem. We have a network, that—cross your fingers—will solve the Traveling Salesman problem.

Now for some remarks. These are not meant to be positive or negative; they are just observations.

First, the network size is quadratic in the number of cities. Next, since we are minimizing a linear combination of the conditions, we are not sure that any one of the energies is actually minimized, and therefore we are not sure that the solution is legitimate. Even if it is legitimate, we are not sure it's optimal. We have a curious situation: Suppose we run this network and get an illegitimate solution. We visit city 5 in positions 3 and 5. How does one translate that into an optimal tour? This example actually happens some of the time.

In the description of the feedback model, we took the non-linearity of the neurons to be hard threshold. When optimizing, it is advisable to soften things because every hard decision creates stability problems. If we choose a sigmoid with a reasonable slope for the non-linearity, we will end up with a solution. It can be proved that if the slope is not too low—higher than the line $y = x$ will do—then the network will always end up in a stable state of, asymptotically, +1's or −1's. We won't get the network stabilizing to a state with values like .7 and −.8. We will always gets +1's and −1's eventually, at least asymptotically, and that helps the convergence greatly.

We need a new set of weights for every problem instance, even if the number of cities doesn't change. If someone gave us a feedback network and told us that we could use it for a computation, we would expect that the problem was embedded in the weights and that the network would take any problem instance of that size as initial values of the neurons and return the solution as the final values of the neurons. But here we have to change the weights in order to load a problem instance.

The choice of A, B, C, and D, is entirely heuristic and absolutely crucial. If we choose them wrong, we end up with a disaster. Even the simplest heuristic algorithm for solving the Traveling Salesman problem will yield a better result than the neural network method, if these constants are wrong. However, it was experimentally observed that the solutions this method gets are very good. When it was tried on $M = 10$—this example was reported—people were very enthusiastic.

But it has been observed experimentally [5], that when M is larger—and never mind 100, 15 is enough—the solution collapses in a ridiculous way. The method hardly ever gives a legitimate solution, and even when it does give a legitimate solution, it gives an inadequate tour. Something is just not working.

5.9   Limitations

We are going to ask a little bit theoretically, whether we stand a chance when we try to solve a hard problem, let's say a hard optimization problem, on a neural network. Maybe we can actually answer the question before we try several experiments. We are going to investigate the network size and the computation time as they relate to the problem complexity [6]. It may be that complex problems need bigger networks, or maybe we need to wait more, or something of this sort. As Chuck Seitz noted, theoreticians love simulations, so we have a simulation here.

First we are going to ask ourselves how quickly the neural network converges to a solution. It has been observed that it converges to a solution extremely quickly—sublinearly. But since we promised theory, we will use pessimistic estimates, which we can prove. We took the case with finite resolution for the weights, which is true for all the experiments that have been run. Nobody uses real numbers as genuine real numbers. If each weight has a finite number of levels, big as they may be, each term in the energy function assumes a finite number of values. The energy function has N^2 terms because of the double summation, and we end up with the energy function having the order of N^2.

We noted that every update decreases E. If something has, at most, order N^2 levels, and every step decreases it, then by the time we get to the minimum of it, we must have not spent more than the order of N^2 units of time. So the time for convergence is at most N^2.

We come to the real question. Is this good or bad? It looks like a good thing to have a system find the solution quickly. Except that it's what you consider to be constant, and what you consider to be variable, because you can always get the solutions quickly by getting wrong solutions. There's nothing to it. Cut the computation when you are tired, and read off the "solution," before it has converged.

Suppose we try to simulate a neural network. We want to simulate a full computation of a neural network of N neurons. We are doing it using a very simple sequential machine, so we have N^2 iterations at most to worry about. We are going to consider that we don't have a built-in threshold of N variables. We are going to consider it variable by variable and add them up, and that takes linear time.

We are going to be very pessimistic. We are going to scan the neurons one by one to look for one to update, and each time we scan them, the last one we look at is the one to be updated. Each update takes N evaluations to compute. Thus, we can simulate the entire computation of a feedback model in order N^4 cycles.

Now let's say that the network does solve a truly hard problem. Simulation will usually tell you that there is no magic, because whatever you can do, we can do. Maybe it will take us longer, but the mystique of having a new system solve a class of problems which we never solved before can always be killed by simulation arguments. The usefulness of the model will depend on whether the simulation overhead is big or small. If it's big, the model may be worthwhile, because the simulation is not that effective. If it's small, then it may not be worthwhile at all.

So suppose we have a genuinely hard problem with proven exponential time complexity. Most of the $\mathcal{NP}$-complete problems are conjectured to be in that class. So we have the time complexity $t(n)$ lower bounded by some constant times α^n. We are using big N for the number of neurons, and small n for the size of input. If we have a solution using a neural network, then give us the neural network that you used, and we are going to simulate it. When we simulate it, it will take us $O(N^4)$ steps. However, we just observed that this problem cannot be solved in less than α^n steps, and by simulating your solution, we were able to solve it in $O(N^4)$ steps. It must be that $O(N^4)$ is at least α^n. If the time complexity of a problem is at least N^{100}, it means that no solution whatsoever will take less than N^{100}. If a solution is found on a different machine that we happen to have simulated in N^{50} steps, then the complexity of the problem could not have been N^{100} because here is a solution which took only N^{50}. This implies that N has to be exponential in n. *So if you ever succeed in solving a problem that is exponential-time using a neural network, then you must have used an exponential number of neurons.* That's why we said that converging to a point quickly was not a good idea, because if the exponential is there, you will either absorb it in the time, or absorb it in the size. In this case you absorb it in the size, which is unfortunate, because you are building this whole network to solve an instance of a problem, and we would rather wait for the solution than fill a room with the network!

Bibliography

[1] J. J. Hopfield, "Neural networks and physical systems with emergent collective computational abilities," *Proc. Natl. Acad. Sci. USA*, vol. 79, pp. 2554–2558, 1982.

[2] Y. S. Abu-Mostafa and J. St. Jacques, "Information capacity of the Hopfield model," *IEEE Trans. Inform. Theory*, vol. IT-31, pp. 461–464, 1985.

[3] R. J. McEliece, E. C. Posner, E. R. Rodemich, and S. S. Venkatesh, "The capacity of the Hopfield associative memory," *IEEE Trans. Inform. Theory*, vol. 33, pp. 461–482, 1987.

[4] D. E. Rumelhart, G. E. Hinton, and R. J. Williams, "Learning internal representations by error propagation," in *Parallel Distributed Processing, vol. 1.* Cambridge, MA: MIT Press, 1986.

[5] J. J. Hopfield, "Collective computation, content-addressable memory, and optimization problems," in *Complexity in Information Theory*, Y. S. Abu-Mostafa (ed.), Springer-Verlag, pp. 99–114, 1988.

[6] Y. S. Abu-Mostafa, "Neural networks for computing?" in *Neural Networks for Computing*. New York: AIP Conf. Proc., vol. 151, pp. 1–6, 1986.

[7] P. Lewis and C. Coates, *Threshold Logic*, John Wiley, 1967.

[8] J. Stephen Judd, "On the complexity of loading shallow neural networks," *Journal of Complexity*, vol. 4, pp. 177–192, 1988.

CHAPTER 6

RICHARD LYON
Apple Computer

VLSI and Machines that Hear

6.1 VLSI Complexity and Area Cost

I would like to make some comments on what Chuck Seitz has observed about VLSI complexity and cost. It's interesting that the people who study VLSI complexity tend to emphasize the AT^2 bound. The reason is, I think, that the question they are addressing is *How does the area grow when you try to make your circuit faster and faster?* Or formally, how does area grow as a function of decreasing T? The answer is that the area must grow as $O(T^{-2})$ or speed^2. If you can make the circuit really fast, T gets very small, and the area required goes up faster than linearly with the speedup.

But if you look at where the AT^2 bound comes from, you can find other components to the area cost. The area required by a circuit to work a problem can be expressed better using three terms, as discussed below:

$$A \geq \alpha/T^2 + \beta/T + \gamma,$$

where α, β, and γ depend on the problem.

The α term dominates and provides the bound $AT^2 \geq \alpha$ for fast circuits. This term comes from all the wires that you need to interconnect things if they are all going to talk to each other in a hurry and solve your problem really quickly; i.e., it expresses the area associated with interconnect wiring.

Consider a binary multiplier. If you ask some of the hard-core engineers about this problem, they may never have studied its complexity. What they may be able to tell you is that the cost of doing a multiply has to do with how many binary full adders you need, which you can work out in any one of several ways. You can use one binary full adder N^2 times, or you can use N^2 of them in an array once each, or you can build a serial pipeline structure or a serial-parallel structure that uses N of them N times. According to this view, if you want to get the job done quicker by a factor of N, you pay for it by using an area that is N times larger. You need that many more full adders to get the job done, and so the area required is, at least, some constant times $1/T$. There are a lot of problems you can look at in this way, and it is a very different model of complexity. Basically there is an area that has to do with the computation elements (the gates) themselves, and leads to the bound $AT \geq \beta$.

For the slow end of the problem spectrum, suppose somebody says *So, if I want to make a really small multiplier, I can use one full adder N^2 times.* When you actually build it, it turns out to be much larger than one full adder, because you need to have shift registers that hold the operands and cycle them around, and so on. It turns out that binary multiplications and FFT's and a lot of other (so-called transitive) problems have this property: *All or most of the output bits depend on all or most of the input bits.* You have to store all the inputs before you can get the first output, and you have to continue to store the inputs until you get the last output. This means that for any organization of the circuit except the purely parallel combinational one you have to have at least N bits of memory in there somewhere, where N is some measure of the size of the problem for multiplying, doing an FFT, or whatever. Thus there is an area associated with memory cells leading to the bound $A \geq \gamma$.

This is all for a fixed problem size. I haven't said how this scales with problem size, but for a fixed problem size, there is a constant amount of area associated with the memory required. The cost that Chuck was talking about is the amount of silicon you need if you want to get your problem done in a set time. You might say that the real cost is the silicon area you need multiplied by how long you use it. You can make a total area-time cost bound by multiplying each of the above area terms by T and adding them up:

$$AT \geq \alpha/T + \beta + \gamma T.$$

This is more or less the cost of building a chip to do a fixed-size problem in time T. The maximum, rather than the sum, is a looser bound because no matter how you do it, you need some wires, you need some gates, and you need some memory. Each one of these bounds is independently correct, and taken together they imply that AT exceeds the sum (at least in practice, if not strictly in theory). Either way you look at it, the way you are going to minimize this cost, if what you are going to do is vary how long it takes to do the problem, trying to get the AT product minimized, is by making the contribution of the α/T term equal to the γT term. The actual work is done by the gates, and this is now a constant.

These are deviations from what some hard-core engineers would say is the ideal measure of what it takes to do the problem, which is the logic gates. You need to support those gates with wires and memory, and you want the overhead of wires to be equal (roughly) to the overhead of memories in order to minimize the cost. That is why you don't want to push things into extremes of either speed or size.

I've had a habit of building bit-serial pipelined multipliers to build signal-processing machines. I often tell people to look at an AT^2 bound and use that as a motivation as to why you might want to build semi-slow multipliers, instead of using the fast-array multipliers you can get off the shelf. I think that if you look at the pipelined multiplier more carefully, you find that the wiring overhead and the memory overhead are roughly of the same order when you use N adders and N timesteps to do an N-by-N-bit multiply.

So, for any particular problem, you can look at how these things go with your architecture, how you trade off time and space and try to equate roughly those support costs in order to get an optimum.

Of course, the cost can actually be a much worse than linear function of area; say, a very high power or an exponential, due to yield considerations. In addition to total area, you need to consider partitioning into reasonable sized pieces of silicon. When you partition across chips, when you put together boards in a system, and so on, you've essentially got a third dimension to work in. This makes the models somewhat less relevant. The models that work on VLSI for wiring complexity, and so on, also work reasonably well on PC boards with any number of layers. But if you're allowing thick wiring mats, then you really need a different model to measure how much it is going to cost to build the thing. This is why models are useful. You scope out a particular domain and describe it by a model. Then you can analyse all kinds of trade-offs.

Chuck mentioned the idea of distributing a computation. If you have an FFT, you can run it sequentially on one very fast multiplier, or you can run it in the same amount of time by using many slower multipliers. If you look at multipliers, a fast multiplier is itself a distributed implementation of multiplication. If you have many elements in the distributed array multiplier, why isn't it a win? The answer is complicated because not everything you do to distribute computations is going to make them run really quickly and efficiently—especially if long carry ripple paths are involved.

Embedding a fast but cost-inefficient parallel array multiplier inside a slow FFT architecture is probably not a winning design strategy. That's why Chuck's idea of distributing the FFT problem itself may be better.

Trade-off spaces are very complicated. When you distribute things and go for massive parallelism, you have to try and figure out how you are going to get proportionate returns for your replication. This is not a simple question. Of course, if getting the problem done in a hurry is worth more to you than the cost of silicon area, then maybe AT is not a good model of your costs.

If you look at the AT cost bound, and you really want to minimize it, it turns out that you want the first and last terms to be equal. That gives you an exact solution for how long it should take to complete the problem, which is $\sqrt{\alpha/\gamma}$. Therefore it doesn't depend on how many gates it takes to work the thing, it only depends on trading off wires and memory and balancing their costs.

If you are within an order of magnitude of this solution, then your design is probably optimized to within the error of your model. People can never really determine these coefficients very well because they depend on particular technologies and particular circuits. For example, how big is a memory cell? It makes a huge difference if you are talking about a one transistor dynamic cell (often on a highly engineered RAM chip) or a two-port eight-transistor register cell in a register file on a processor chip. These coefficients can all change by more than an order of magnitude depending on how you design your circuits and layouts. The models give you a weak guidance, but it can often be a very useful guidance.

In his thesis, Clarke Thompson took an extra parameter P to represent how many problem instances something was working on at once [5]. You can then make a nice model based on AT/P (which is what the cost per problem is to us). When you start getting into things like array multipliers that have internal pipelines, they can become very efficient again. But more typically P is one or two, because there is a big memory cost in the pipeline latches you need to get it much higher, which can complicate things. The higher latency of this class of AT-efficient pipelined circuits is another type of cost that may be a problem in the architecture one level up.

In summary, there is good reason to study VLSI complexity theory and then to go ahead and interpret that theory in terms of real engineering practice—but beware of trying to apply interpretations that may be too simplistic for the problem at hand.

6.2   RAMs and Circuits

6.2.1   Analog Circuits

One of the questions that Yaser Abu-Mostafa asked in Chapter 5 was *What is analog good for—what does it buy you in neural networks, etc.?* That's not a question I am going to answer, but I want to discuss analog. Yaser pointed out that in neural networks, there are various ways that analog values could help make systems more stable or robust, or whatever. I think that may ultimately be important. Probably it is important in affecting the constant factors from the point of view of the theoretician, but it doesn't change the order of anything.

To a practicing engineer, however, the differences between analog and digital can be very important. The differences can sway you one way or the other. I am not saying that one is always better than the other, but for example, consider the cochlea chip discussed in the next section. The fact that the circuit is analog means that we get essentially the entire function of a cochlea on one modest-sized chip that has a current drain of about only a few microamps. That is an awful lot less power than it takes to do a similar job with a digital implementation. So that's one way that the difference may affect you. But now let us a consider a way that analog effects can have an important impact in digital circuits, namely in random-access memory cells and decoders.

6.2.2   RAM Cells

Consider the RAM cell variations shown in Figure 6.1. I have borrowed the left-hand column of this figure, showing a reconstructed evolution of the RAM cell, from a recent text by Glasser and Dobberpuhl [1]. They illustrate a succession of random access memory cell designs in nMOS. The right-hand column shows some alternate designs, some of which use CMOS.

They start with the six-transistor design (a): cross-coupled depletion-load inverters and a pair of access transistors, which are switched by a word line to connect the bistable element onto the bit lines. By forcing those bit lines, it is possible to set the state of the bistable. By sensing those bit lines with the access transistors enabled, the state can be read back. If you're careful you can read it without rewriting at the same time, and it holds its state statically.

The pull-up devices (depletion-mode transistors) take up a lot of space because they have to be long and skinny. They also have DC power consumption, so they cause space problems and power problems. If you get rid of them, you're left with a design (b) which works more or less the same way, except that it has no connection to a power supply. It holds its data dynamically, as the capacitance of the gates allows it to remember that one of the transistors is on, and the other one is off; but there is nothing to force it to stay in that state. It is not continuously self-restoring, but you can still use it as a memory element. It is dynamic, but if you're clever when you read it, you can cause it to restore itself (or if you're not clever you can cause it to erase itself before you can read it!). You can do a refresh on read, or you can read it and rewrite it occasionally.

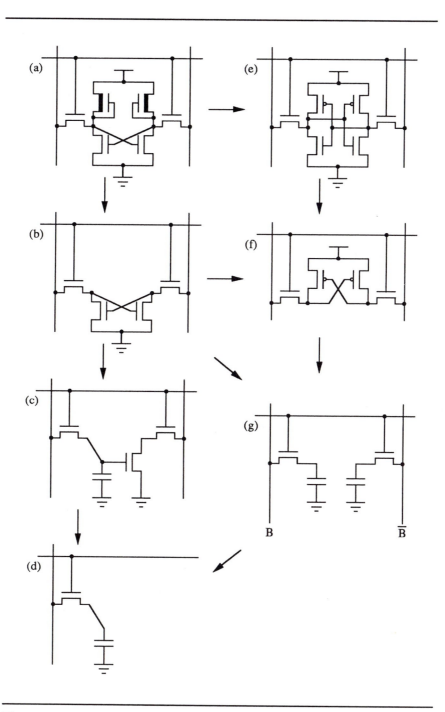

Figure 6.1   Evolution and side-lines in RAM-cell development.

There are well-known ways of using dynamic memory cells. If you get rid of the pulldown device on the left in (b), you get a three-transistor variant (c) of the same idea. That extra transistor in the cross-coupling is not doing anything active, and if you get rid of it, the cell works the same way, more or less. It holds its state on a gate capacitance, and by turning on the read device, you can read the state out onto a bit line. The cell (c) is drawn with two bit lines, one of which is used for writing, the other for reading. It can also be built with the two bit lines merged into one, making it a little smaller. There are several variations on that kind of design.

The three-transistor cell has gain from input to output through the storage device, but it can't do a self-refresh because the polarity of the signal that you read out is the opposite of the polarity of what you write. If you want to refresh it, you have to execute a complete read cycle and a complete write cycle; that is, you precharge the bit line, read the data out, invert it, and send it back in again. So it's a little bit more of a nuisance to refresh.

The final step in the traditional RAM cell evolution, the one-transistor design (d), does not have gain and reads with the same polarity that you write. If you write a high voltage, you'll sense a (very weak) high voltage when you read it. With a clever sense amplifier, you can cause that to be self-restoring on read. It's not the cell that restores itself, but externally the sense amp can provide the gain to refresh the cell on a read, much more quickly than doing a complete read followed by write cycle.

The four-transistor nMOS dynamic cell (b) is often modified by adding high-resistance polysilicon pullup resistors to make it static. Since this is not an option in standard logic processes, we don't discuss it further.

This is the classic evolution of RAM cells. By the way, these cells are all still in use, and account for probably more than 99 percent of all circuits in use today (though what they do in their arrays is usually trivial and boring). In logic chips you might find register files made with static cells because on a processor chip you do not generally have any guarantee about how often cells will be read and rewritten. Four-transistor dynamic cells are sometimes used in processor chips because you can arrange that you access them often enough to refresh them. If you care to do that, you can save some area.

The three-transistor cell (c) is one I have used on several signal processing chips where it is easy to arrange that your algorithms are going to access data over and over, say 20,000 times a second for doing digital filters with a 20 kHz sample rate. You can forget about the fact that it's dynamic because you are never going to require it to remember any data long enough. If you can get away with a system with that kind of constraint, you can use dynamic cells.

The final one-transistor cell is the one that's used on the really high density DRAM chips. It is not used on standard logic chips much because it's a little tricky to build. You typically use a special memory process, and a lot more extra support circuitry is put around it to make good sense amplifiers, and so on.

But there are other variations you can do if you have CMOS. The standard static memory in CMOS is a six-transistor device (e) where the pull-ups are active p-type devices (instead of being depletion loads which waste lots of power). This cell has no DC power dissipation when it's holding data statically.

The six-transistor static cell is pretty popular. It has the area disadvantage of six transistors, two of which are a different type from the others, so they have to be in a well. What you might do, now that you have the six-transistor CMOS design, is to get rid of the two lower cross-coupled devices, leaving the two upper devices of the other type (f).

This four-transistor cell has several nice properties. One is that you can build it in a standard CMOS logic process, which was my original motivation for designing it [3]. I wanted something that was a little more compact than the standard six-transistor cell, one that I could use on a logic chip that I could get fabricated through MOSIS.

The other thing you can do from here, or from the nMOS four-transistor cell, is to make a variant of the one-transistor cell that is symmetric with two capacitors and two transistors (g). You can think of this as being part of the evolution that got us to the one-transistor cell, if you like.

When you first look at this two-transistor cell, it doesn't have any obvious benefit. You are storing data redundantly (in complement form) on two capacitors. But it has the nice feature that you can sense read data differentially. You don't need dummy reference levels to sense relative to, and you also get twice as much differential voltage, because you stored the same data in both directions on two capacitors. So this is a variant on the one transistor design that you could actually imagine using on a standard logic process. You can get more than twice as much signal, and signal-to-noise ratio, out of this kind of cell.

Of all the choices implied by the variety of RAM cells, the static versus dynamic issue is often the toughest. I think you are always going to have a much smaller cell if you can use a dynamic cell, but that has implications at the system level where there is always a trade-off between the density and the complexity. Eventually, it may be the case that dynamic memories won't really be a feasible way to put a lot of memory on a chip, for the following reason: as devices scale down and down, it takes relatively more capacitance area to get a good dynamic cell to work. It may be that the static cells get smaller and catch up again. Then again, they need capacitance to be robust against alpha particles. It's not really clear, therefore, what will happen eventually, but there will be increasing difficulties in making robust dynamic designs in very small feature sizes.

Other kinds of memories that I'm not going to talk about, like ROMs and CAMs, and so on, have even more choices about circuit forms, but even within RAMS, the field is wide open. You can make any kind of circuit you can imagine. I think CMOS hasn't been as well explored as nMOS in a lot of ways. I'm going to illustrate two circuits that I have come up with, just because I needed some memory. One is the RAM cell and the other is the address decoder circuit.

The four-transistor cell is redrawn slightly differently in Figure 6.2. I haven't said yet whether it's static or dynamic, and the reason is that it's in-between. It's got complementary bit lines (bit and bit-bar), cross-coupled devices, and a word line which, if high, turns off the access transistors. That leaves a dynamic cross-coupled circuit, just as in the nMOS four-transistor cell, effectively disconnected from the bit lines. With just the two cross-coupled transistors, there's no current and no gain, so it's just dynamic. When the word line is low, it couples the cell to the bit line through the access transistors, and you can read the cell.

But the interesting thing about this circuit is what happens when you exploit the analog properties of these access transistors and turn them on some intermediate amount—don't turn them on hard, and don't quite turn them off. The nice thing about MOS transistors is that you can control the currents through them over many orders of magnitude in the subthreshold region. They are very well controlled down to currents of one picoamp, or thereabouts. You can set the word line at a voltage that makes these transistors act as current sources with very small currents. A current source is like a high impedence load; you have very high gain cross-coupled inverters here, so it's static. If you hold the word line at some intermediate voltage, you get a variety of choices about how to use this cell. You can use it dynamically and you can use it statically. That is, you can use it statically while the bit lines are held at VDD. I.e., if you connect the cell

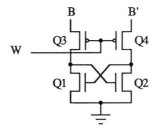

Figure 6.2   A new 4-transistor sort-of-static RAM cell.

to a power supply by precharging the bit lines, and you bias the word line
correctly, then it's a static cell. While you're reading and writing, the bit
lines won't be held at the power supply voltage, and so it will be a dynamic
cell. I call it a sort-of-static cell, not to be confused with pseudo-static or
quasi-static!

Figure 6.3 shows how we bias the cell up to be static. I'll call the cross-
coupled devices Q1 and Q2. I learned after I drew all these figures that the
Q1, Q2 terminology is usually reserved for bipolar devices and is not used
for MOS devices. I'm supposed to use M1 and M2, but in subthreshold

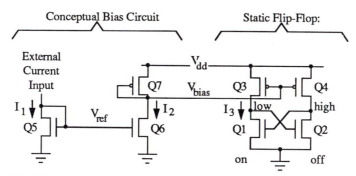

Q6, Q7 act as word-line driver in unselected state

First current mirror Q5, Q6:

$$I_2 = a_1 \cdot I_1 \text{ , for } a_1 = \frac{\text{Width (Q6)}}{\text{Width (Q5)}} \cdot e^{(V_{T_5} - V_{T_6})/V_T}$$

Second current mirror Q7, Q3:

$$I_3 = a_2 \cdot I_2 \text{ , for } a_2 = \frac{\text{Width (Q3)}}{\text{Width (Q7)}} \cdot e^{(V_{T_7} - V_{T_3})/V_T}$$

$$\therefore I_3 = a_2 \cdot a_1 \cdot I_1 \qquad\qquad \text{where } V_T = n\frac{kT}{q} \approx 40mV.$$

$$I_3 = 1pA \ (10^{-12} \text{ A}) \text{ is easily attained.}$$

(All threshold values V_{T_i} are magnitudes of enhancement- mode
thresholds of Qi.)

Figure 6.3   A bistable cell biased for low power static operation.

these devices look more like bipolars than MOS anyway, so we can get away with the notation.

So suppose we've got the cell in a state where Q1 is turned on and Q2 is turned off. That means there's a current through Q1 that's pulling its drain node low. There's no current through the other leg, so Q2's drain node should be high. The complementary state goes through the same way.

Let's look at the word-line bias. We're going to put a bias voltage on the gates of the access devices Q3 and Q4 that connect to the bit lines. For the sake of telling you how this cell works when it's static, I just say the bit lines are connected to VDD. When the memory is in operation, the bit lines are connected to VDD through a pre-charge transistor, and the bias voltage is set through the word-line driver circuit.

For now, forget about the word-line select driver and view the bias function conceptually, as when the cell is not being accessed onto the bit lines. That is, the access devices are almost turned off. The word line is held at a bias voltage that's just high enough to be somewhere near a threshold below VDD. That voltage can be established by a current mirror trick. Referring to Figure 6.3, Q7 and Q3 form a mirror where the current I_2, being pulled down by transistor Q7, pulls the gate down just low enough so that enough current comes through to set the same voltage on Q7's gate, and that makes the same current (or the same current scaled by the ratio of the widths of those two devices) go through the cell.

There is another current mirror, the pair of devices Q5 and Q6, which makes I_2 equal to I_1 times the ratio of the width of those devices. Therefore you get the current I_3 inside the memory cell to be equal to these two constants that are the width ratios of the transistors times some externally supplied current that you can use to program how much current you want in your memory cell. If you don't supply any external current, and you let this gate voltage drift down toward ground until there's no current, then your system will be dynamic. If you put about a microamp in as I_1, and you've got a 10:1 ratio to scale down to a tenth of a microamp as I_2 and another 10:1 ratio to scale that down to a hundredth of a microamp as I_3, then every cell will have 10^{-8} amps running through it. If you have, say, 10^4 cells—10K bits—then you've got 10^{-4} amps on your whole chip, or 1/10 of a milliamp running through the memory cells. Thus, you can control the power dissipation of a whole array of these cells by one external programming chain. Of course there are other factors I haven't noted, as shown in Figure 6.3, such as exponentials of the differences of the thresholds of these devices. If the thresholds are not close to each other, these factors become relevant. One hopes that the thresholds are fairly well matched, so that currents are off by a factor of only 2 or a factor of 4, or some relatively small number rather than orders of magnitude that you can lose control over.

The bias current needs to be adjusted to be comfortably larger than the junction leakage current, at the temperatures at which the chip is expected to operate. At room temperature, I think typical leakages are much less than that (10^{-14} A), so 10^{-12} A would work at room temperature. But if you want to run this chip hot, you had better program enough current per cell to make it reliably static. You can imagine the required current being the same as the actual current in a six-transistor cell, just enough current to replace leakage. What you would actually have to do here is program it for several orders of magnitude more than that, in order to be on the safe side, so the power dissipation of this design is on the order of, say, 10 or 100 or 1,000 times as much as a six-transistor cell.

This power level is probably still a lot less than what you typically get with the four-transistor static cell that uses high resistance polysilicon resistors as loads. On a power dissipation scale this new cell is intermediate between a four-transistor (memory process) high resistance poly cell and a six-transistor cell. It's also intermediate in terms of area—it has two transistors fewer than the six-transistor static, but it's bigger than the cell with poly loads because it has two different kinds of devices and it needs a well clearance. In both area and power, it's intermediate between what you can do with a memory process and what has been done on a logic process. I think it's an interesting point to have available to you in a standard logic process.

6.3   Analog Computation in Hearing

In this section I describe the analog VLSI implementation of the first part of a model of hearing: an electronic *cochlea*, or inner ear.

The first question we have to answer in building VLSI hearing machines is *How does hearing work?* Once that question is answered, we can attempt to build machines that do the same thing. That is, we first try to come up with a translation of what's going on in hearing into something that resembles a computer science problem or algorithm (although we are not going to cast it in exactly that form), and then we try to implement it. We want to build machines that can hear as well as we do. Ultimately, we want to be able to build speech recognition devices and be able to talk to our computers. I can't say we are very close to that, but we have some pieces along the way.

The question of how to implement a machine sensory system is interesting in itself, independent of the underlying models. Do you use digital techniques or go analog? For years when I gave talks about what I was doing, there was agreement that a digital implementation was the obvious smart and modern solution, for lots of good reasons. But I have recently been involved with some very large scale, laboratory analog work that has been evolving in Carver Mead's lab at Caltech. Now I find there are interesting reasons for going analog, as well! Since digital approaches have become fairly well known, this section concentrates on an alternative *analog* approach to implementation of the cochlea models.

The embedding of the problem into the analog implementation is very direct. Wave propagation in the physical system is modeled and implemented by spatially discretizing the biological cochlea and then building the little pieces of the system as simple circuits that hook up to their immediate neighbors. Everything is directly hard wired for its function and everything is very local. There are no switching networks, no multiplexing. The data representations are fairly trivial. Physical variables in the cochlea, like pressure and displacement, are represented by physical variables in the circuits, like voltage and current. With this approach we build systems that are incredibly parallel and efficient.

An interesting side effect is that the energy cost of implementing the model this way is amazingly low: on the order of six orders of magnitude less power dissipation than an equivalent digital implementation. That is one possibly important payoff when using an analog approach.

Figure 6.4 is a sketch of the inner ear. It consists of a coiled structure known as the cochlea, plus the semi-circular canals that you use for knowing how your head is oriented. There is fluid in ducts in the coiled up cochlea, and when the eardrum presses on the bones of the middle ear, the stapes bone presses on the fluid, which moves around inside and bulges back out through the round window. So we have a closed fluid system. Mounted inside one of the ducts is the organ of Corti, which holds hair cells that detect the fluid motion. So it is just a physical transducer. It takes sound waves and converts them to a complicated pattern of fluid motion detectable by hair cells. Each hair cell hooks up to several dozen auditory neurons, which send signals to the brain about what is going on. That describes just the most peripheral level of hearing. It is the only level I will say much about, but there is a lot more that goes on later, farther into the brain.

6.3.1 VLSI Implementation

The cochlea model chip is just a cascade of simple analog filters. We design each of these filters to model a small section of the cochlea transmission line. We would like to be able to put a signal in, where it says input on the floorplan (Figure 6.5), have it serpentine through the filters, pick up taps at various places, and look at how the system responds to sounds. We would like the outputs of these taps to look a lot like what really goes on in the cochlea at selected places. We have done all the mathematics that says exactly what these transfer functions need to be. You can derive it from the hydrodynamic model of the cochlea and, to first order, what they look

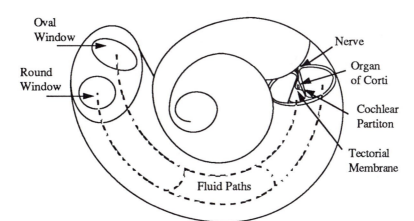

Figure 6.4   A sketch of the cochlea, or inner ear, also known as the membranous labyrinth.

like are delay lines, with increasing delay as the signal propagates through the filter cascade. Each one of the stages is a little different in the sense that it has a different delay. But they are all self-similar in the sense that there is a simple scaling relationship between them. Each one wants to be slower, by some constant factor, than the one before it, so there is an exponential or geometric progression of parameters.

The second-order behavior, besides the delay, is an attenuation of high frequencies. That high-frequency loss is easy to achieve; in fact, you can't do otherwise. If you build any kind of a delay stage, it's going to roll off at high frequencies. Then the third-order operation that it needs to perform is to adjust the middle range. You want to be able to control how much gain it has for frequencies that are near its characteristic frequency (roughly the reciprocal of its delay). You'd like to be able to have a little more gain or a little less gain. You would like to be able to control that gain to adapt the overall gain of the model in response to the loudness of incoming sounds. This nonlinear adaptive aspect of the model is very important, and is discussed elsewhere [2].

We are going to build these filters out of simple analog continuous-time circuits. We could build this on a digital computer as well, but the chip we've built is massively parallel—it uses no less than $N = 480$ filters!

430   Lyon

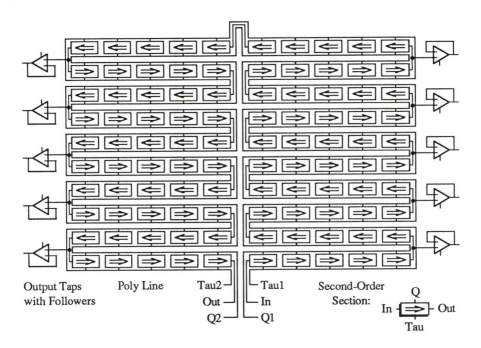

Figure 6.5   Floorplan of analog VLSI cochlea chip.

Each one of those filters requires three amplifiers, so there are almost 1,500 amplifiers in total. That's fairly parallel for a one-chip system.

Figure 6.6 shows circuits for differential amplifiers, for which we use an op-amp-like symbol. These are really what we call transconductance amplifiers, which act like op-amps with certain well-controlled limitations. Like an operational amplifier, they have very high voltage gain, but unlike an op-amp, they have relatively limited current output. In fact, the current output is going to be some well-controlled transconductance times the differential input voltage—transconductance means output current per input voltage. When a voltage difference is applied across the inputs, the differential source-coupled pair splits a constant bias current into unequal parts.

The current mirror arrangement makes the currents in the top two devices equal, and the output node takes a difference, yielding a net output current. So if the inputs are balanced, the output current is zero. The wide-range amplifier circuit is a little bit fancier and reflects things around a few more times to get an output that can operate over a wider range of

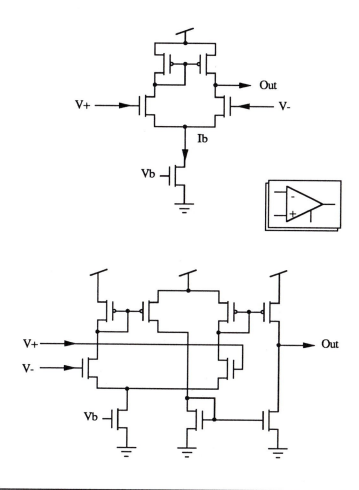

Figure 6.6 Basic and wide-range CMOS transconductance amplifier circuits.

voltages. It also has a little bit higher voltage gain. For these reasons we have used it in some of our cochlea test chips.

The transconductance amplifiers are our basic building blocks. When we build filters out of them, they look a little bit different from the kinds of filters we would build out of op-amps, because they don't need resistors to go with them. The well-controlled transconductance gives you the impedence you need to combine with a capacitor to make a time constant.

Figure 6.7 is a first-order filter circuit. This gives you the basic property we need in the cochlea, which is a delay with high-frequency attenuation. You put a voltage in the input, and if you wait long enough you'll see it at the output. The time constant is going to be C/G, that is, capacitance over conductance, or RC, where $R = 1/G$ is the effective resistance of an output as a follower when it has a transconductance of G. There is a bias terminal that controls the bias current in the amplifier that sets the transconductance. The bias current is going to be an exponential function of the bias voltage when you operate these devices in a subthreshold region.

To set the time constant of this kind of circuit to be in the audio band, from say 20 Hz up to 20 kHz, you need transconductances that are very small to get resistances that are very large, so that you can use modest-sized capacitances. When you operate at such low currents, you are well below the normal currents at which MOS devices operate in digital systems. You are in a subthreshold region where the currents are exponential functions of gate voltage. Typically, we operate on the order of 10^{-10} to 10^{-7} amps for the audio range.

A better filter stage for modeling the cochlea is a second-order filter, shown in Figure 6.8. I've now labeled the transconductance bias terminal the τ knob here because that sets the time constant of these first-order stages. We hook up an amplifier with positive feedback in the feedback

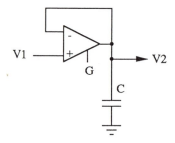

Figure 6.7   A follower/integrator circuit.

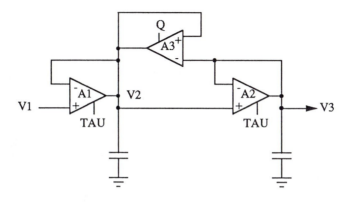

Figure 6.8   Second-order filter circuit.

path, which will tend to boost the middle frequencies and make it like to ring a little bit. It has a control on its bias, which we will call the Q knob, because it will set the Q of that second order system (Q being the familiar term describing the amount of resonance in a second-order system). Changing the bias on the Q terminal doesn't affect the delay much at all. It doesn't affect the gain for low frequencies which is unity, and it doesn't affect the gain for high frequencies, which is determined by the roll-off of these RC's. But it does affect the gain in the middle, near the corner, between low and high frequencies. This gives you that extra degree of freedom you need to adjust the flatness and the gain of this filter to make a good model of the cochlea.

6.3.2   System Response

The test chip looks like Figure 6.9, just like the floor plan I showed you that had only 100 stages because that is about how many I could fit on a figure.

The transfer functions can be computed by taking the transfer function of a simple second-order section from any textbook [5]. We've drawn it in Figure 6.10 for three different values of the Q setting. The lower curve is 0.7, which is just below the point known as maximally flat; that is, 0.707 or $1/\sqrt{2}$ is a Q for which the transfer function stays as high as it can without going above unity. Then we can put slightly higher Q's and get a little bit of gain out here in these middle frequencies, and those are Q's of 0.7, 0.8 and 0.9.

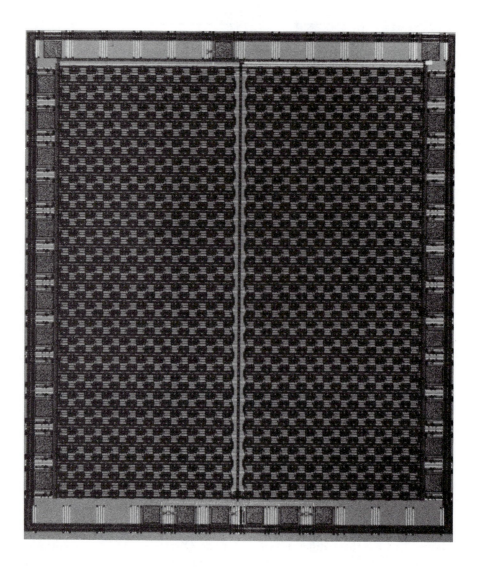

Figure 6.9   Photomicrograph of 480-stage cochlea test chip.

We are not going to operate this thing in a high-Q region, we are going to operate it in a nearly flat region. When you cascade a whole bunch of those together—that's like putting in the dotted lines for the $Q = 0.9$ case, and adding all those up on the log-log plot or multiplying all the transfer functions together. Then you get the curves on the bottom of Figure 6.10; notice the different scales. At low frequency it's unity gain, 0 dB. When you get out into the middle, you might have something like 40 dB of gain, even though the Q is extremely low, and so you get this pseudo-resonance. It is quite wide, maybe on the order of an octave wide, but it's got a lot of gain. It is very unlike what you'd get by making a resonator with that much gain because a resonator with that kind of gain would have a very narrow response, a very high-Q narrow response. By changing the Q over a very narrow range from 0.7 to 0.9, you can vary the gain by 40dB or more.

Here, then, is an effect that can be used as a wide-range gain control to adapt to a wide range of input signal levels, which is one of the most amazing things that your ear is able to do. You can deal with very quiet sounds and very loud sounds.

These high-order filter cascades can be described by plots of their poles in the s-plane, which are just the collection of all the poles of the cascaded sections up to the output tap in question. This kind of pole arrangement is a very unusual way to build high-order filters, something you won't find in any textbook. When you try to build those with non-ideal MOS devices, your actual bias currents and transconductances will vary all over the map—this effect is usually termed threshold variation, though that's a strange notion when operating in subthreshold.

In simulations of random threshold variation, we have assumed that the current in a transistor can vary over a range of a factor of 2, which is about what you'll find in typical digital logic processes—sometimes worse, sometimes better. Then one of these second-order sections can have an effective α parameter, which is a ratio of feedback gain to feedforward gain, that varies by a factor of 2 above or below nominal, depending on whether the forward gain is low and the feedback gain is high, or vice versa. An α parameter is a measure that essentially tells you the angle of the pole. That gives you the two dotted lines in Figure 6.11, and it says all the poles are going to be between those limits. We just simulated random numbers and made a scatter plot of where the poles really end up for the circuit we built, for the nominal conditions under which we tested the chip. In this case, the frequency ratio from one stage to the next is supposed to be around 1.0139. That is a 1.4 percent change from one stage to the next. In 480 stages that will give you three orders of magnitude and cover the range from 20 Hz to 20 kHz. What you find is that the characteristic frequencies of the stages don't even end up being monotonic. Some of them get higher from one stage to the next. Some get lower by a factor of 2 even. It turns out to be nothing like the ideal thing that we want to design. The Q's are going

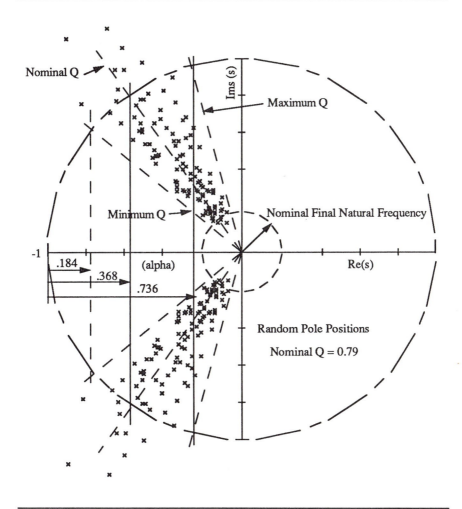

Figure 6.11   A typical s-plane pole plot of a cochlea filter cascade.

to be all over the place. But the nice thing is that you cannot tell from the collective response when you measure the overall transfer functions. Murphy's Law didn't apply this time.

We measured the transfer functions, as shown in Figure 6.12. This was set up with some modest gain like a Q of 0.79. We measured a couple of different taps on the chip. The dots are measurements made on a digital voltmeter, using a fixed amplitude input sine wave and changing the frequency. The curves are amazingly close to what the theory says they

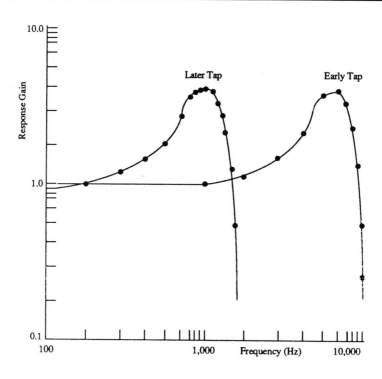

Figure 6.12   Measured frequency responses at two cascade output taps.

should be, and so this wide dispersion of Q's and center frequencies doesn't mess things up too much. We were very lucky in this case.

Here is one way to explain the effective composite response obtained when you add one more stage. It doesn't really matter whether that stage's characteristic frequency is a little high or a little low because it has roll-off. The roll-off starts gradually, but in the right neighbourhood, and no matter whether it starts too high or too low, it is going to reduce the effective bandwidth of that next stage. So the effective bandwidths of the output taps, going from one to the next, are going to be monotonic, even though the increments between them are not. That helps, but the gains are going to be fairly variable, and it would be nice to have all these stages self-adjust. We need them to self-adjust anyway, to adapt to loudness variation, so the ability to adjust these filter stages might solve two problems at once.

Besides the frequency response, we did step responses, adjusting things a couple of different ways; see Figure 6.13. If you adjust the Q to about 0.69, roughly the maximally flat condition, or just below it, you get a step response that doesn't ring much at all. If you crank it up to about 0.79, which is where we measured that transfer function, you get a couple of cycles of ringing. These actually agree pretty well with the kinds of responses people are able to infer from the living system using the reverse correlation measurements, and so on. A healthy cochlea will have a couple of cycles of ringing if measured at a fairly low signal level. If you overdrive them, or you treat them with toxic drugs first, or let them get oxygen starved, or one thing or another, they tend to revert to a more dull kind of a condition with less gain, less ringing—it looks more like a passive or "flat" system.

An appealing feature of these models is that their implementation is relatively trivial—it's just a cascade of simple filters. Essentially you design one and you hook them all together, and the only thing that needs to be different between them is an exponential scaling of the currents or the time constants, which can be done simply by biasing them with a linear voltage gradient.

We can get an exponential current gradient or exponential time constant change simply by using a long resistive polysilicon wire that snakes along

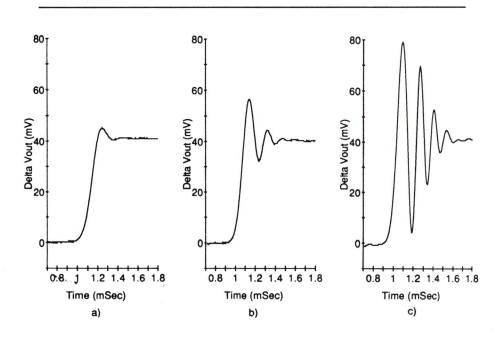

Figure 6.13   Measured cochlea chip step responses for various Q settings.

with the sections. When you put voltages on the wire's two ends, it creates an exponential time constant gradient through the whole design. This incredibly trivial array of simple processing elements with nearest neighbor communication solves a large part of the hearing problem that would have been difficult to solve algorithmically, but is not so hard to build an analog machine for it.

All the other interesting levels of the auditory nervous system are available to stimulate future work in this area. Circuits that sense the output of the cochlear filters and provide nonlinear adaptation are our current focus, and will integrate right into the parallel analog processing array. Later stages to model higher levels of the nervous system may be built in the same array chip, or may be separated by a communication channel analogous to the cochlear nerve.

Bibliography

[1] Dobberpuhl, D. W., and L. A. Glasser. *The design and analysis of VLSI circuits.* Reading, Mass.: Addison-Wesley, 1985.

[2] Lyon, R. F., and C. Mead. An Analog Electronic Cochlea, *IEEE transactions on acoustics, speech, and signal processing,* 36(7):1119–1134, 1988.

[3] Smith, R. J. *circuits, devices, and systems 2nd Ed.* John Wiley & Sons, Inc., 1971.

[4] Sproull, R. F., and I. E. Sutherland. *A theory of logical effort.* Tech. Rep. SSA 4759, Sutherland, Sproull, and Associates, Inc., 1987.

[5] Thompson, C. D. *A complexity theory for VLSI.* Ph.D. Thesis. Computer Science Dept., Carnegie Mellon University, August 1980.

CHAPTER 7

BRYAN ACKLAND
AT&T Bell Labs

Knowledge Based VLSI Design Synthesis

Introduction

This chapter describes some of our experiences in applying knowledge-based programming techniques to the synthesis and verification of VLSI (very large-scale integration) designs. The following sections deal with two projects for the automatic synthesis of VLSI circuits and layout that are actually going on in our area at AT&T Bell Laboratories. The two projects are named *SYNAPSE* and *CADRE*, and they occupy complimentary positions in our simple model of the design process, as shown in Figure 7.1. SYNAPSE [4] is a tool developed by P. Subrahmanyam for synthesizing circuit structure from high-level behavioral descriptions. CADRE [1] is a project in which we are attempting to convert structure into full custom physical layout. Ultimately, we'd like to be able to hook these two together and thus span the whole design process. We recognize that it is not possible to separate design neatly into two pieces, with no feedback across the structural boundary. But for now these two problems are difficult enough in their own right that we'll settle for these partial solutions, and then later look at the problem of merging the two tools.

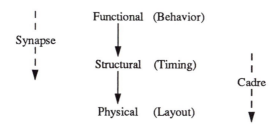

Figure 7.1   Synthesis tools SYNAPSE and CADRE.

7.1   SYNAPSE—Objectives and Techniques

The primary goal of SYNAPSE is to synthesize VLSI systems from very high-level behavioral specifications. These specifications consist of function plus constraints on timing, pin-out, area, protocols, etc. The output is a VLSI architecture that meets these specifications.

A second goal is to make the design process more robust by producing designs that are provably correct. This can be done using representations and techniques that are amenable to formal machine verification. Recognizing that such a system will be substantially different from the CAD tools suites we have available today, a third goal is to provide a CAD platform on which we can evolve gradually toward this rigorous style of design correct by construction.

A popular model, which shows some of the complexity of the design process, is shown in Figure 7.2. This model has three separate dimensions in the design space: behavioral, structural, and physical. And along each of these dimensions are different levels of abstraction. In the structural dimension, for example, we can have a high-level representation such as register transfer, or we can have a low-level representation such as transistor connectivity. The figure gives examples for the other two dimensions. The overall design process begins, therefore, high on the behavioral axis and moves, through a series of transformations, to a point low on the physical axis that represents the physical mask description we will hand off to manufacturing.

We model design as a series of transformations between different representations. But the goal here is to perform these transformations in a rigorous, provably correct fashion. Unfortunately, this goal is not well supported by the representations and tools in use today. Each CAD tool or

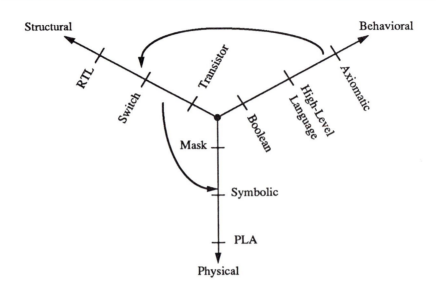

Figure 7.2   Y-Chart of design representations.

system has its own set of concepts, terminology, semantics, and idiosyncrasies. In order to build rigorous, automatic design tools we first need to develop formal descriptions of desired circuit behavior and structure. Second, we need formal descriptions of the primitive components of each of the representation levels together with rules that describe how they may be combined and what the resulting combination represents. Third, we need verifiable techniques for transforming from one representation to another.

Now this is, of course, a huge undertaking, given the limited resources of one or two researchers. Subrahmanyam has decided, therefore, to focus on pieces of the problem, on particular design tasks in which he can study methods of representation and transformation. This has included some high-level and some low-level projects. At the high level, he has developed techniques for converting recurrence relations into a systolic array structure. I'm going to describe a low-level project: converting Boolean expressions into transistor-level circuit connectivity [5].

7.1.1   Leaf Cell Synthesizer

Suppose we have organized our design into a structural hierarchy consisting, at the lowest level, of a number of leaf cells. Some of these leaf cells will be represented, at the behavioral level, as combinational logic functions. We wish to create, in a provably correct fashion, a technology-specific transistor schematic that will perform the required function.

The first step is to represent, in a formal way, the available signals and the desired behavior. The second is to find a representation for the primitive elements in the available technology and rules on how they may be combined or connected to form a circuit. The third step is to discover a transformation mechanism that can take us from one representation to another. We will examine these steps one at a time by way of the simple example of a logic inverter. This is not to say that the synthesis of an inverter is, by itself, an interesting problem; rather, it is an instructive example. Later I will show you a more complex circuit synthesized by this system.

7.1.2   Representations of Desired Behavior

A simple way to represent circuit behavior is in the form of a truth table. Figure 7.3 shows a truth table describing the desired behavior of our logical inverter in terms of an input signal, *a*. This behavior is formally represented as a predicate of type *circuit*, in which the first parameter is the name of the circuit, X, followed by a list of the outputs obtained as we increment our way through the truth table. Our available input, *a*, is simply a special circuit (an input wire) whose output is known to take on the values *[0,1]*.

7.1.3   Representation of the Circuit Technology

Suppose we are using a CMOS technology. One of the primitives we can use in this technology is a "grounded N transistor". This is an N device with its source connected to *VSS*, as shown in Figure 7.4. Its behavior is represented by a function that maps a Boolean input (gate) into a value

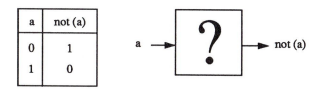

circuit (X,1,0)
input signal : circuit (a,0,1)

Figure 7.3   Logical inverter—desired behavior.

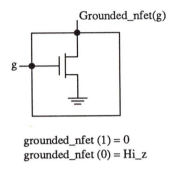

Grounded_nfet(g)

g

grounded_nfet (1) = 0
grounded_nfet (0) = Hi_z

Figure 7.4   Circuit primitive—grounded N transistor.

on the output (drain). If the input is 1, the output is 0. If the input is 0, the output takes on an unknown value we call *HiZ*. There are similar descriptions for other primitives in the technology, such as P transistors, N transistors that are not connected to *VSS*, etc.

In addition to these primitives, we need some composition rules for joining them. One of these is shown in Figure 7.5. Suppose we have two subcircuits called *wire1* and *wire2*, and we wish to connect their outputs. Then we will create a new circuit (called *connect(wire1,wire2)*) whose behavior is described by a new predicate based on the *join* function. *Join* is a function that maps two logical values into a third "joined" value, as shown in Figure 7.6. If the two input signals are equal, the output simply takes on that value. If one of the inputs is *HiZ*, the output takes on the value of the other. If the two inputs are different, we get a short circuit that is flagged as an error.

Now we can see how these formal representations allow us to derive and prove circuit behavior. Suppose I connect an N and P transistor in the classical CMOS inverter structure, as shown in Figure 7.7. We have a circuit that can be described by the predicate:

$$CIRCUIT(connect(pfet-vdd(a), nfet-vss(a)), join(1, HI-Z), join(HI-Z, 0))$$

Now substituting for the semantics of *join* gives

$$CIRCUIT(connect(pfet - vdd(a), nfet - vss(a)), 1, 0)$$

This describes a circuit that, in fact, has our desired behavior. And so we are able to show rigorously that our connection of a single N and P device yields the behavior of a logical inverter.

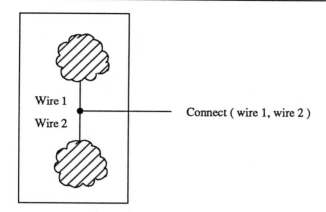

$$\text{Connect (wire 1, wire 2)}$$

If CIRCUIT (wire 1, r1, r2) & CIRCUIT (wire2, s1, s2)
then CIRCUIT (connect (wire1, wire2),
                    join (r1, s1), join (r2, s2))

Figure 7.5   Composition rule connect.

join (x, x) = x; { Identical signal values }

join (x, Hi Z) = x; { An electrically isolated node }

join (x, y) = a_short; { x ≠ y }

Figure 7.6   Join function.

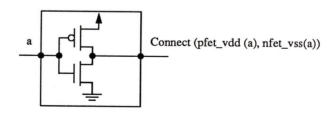

Connect (pfet_vdd (a), nfet_vss(a))

Figure 7.7   Inverter circuit.

7.1.4   Transformation Technique

So far I have shown how the behavior of a particular circuit structure can be derived. Our next step is to find a technique for automatically converting desired behavior into circuit structure. The tool we use is one that can efficiently search through different possible connections looking for a solution. It is a theorem prover written in Pascal called *ITP*. We provide as input a number of axioms that describe all the information about our circuit primitives, the input signals available, and our composition rules. Then we make the assertion that no circuit with the desired behavior exists. The theorem prover will attempt to refute our assertion by finding a composition of primitives that does indeed exhibit the desired behavior. What we obtain is not only a circuit, but also a proof that this circuit behaves as required.

Our leaf cell layout tool has been used successfully to design a number of simple circuits. One of the more interesting is an exclusive-or gate. When we first attempted to synthesize this circuit using a rather simple model of N and P transistors, we obtained some rather naive circuits in which P transistors were used to drive a 0 level and N transistors were used to drive 1. Those familiar with CMOS will know that this is poor circuit design. And so we had to introduce more sophisticated descriptions of the primitive components that included the concept of strong and weak levels. Once we did, it produced the circuit shown in Figure 7.8. The interesting point here is that from a human design perspective, this is a nontrivial solution.

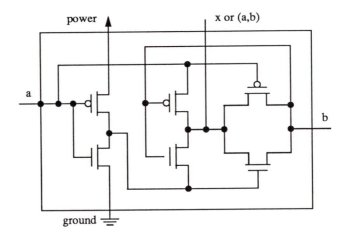

Figure 7.8   Synthesized EXOR cell.

Rather than work from logic primitives (as we tend to do), the theorem prover looked for an implementation that was optimum at the transistor level according to some cost function (in this case transistor count).

The time taken to discover this solution is about ten or fifteen seconds. Obviously, as the complexity of the circuit increases, the search time will increase exponentially without some form of clever pruning. Nevertheless, it demonstrates how formal verification tools can be used to reliably synthesize novel circuit structure from function.

7.2   CADRE—Custom Layout Synthesis

The second piece of work I will describe is a system called CADRE. Here the goal is to convert structural specifications into full custom layout. Input to the system is a hierarchy of structural blocks, each consisting of subblocks and their interconnections, as shown in Figure 7.9. At the lowest levels in the hierarchy are leaf cells (see Figure 7.10). Additional constraints specify area, speed, etc. Output from the system is a full custom symbolic layout. What I mean by full custom is a layout in which we have a floorplan of abutting leaf cells whose physical interface (shape, port positions) has been customized to the particular circuit under design. Such layouts are typically built around regular patterns of control and data flow. They are

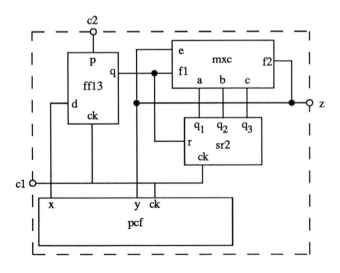

Figure 7.9   Structural composition cell.

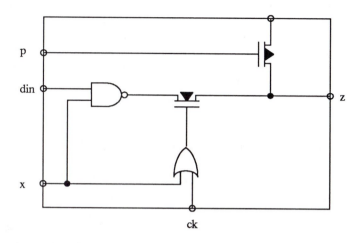

Figure 7.10   Structural leaf cell.

also normally considered to be within the province of expert human layout designers.

7.2.1 Cooperating Agents

If we examine this problem, we can identify some attributes which are, in fact, common to a broad class of design situations. First, it's a problem in which there is a strong notion of state; the order in which we decompose and attack the various subproblems is very important. Second, there is a need for backtracking. Even well-organized designers who adopt a disciplined top-down design strategy discover conflicts that lead to modification and redesign. Third, we need to bring together knowledge from many different sources in order to arrive at a solution. We need to understand, for example, floorplanning, transistor layout, routing, and performance optimization.

Based upon these attributes, we developed an architecture (see Figure 7.11), in which there are a number of expert agents, each of which knows about a particular aspect of the custom layout process. There is, for example, an agent that knows how to generate floorplans, and one that understands routing. These various agents are hooked together with a central coordinating agent known as the manager.

This type of architecture has several advantages. First, the modularity allows us to build these agents somewhat independently of each other. The person working on the floorplanner can concentrate on floorplanning problems without understanding all the intricacies of the design process.

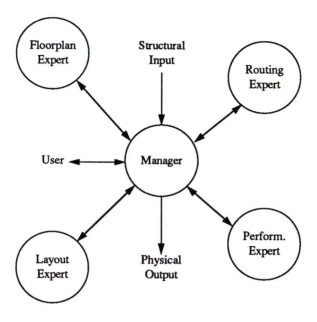

Figure 7.11   Cooperating expert agents.

Second, we can build these agents using different programming techniques. Some, for example, may be rule-based, others algorithmic; or we can even insert a human agent for some of the more difficult parts of the problem. A third advantage is that it allows us to build, initially, a minimal system and then subsequently expand it by adding more agents to deal with more subtle aspects of the design process.

The disadvantage of this architecture is the burden it places on the central manager. The manager is responsible for assigning tasks to the agents and interpreting their responses. More importantly, the manager ensures that the various agents work together in a constructive fashion. The manager provides the overall design strategy.

If we take a look at the way in which human designers tackle this problem, we see they use a mixture of top-down and bottom-up techniques. A designer may start off, for example, in a top-down mode until he or she reaches some critical cell. At that point, the designer jumps to the leaf cell level to gain a better understanding of what a good physical floorplan will look like. CADRE also uses a mixture of these two approaches, albeit in a more organized fashion. We begin the design process at the root of the structural hierarchy, first producing a floorplan of that node. This creates external constraints for the children of the root. We now proceed to floorplan these cells. We continue in this top-down fashion until we invariably design ourselves into a corner, i.e., we finish up trying to design a cell to impossible physical specifications. At this point, we backtrack to the parent of the offending cell, carrying information on how the constraints might be relaxed in order to remedy the problem. In redesigning the parent, we may find we need to backtrack still higher in the hierarchy. Eventually we resolve the problem and proceed back down the tree.

At any one point in the design process, therefore, we will be designing a cell subject to external constraints that have been passed from above and internal constraints that have been propagated from below. Let's take a look at some of the individual agents and see how they deal with this demanding design environment.

7.2.2   Floorplanner

The input to the floorplanning process, as shown in Figure 7.12, consists of a set of submodules and their interconnections together with various constraints on area, aspect ratio, port positions, etc. The output is a rectangular subdivision of the silicon surface in which we attempt to minimize area by maximizing the connectivity between adjacent cells in the floorplan.

A number of algorithmic solutions have been proposed for this problem based on graph theoretic techniques. These have the desirable property

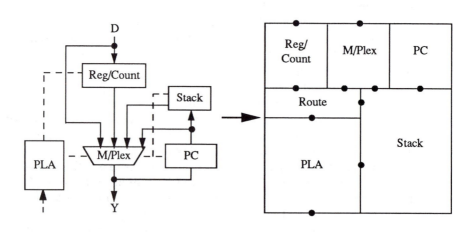

Figure 7.12   Floorplanning agent.

that they work as well for large problems as they do for small examples. In addition, they are predictable in terms of quality of layout and execution time. They do have the disadvantage, however, that they are rather inflexible in terms of the constraints they can process as part of the input specification. This, in turn, is limited by the nature of the model on which the algorithm is built. In the CADRE environment, where our understanding of the process and the nature of the constraints to be propagated is continually changing, this inflexibility can be a problem.

A rule-based approach would seem to offer more flexibility in dealing with this changing environment. The disadvantage is that human designers are only good at floorplanning relatively simple structures of five to twenty submodules. So how can we develop rules for more complex problems? Fortunately we do not have to. In the CADRE environment, we rely heavily on the hierarchical partitioning of the problem to keep the complexity of each module within bounds. We use a mixture of rule-based and algorithmic techniques to achieve a flexible agent that gives good performance for moderately sized floorplanning problems.

If we study the way a human designer attacks a floorplanning problem, we see, in general, two phases. The first is a topological planning phase in which the designer experiments with different relative placements, concentrating mainly on communication and topological packing. Once a good topology has been selected, the designer then determines exact geometrical dimensions to satisfy constraints such as area and communication bandwidth. Our floorplanner [6, 7] similarly splits the process into two phases. The first phase, topological layout, is rule-based because this is where we need the greatest flexibility. The second phase, geometric realization, is concerned with meeting geometric constraints and uses a fairly straightforward graph minimization algorithm.

Topological layout is performed with the aid of a special graph representation called the rectangular grid graph. An example and its corresponding floorplan is shown in Figure 7.13. In this graph we represent not just the connectivity between the modules but also their adjacency. Note that each node has four sides representing the four sides of the actual submodule. In addition, four special nodes represent the external boundary of the module. A number of rules govern how connections may be placed on the graph. These ensure the existence of a corresponding rectangular topology.

The topological planning process is one of placing submodules on the grid graph so as to "best" satisfy connectivity constraints. This placement is performed according to a number of rules. Like many design problems, however, this placement is a task in which there is a strong notion of state; there are subtasks that need to be performed in a definite order. We begin, for example, by focusing on the principal interconnects, buses. Then we examine the modules connected to each bus. Once modules are placed, we must satisfy all their connectivity. For each of these subtasks there is a rule

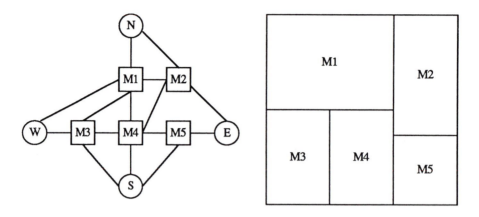

Figure 7.13   Rectangular grid graph.

set. In addition to these rules, however, we need some form of controller to activate them at appropriate times. This control is provided by regular procedural code; a flow chart is shown in Figure 7.14. Each rectangular box represents a rule set applied to one subtask.

In addition to these rule sets and a procedural controller, we need assistance from some algorithmic operators. The human floorplanning process is visually oriented and relies heavily on pattern recognition. Simple tasks like testing for line intersection, which are trivial for the human, must be performed in code. To implement these using rules would be hopelessly inefficient, and so we have a number of algorithmic routines on which the rules call either to test the layout or perform some operation. The overall structure is shown in Figure 7.15. At the highest level is the procedural controller. It activates rule sets that in turn invoke a number of algorithmic operators.

Figure 7.16 gives an example of one of the rules from the *place-module* rule set. In English, it says that if I have already placed two modules m1 and m2, and I am about to place a module m3 that is connected to both m1 and m2, and I know that m1 is considerably larger than m2 and m3, then I do not want to place m3 in a position such that m1 is orthogonally adjacent to m2 and m3. If I do, the result will be a floorplan with extreme aspect ratios, as shown in Figure 7.17.

Figure 7.18 shows an example of a rule from the *route-module* rule set. This rule says that if I have placed modules such that their interconnection

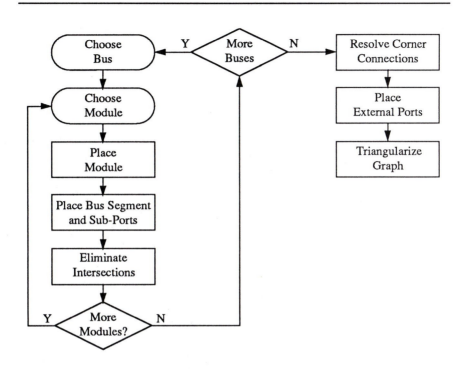

Figure 7.14   Control of planner.

leads to a cross-over, I need to create some space (possibly by inserting a grid line) so that I can subsequently insert a routing module.

Once we have determined a suitable rectangular topology, we proceed with the second phase of determining the actual geometric dimensions required to meet constraints on area, aspect ratio, port position, connectivity, etc. As I mentioned earlier, this is a graph-based algorithm, and I don't intend to go into it fully here. Basically, we assign variables to the positions of all the edges of the modules. The various constraints can then be cast as arcs in a constraint graph, as shown in Figure 7.19. There are, in fact, two graphs—one in the X direction and one in Y. Because of the inclusion of quadratic constraints (due to area), there is no reasonable technique for arriving at an optimal solution. Consequently we use an iterative technique that seeks to minimize area by modifying the aspect ratio of submodules that are on the critical path in one dimension and not the other. Our experience has been that this technique gives a good solution, usually close to the optimal.

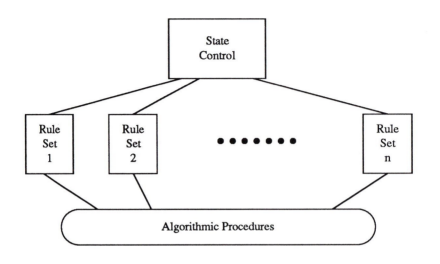

Figure 7.15 Overall structure of planner.

IF

M2 and M1 are placed Adjacent
and M3 is connected to both of them
and M1 is substantially larger than M2 and M3

THEN

Add a strong Penalty to grid locations that cause
M2 and M3 to be orthogonally adjacent to M1

Figure 7.16 Place module rule.

7.2.3   Leaf Cell Layout Agent

Now I want to discuss the agent that is responsible for detailed transistor layout within a leaf cell. Input to this agent (known as EXCELLERATOR [3]) is a description of transistor connectivity together with constraints on port positions, aspect ratio, etc. An example is shown in Figure 7.20. Output is a custom symbolic CMOS leaf cell layout, as shown in Figure 7.21.

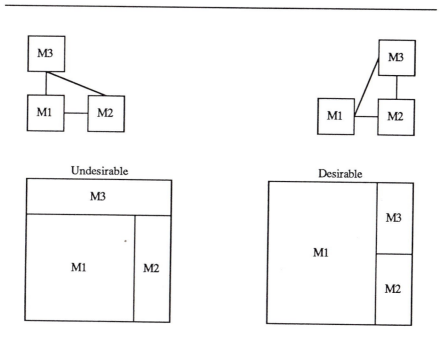

Figure 7.17   Resulting layout.

Layout is produced in the gate-matrix style. That is, we lay out rows of transistors, trying to maximize vertically common gate connections and diffusion source/drain overlap. We are not restricted, however, by the routing conventions of true gate-matrix layout.

Before getting into the details of EXCELLERATOR, I want to give some background on the way in which it was developed. Our initial solution to the problem was a system called TOPOLOGIZER [2], a rule-based system built around the "redesign" model. It provided good solutions some of the time, but not always. It would occasionally get stuck in poor local minima. From our experience with that tool, however, we were able to develop EXCELLERATOR. This tool is also modeled on techniques used by human designers, but is implemented, for the most part, using algorithmic code. This scenario is an excellent example of the way in which knowledge-based programming techniques can allow us to begin exploring poorly understood problems, and then, having gained some understanding, move on to a more efficient solution.

Like the floorplanner, EXCELLERATOR divides its task into two phases: The first is transistor placement; the second is signal routing. In the first

IF

There is a X-type intersection.
and Grid graph width is greater than height

THEN

Insert new column

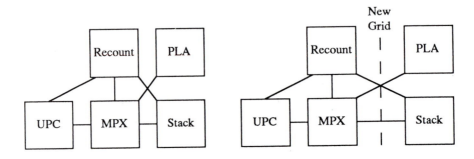

Figure 7.18   Routing rule.

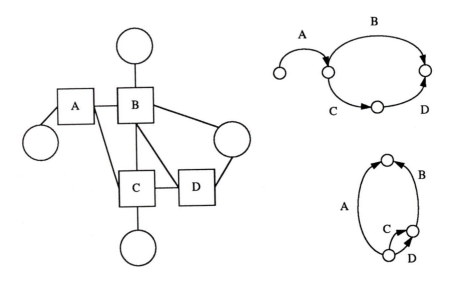

Figure 7.19   Geometric solution by constraint graph.

Cellname:		Transistors:			
bit-equal-cell		nfet	p2	input1	vss
Constraints:		nfet	p2	input2	vss
aspect X10 Y8		pfet	p2	input2	p1
port input1 north		pfet	p1	input1	vdd
port input2 north		nfet	xout	p2	vss
port out east		nfet	xout	input1	n3
		nfet	n3	input2	vss
		pfet	xout	p2	p3
		pfet	p3	input1	vdd
		pfet	p3	input2	vdd
		nfet	out	xout	vss
		pfet	out	xout	vdd

Figure 7.20   Input specifications to EXCELLERATOR.

phase, transistors are initially placed on a grid according to their gate signal. All transistors with a common gate are placed on a single vertical column, as shown in Figure 7.22. Transistors are then swapped and flipped so as to cause matched source/drain regions on adjacent columns. Once matched, columns are merged as shown in the figure. Some columns are split in two to satisfy aspect ratio or vertical height constraints. This matching operation is carried out using exhaustive search. Although this

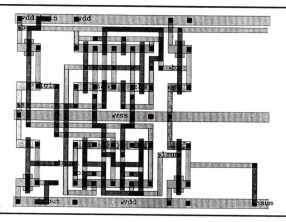

Figure 7.21   Symbolic output produced by EXCELLERATOR.

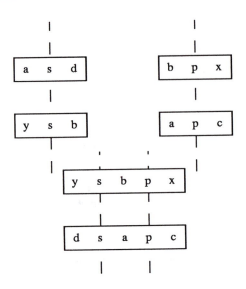

Figure 7.22   Leaf cell transistor placement.

would be prohibitively expensive in the general case, it is adequate for the number of objects being manipulated in relatively small leaf cells.

The second phase, signal routing between the transistors and external ports, is the more interesting, and so I will spend more time on this part of the problem. It is worth noting that this is a much more difficult task than the classical switchbox or channel routing problem. First, we have multiple routing layers. Second, we have our connections spread out across the 2-D surface and not just on the boundary. Third, there are obstacles of arbitrary rectilinear shape, over which some routing layers are allowed to pass. Finally, there is no preferred direction for routing any one layer.

Figure 7.23 shows the overall control of the routing process. We start out with an initial spacing between the rows of transistors based on our estimate of the routing requirements of the cell. Then we select connections, one at a time, and attempt to place them in the layout. Selection is based on a number of rules that prioritize wires on the basis of length, layer, function, etc. During the routing process, we may rip up and reroute wires that have already been placed. If no solution can be found for a particular net, then we insert a new grid line to allow the process to proceed. We continue in this fashion until all routes have been completed.

Routing is performed using a recursive rip up reroute scheme based on

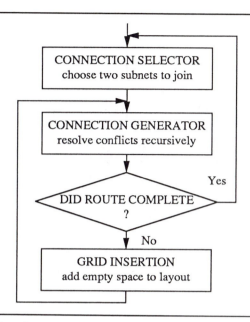

Figure 7.23   Flow chart of routing operation.

an A-star maze router. Maze routing is a technique for finding the optimal path between a source and a destination by assigning costs to each cell in a routing matrix, as shown in Figure 7.24. We start at the source and move out, recursively searching the matrix, adding one each time to our cost function. If there are two paths to a cell, we choose the lower of the two costs. Once we reach the destination, we trace a path back to the source as a sequence of monotonically decreasing costs. The problem with this technique is that it is very time consuming because of the large amount of search required.

In A-star routing, we add a second component to the cost function, which is an estimate of the cost of routing from that cell to the destination. The estimate used in EXCELLERATOR is simply the Cartesian distance between these two points. This helps direct the search process and limit the number of cells visited. Now we extend the process to 2-1/2 dimensions to include routing on different layers, as shown in Figure 7.25. We also assign costs to the different routing layers and vias and also other items like corners.

Another problem with maze routing is that it greedily consumes available space according to the order in which the routes are attempted. We need a technique to deal with being unable to complete a route because its path is blocked by previously routed nets. One technique that has been used

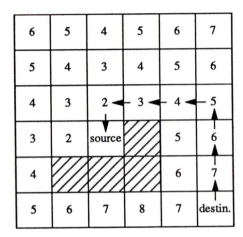

Figure 7.24   Maze routing.

previously is *rip up and reroute*. The idea is to rip up the blocking net, put it back on the "to be routed" stack, and then complete the current net. There are, however, a number of situations in which this can lead to cyclic reroutes as shown in Figure 7.26. There are two routes to be made here. Whichever one we make first blocks the other, so we cycle back and forth ripping up one and routing the other.

The approach taken in EXCELLERATOR is different in that we do not discard the blocking net. Rather, we suspend the current net at the point where the blockage occurs and then try to reroute the blocking net in such a way as to avoid the conflict cell. Once the blocking cell has been rerouted, we continue with the current net, as shown in Figure 7.27. In modifying the blocking route, we incur some cost. In the example shown, the cost of the blocking route increases from 20 to 30. This increase (10) is added to the maze cost of the current net at the conflict point. Remember that each point in the maze represents only one possible path. We may consider, therefore, a number of different reroutes of different nets before arriving at the minimum cost solution.

This rerouting can, in fact, be performed recursively. In other words, we may suspend the rerouting of a net to investigate the rerouting of a secondary blockage. An example is shown in Figure 7.28. Results have shown that this scheme works well up to two or three levels of recursion. Beyond that, the process becomes intolerably slow as the search space grows exponentially. There is an interesting parallel here to the way human

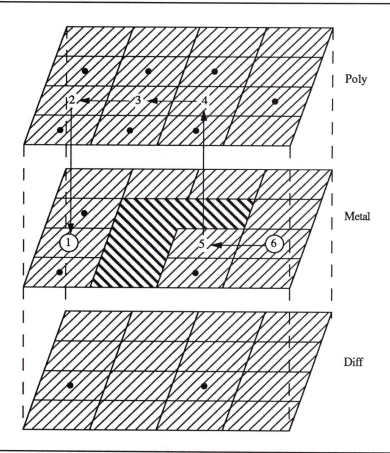

Figure 7.25   A-star maze routing in 2-1/2 dimensions.

designers behave. The human will mentally rip up and reroute to achieve a solution and will, in fact, do this recursively if necessary. Beyond about two or three levels of recursion, however, the designer is not able to remember all the relevant information. Nevertheless, this type of exploration is sufficient to achieve good results.

How well does the program work? We have used this tool to generate layouts for a number of different leaf cells ranging in size from 10 up to about 50 transistors. The areas of these cells are comparable (within 20%) to those achieved by experienced human designers. The automatic layouts are usually smaller than the human ones.

versus

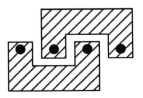

Figure 7.26 Rip up cycling.

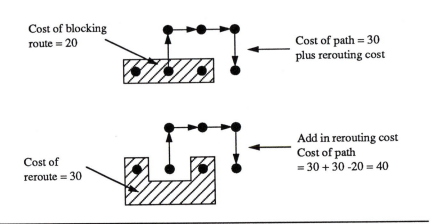

Figure 7.27 Rerouting cost.

7.2.4 Design Manager

Finally, I want to take a look at the manager. This agent is in many ways the key to the whole system. It is responsible for organizing the operations

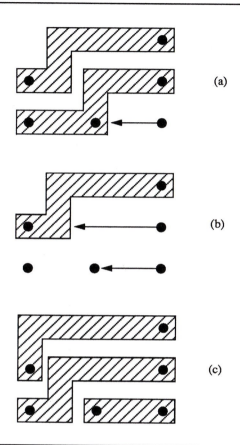

Figure 7.28   Two-level recursion.

of the individual design agents into some sort of overall design plan. To achieve this goal, the manager must

- Maintain a design database: Organize the myriad pieces of design data into a structure that facilitates analysis and modification by the various design agents.

- Assign tasks to the agents: Retrieve information from them; act as an interpreter from one agent to another.

- Maintain consistency: Propagate constraints where appropriate; know which pieces of design data go together to make a valid partial design.

- Interact with the user: Inform the user of what is happening and why; allow the user to make suggestions and/or override.

- Maintain a design history: This is used for explanation purposes and also for aborting unsuccessful subtasks in a clean, predictable fashion.

- Make design trade-offs: Choose between competing strategies; know how far to backtrack when a conflict occurs.

A full description of the manager and its implementation is beyond the scope of this chapter, but I will look briefly at the different types of knowledge the manager needs to maintain and manipulate in order to perform the tasks mentioned above. Design knowledge can be broken down into three classes. The first is problem-specific design knowledge, which includes specifications, layouts, net lists, history, etc.—in other words, data which describe the current state of a particular design. The second class is domain-specific knowledge, information that tells us how to design a VLSI chip successfully. Some of this is contained in the design agents, and some resides in the manager—knowledge about how and when to invoke a particular agent, for example. The third class is control knowledge. This is meta-knowledge that tells us how and when to apply different design strategies, when to communicate with the user, what preconditions need to be satisfied before a particular action may be invoked, etc. If we take a look at just one of these, problem-specific knowledge, we can see just how rich and complex this information space might be.

Problem-specific knowledge, remember, is the data used to describe the actual chip we are building. Traditionally, designers like to manage this data using hierarchy, but at least three orthogonal hierarchies can be applied. Each views the design process from a different perspective.

- Architectural Hierarchy. This describes modules in terms of submodules and interconnection. It represents the logical structure of the circuit.

- Context Hierarchy. This relates modules according to the physical context in which they are placed. An adder cell, for example, may be used a number of times within an ALU such that each instance has an identical physical environment. The same adder may be used elsewhere in the circuit, however, with quite different physical interface constraints.

- Refinement Hierarchy. The layout of a particular cell may be redesigned a number of times as we iteratively backtrack through the design space. We may have to keep a number of versions together with design history information.

The structure we choose to represent these different types of information must be able to answer questions like, "What subset of information represents the current state of the design?" and, "If we backtrack, what information should be retained and what should be discarded?"

Thus one of the most important tasks we have faced in building this manager is developing a suitable data representation. What we have developed is an object-based representation consisting of four principal sections, as shown in Figure 7.29. The module section is an object-based description of the problem-specific data. It describes cells, floorplans, specifications, etc. These objects are organized into a rich classification hierarchy in which inheritance between objects is used to simplify and reduce the amount of data actually stored.

Attached to these objects are operators (or messages). These are domain-specific knowledge in the form of executable code. Some of them are pre-processors for the external agents (e.g., floorplanner). Operators also make design decisions, perform transformations, propagate constraints, etc. The kernel is the control section that drives us toward a solution. It is a procedural object that chooses what to do next when the design is in a stable (all constraints consistent) state.

Finally, the tasker is a set of objects that provides control over how we

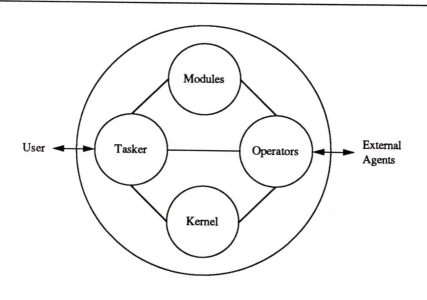

Figure 7.29   Internal structure of manager.

perform design operations. The design process is divided into a number of atomic tasks. For each task, the tasker checks preconditions, selects an operator, and then propagates the result. It provides an interface to the user, both synchronous (ask the user) and asynchronous (interrupts). It also transactionalizes the design process, generating check points to which the manager can return in the event of an interrupt or backtrack.

The status of the manager is that we have implemented much of this code using KEE. This software package provides us with objects, inheritance demons, rules, etc., together with a graphics-based entry and debugging interface. It is yet to be seen, however, whether this class of commercial system can really cope with the amount of data required to solve this type of problem.

7.3   Future Directions

I want to end this chapter with a quick summary of what I feel are the key issues to be resolved in order for future progress to occur in this area. The first is that we need to develop good models for storing and manipulating design knowledge. The simple rule-based production system model is not sufficient for dealing with design problems. A number of attempts have been made to integrate control and design data into rule-based systems. Examples are the redesign model, blackboard systems, and a host of other ad hoc solutions. Progress in this area will be slow until we discover a general model around which we can build good software development tools.

The second area that is going to be important is learning. If we look at those expert systems that have gone beyond laboratory experiments to become useful tools, we find that each has taken many person-years to develop and refine to this level of operation. This presents a real problem in a fast-moving field such as VLSI design. How can we build systems in time for them still to be useful when they are complete? How can we build systems that can adapt to changing technology and design styles? The answer lies in the ability to acquire knowledge automatically and update design rules and strategies. The subject of machine learning, however, is still an open research topic in the AI community.

The third area is the need for computer architectures better suited to the execution of rule-based, knowledge-based programming styles. One obvious improvement that can be made is to apply parallel processing techniques. Many of the systems that have been built, including CADRE, use multiple agents. Much of the searching of the solution space for design alternatives could be performed in parallel. Finer-grain opportunities also exist within the individual design operations. Progress in this area is needed to yield systems that produce good designs on a modestly priced machine in a reasonable amount of time.

Bibliography

[1] Ackland, B., et al. "CADRE – a system of cooperating VLSI design experts," *IEEE Intl. Conf. on Computer Design,* pp. 99–104, 1985.

[2] Kollaritsch, P. and Weste, N. "Topologizer: an expert system translator of transistor connectivity to symbolic cell layout," *IEEE Journal Solid State Circuits,* 20(3):799–804, 1985.

[3] Poirier, C. "Excellerator – automatic leaf cell layout agent," *IEEE Intl. Conf. on Computer Aided Design,* pp. 176–179, 1987.

[4] Subrahmanyam, P. "SYNAPSE: an expert system for VLSI design," *IEEE Computer,* 19(7):78–89, July 1986.

[5] Subrahmanyam, P. "LCS – a leaf cell synthesizer employing formal deduction techniques," *24th. Design Automation Conf.,* pp. 459–465, 1987.

[6] Watanabe, H. and Ackland, B., "Flute – a floorplanning agent for full custom VLSI design," *23rd. Design Automation Conf.,* pp. 601–607, 1986.

[7] Yu, M-L. "Fork – a floorplanning expert for custom VLSI design," *IEEE Intl. Conf. on Computer Aided Design*, pp. 34–37, 1987.

Index

accepted traffic, 170
adaptive routing, 151, 163, 164
address allocations, 230
address decoder circuit, 424
address maps, 225-228, 254-260, 337,
    362-363, 367-369
algebraic straightline programs, 120-124
all-to-all broadcasting, 228, 242, 281,
    301-303, 309
all-to-all personalized communication,
    236, 280-281, 284, 303-309, 326,
    333, 346-347
ALT(log n, expolylog n), 115, 119,
    125
alternating Turing machines (ATMs),
    87-88, 111, 115, 118-119, 125
Ametek Series 2010, 30-47
arbiters, 189-192, 194
arbitrary-delay communication, 57
architecture of neural networks, 391,
    396, 398
area-time product, 4, 416-419

balanced cyclic reduction algorithm,
    243
bandwidth, 148, 150, 165, 169,
    200-202, 204, 208
barrier synchronization, 57, 64-65
BBN butterfly, 147, 158
Bernoulli trials, 395-396
binary multiplier, 417
binary n-cubes, 25, 38-41, 58-60, 251

binary-reflected Gray code, 264-268,
    272-274, 290, 322-325
bin packing, 131-135
bisection width, 143, 145, 148-149
blocking network, 185
broadcasting, 281, 283, 295, 297-299,
    301-303, 309
buffered communication, 57, 156-157,
    165, 170, 218
butterfly, 144, 147

Cannon's algorithm, 228
census functions, 102-104
c-future tags, 211-212
Chandry-Misra concurrent shortest path
    algorithm, 53-54
channels, 144, 151, 153, 155-165, 169,
    172-177, 180-182, 189-190, 196, 199
circuit-value problem, 99-100, 116
circular buffer, 208
clock driver tree, 194-195
clock skew, 194-196
cochlea, electronic, 427-440
cochlea, hydrodynamic model of, 429
cochlea model chip, 420-440
communication networks, 143, 169-184
commutator, 73
constant-wire-bisection assumption, 41
consumption assumption, 172
contour-path technique, 110
Cosmic Cube, 24-28, 34, 36, 41, 46,
    57, 64, 68

Cosmic Environment, 47
CRCW-PRAM, 94-96, 362
CREW-PRAM, 94-95, 97
crossbar network, 185
cut-through routing, 34, 36-38

dag (directed acyclic graph), 110,
    120-121, 123, 125, 128
data parallel programming model, 225
degree of a node, 145
Delta rule, 401-403
Denelcor, 158
design manager, 451, 464-468
desperation routing, 158
detailed balance, 239
deterministic routing, 151
diameter of a network, 92-94, 145-146,
    148-150, 160, 165, 169
differential amplifiers, 430
Digital Orrery, 17-18
dimension permutations, 243-244,
    333-352
direct-connect routing, 30
directed acyclic graph (dag), 110,
    120-121, 123, 125, 128
direct networks, 146, 148-150, 154,
    165, 169, 178-184
distributed discrete-event simulation, 67,
    77, 79
divide-and-conquer methods, 227-228,
    241, 259
Dongarra and Sorensen algorithm, 228,
    242
doubling tree, 60
dual paths transpose (DPT) algorithm,
    327-328, 346
DRAM cells, 422-424

e-cube routing, 151-152, 163-164, 179,
    182, 188, 199
embeddings based on partitioning, 272
EREW-PRAM, 94-96, 108
EXCELLERATOR, 456-464
exponential current gradient, 440

fast Poisson solvers, 277
feedback networks, 390-394, 404-412
feedforward networks, 390-391,
    396-399

fine-grain, 3, 20, 48, 52
finite element method, 233-234
first-order filter circuit, 432-433
flits (flow control digits), 35-38,
    152-158, 160, 165, 175-176,
    185,189-190, 192, 199
floorplanning, 449-450, 452-456
flow control, 145-150, 152-158, 165,
    169-170, 173, 184-185, 189-191, 199
flow-control hierarchy, 34-35
Fluent abstract machine, 245, 364,
    369-372, 376-378
flux of the network, 94
full-cube, nonseparable dimension
    permutation algorithms, 351-352
future tags, 210-211

Gaussian elimination, 71-72, 125-131,
    227, 234, 236, 258
geometrical-design-rule checking, 79-80
grain size, 3, 23, 205, 218
graphics pipeline, 74-76
Gray code encoded matrices, 330
Grosch's Law, 14

Hamming distance, 226, 274, 339
Hebbian rule, 394
heterogeneous computers, 23
homogeneous computers, 23
hypercube, 3, 21, 25, 30, 69, 89-91,
    251
hyper-pyramids, 308, 310-321, 316

ILLIAC IV, 91
indirect networks, 146-148, 169,
    177-179, 182
in-forest, 104
ITP, 448-449

Jacobi's method, 226
J-Machine, 199, 204, 206, 214, 216

k-ary n-cubes, 146, 148-152, 163,
    165-166, 169, 179, 183
KEE, 468

latch, 184-185
lattice-gauge theories, 236-241, 243

line compression, 271
linear code decoding, 405-406
linearly separable functions, 391
LINPACK, 69, 71
list-ranking problem, 107-108
load balancing, 20, 28, 76
low-dimensional networks, 166-168

matrix multiplication, 69-71, 75
matrix transposition, 242-243, 321, 325-327, 330-334, 347
matrix transposition with Gray code to binary code conversion, 330-333, 346
maximum likelihood decoding, 405
Metropolis algorithm, 238, 240
MIMD (multiple-instruction, multiple data) computers, 18, 21, 54, 62-64, 74
misrouting, 157, 158
Moler, Cleve, 69, 71
Monte Carlo methods, 66-67, 238-241
Mosaic A chips, 48
multigrid method, 241-242, 308
multilayer neural networks, 398, 402
mutual-exclusion circuit, 11-14

Navier-Stokes equation, 231-233, 243, 246
n-body problem, 72
NC, 98-100, 114-115, 117-119, 124-125, 130-131, 134-135
nearest neighbor search problem, 405-406
n edge-disjoint spanning binomial trees, 229, 281, 292-297, 299-301, 303-304, 307-308
n rotated Hamiltonian paths, 289-290
network design frame (NDF), 188-191, 194, 197-198
network saturation, 144, 170, 182

oblivious routing, 151
Omega network, 22
one-dimensional partitioning of matrices, 255-256, 326
one-to-all broadcasting, 281, 297-300, 309
one-to-all personalized communication, 281, 283, 299-301, 309, 326, 340, 352

optimal parallel algorithm, 107, 109-110, 134
organ of Corti, 428
out degree, 145

packet buffers, 158
packets, 35-36, 39-42, 44-45, 152-158, 160-163
parallel computation thesis, 96-97, 99
parallel-prefix computation, 105-106
path doubling, 98, 104-105, 107-108, 110
P-complete algorithm, 125-127, 131-135
P-complete problem, 99-100, 116, 125-128, 131-135
perfect shuffle, 92
physical transducer, 428
planar-monotone-circuit value problem, 116-118
plaquettes, 237, 239, 241
$PM2I$ networks, 236, 243-244
power-delay product, 7, 9, 14
PRAM (parallel random-access machine), 88, 94-98, 101-104, 109, 112, 121, 123-125
process group, 33

quantum chromodynamics (QCD), 228, 236, 241, 260
quantum electrodynamics (QED), 228, 236, 241
queue stage, 175-176, 185-186

RAM cells, 420-427
Reactive Kernel, 46-47
regular computation graphs, 225
routing-mesh, 30-31, 34
routing methods, 144, 151, 169, 185-189, 199
routing relation, 150-151, 160-161, 165, 175, 185
RP3, IBM, 87, 89, 147, 165

second-order filter circuit, 433-434
segment-based memory management, 215-216
self-timed circuit, 11
self-timed router, 191-194, 199
service time, 172-179, 181

shortest path in a graph, 52-53, 67
shuffle-exchange, 92-94, 96, 101, 109, 125
single path transpose (SPT) algorithm, 327-328, 346
SRAM cells, 420, 422-424
spanning balanced graph, 286-289, 297, 299, 301, 303-304, 307-308
spanning balanced *n*-tree, 281-282, 286-289, 299
spanning binomial trees, 282, 285-288, 290-293, 297, 299-301, 303-304, 306-308
stable dimension permutations, 335-341, 345-351
stable state, 393-396, 404-407, 411
step embedding, 268-271
stiffness matrix, 233-234
store-and-forward networks, 36-38, 153-154, 158-159, 168-169
structured buffer pools, 158-160
successive overrelaxation (SOR), 226
sum-of-outer-products algorithm, 394, 406
switching energy, 7, 9-10
symmetric successive overrelaxation (SSOR), 226
synchronization, 56-57, 63-65, 190-196, 199-200, 205-207, 210-213, 218
synchronization tags, 210-213, 218
synchronous routers, 191, 194, 196, 199
systolic algorithms, 67

Thompson, Clark, 4, 6
threshold functions, 391, 393, 397-398
threshold logic, 391
time-on-target, 60

topological layout, 453
topological sorting, 110
torus networks, 40-41
torus routing chip, 141, 151, 185, 189, 193
transconductance amplifiers, 430
Traveling Salesman problem, 406-411
TRC chip, 141, 151
2-cycle signaling convention, 192-194
two-dimensional partitioning of matrices, 255-256, 259, 326-327

Ultracomputer, 87-89
unbounded-slack communication, 57
unbuffered communication, 56-57
unification, 99
universal-interaction technique, 72

virtual channels, 155, 160-165, 169, 181, 193
virtual cut-through, 154, 156, 173
virtual processors, 225, 254, 257-258, 277-278, 280, 321, 326

weakly synchronized execution sequences, 65
Wide Angle Wave Equation, 235-236
wide-range amplifier circuit, 431-432
wide-range gain control, 436-437
wire delay models, 247-249
wiring complexity, 150
wormhole routing, 30, 34, 36-38, 153-156, 159, 161-163, 165, 168-169, 170, 173
wormlock, 161

zero-slack communication, 56-57, 63